D1035488

TECHNIQUES OF ELECTROCHEMISTRY

TECHNIQUES OF ELECTROCHEMISTRY

Volume 3

Edited by

ERNEST YEAGER

Chemistry Department
Case Western Reserve University
Cleveland, Ohio

ALVIN J. SALKIND

Department of Surgery
CMDNJ-Rutgers Medical School
Piscataway, New Jersey

A Wiley-Interscience Publication

JOHN WILEY & SONS, New York • Chichester • Brisbane • Toronto

Library of Congress Cataloging in Publication Data (Revised)

Main entry under title:

Techniques of electrochemistry.

Includes bibliographical references and indexes.
1. Electrochemistry. I. Yeager, Ernest B., 1924– ed. II. Salkind, Alvin J., 1927– ed.

QD553.T4 541'.37 73-37940
ISBN 0-471-02919-x (v. 3)

Printed in the United States of America

10 9 8 7 6 5 4 3 2 1

CONTRIBUTORS

Theodore R. Beck, Electrochemical Technology Corporation, Seattle, Washington

Ralph Brodd, Union Carbide Corporation, Consumer Products Division, Cleveland, Ohio

Dodd Carr, International Lead-Zinc Research Organization, New York, New York

James Doe, ESB Technology Center, Yardley, Pennsylvania

Gerald Halpert, National Aeronautics and Space Administration, Goddard Space Flight Center, Greenbelt, Maryland

James Hoare, General Motors Corporation, Research Laboratories, Warren, Michigan

A. Kozawa, Union Carbide Corporation, Consumer Products Division, Cleveland, Ohio

Mitchell La Boda, General Motors Corporation, Research Laboratories, Warren, Michigan

Irving Miller, University of Illinois at Chicago Circle, Chicago, Illinois

Jacques Pankove, RCA Corporation, Sarnoff Research Center, Princeton, New Jersey

Alvin J. Salkind, College of Medicine & Dentistry-Rutgers Medical School, Piscataway, New Jersey

Dennis Turner, Bell Telephone Laboratories, Murray Hill, New Jersey

PREFACE

Electrochemistry and electrochemical techniques are generally regarded as a specialty, rather than a basic discipline. Most of the researchers and practitioners using electrochemical techniques were trained in some other field, usually chemistry, biology, physics, electrical engineering, or chemical engineering. As a result, there has existed for some time a need for a compilation of the various standard techniques.

The chapters in this book, and in other volumes in the series, are all written by authors recognized for their work with electrochemical techniques. The material is directed to the nonelectrochemist as well as to the electrochemist who wishes to gain insight into techniques with which he is not very familiar. The background knowledge required of the reader is that ordinarily obtained in undergraduate courses in physical chemistry and instrument analysis. The emphasis here differs substantially from that in the Advances in the Electrochemistry and Electrochemical Engineering series, now edited by Charles Tobias and Heinz Gerischere. We refer the reader to this excellent series for reviews of a rather sophisticated nature, principally directed to the advanced technical student and the experienced electrochemist.

This volume is devoted mainly to industrial and applied techniques. In Volume 1, techniques relating to electrode processes are discussed. Volume 2 was devoted mainly to techniques relating to studies of electrolytes.

We thank the contributors to this volume, and this series, who gave generously of their time in helping to make this third document of the techniques of electrochemistry. We hope that our colleagues will bring to our attention any errors of omission or comission that have occurred.

Ernest Yeager
Alvin J. Salkind

Cleveland, Ohio
Piscataway, New Jersey
January 1976

CONTENTS

TECHNIQUES OF ELECTROCHEMISTRY

I. INDUSTRIAL ELECTROCHEMICAL PROCESSES

Theodore R. Beck
Electrochemical Technology Corp.,
Seattle, Washington
and
Department of Chemical Engineering
University of Washington
Seattle, Washington

1. Introduction

Industrial electrochemistry provides a variety of products used by modern society; aluminum, magnesium, sodium, chlorine, sodium hydroxide, chlorates, perchlorates, adiponitrile, tetraalkyllead, and electrorefined metals to mention a few. Increasing use of electrochemical processes is expected as electrochemical tehcniques become better and more widely understood and as the economics of electrochemical processing improve through use of larger and more efficient cells. Continuation of the decrease in cost of electrical energy over the long term relative to other forms of energy or to chemical oxidizing and reducing agents would add further impetus to use electrochemical processes.

Industrial Electrochemical Processes

Many types of electrochemical techniques are used in the electrolytic industries:

- Laboratory and pilot plant studies
  - Exploratory
  - Directed
  - Scale down (reverse of scale up)
- Cell and process design
  - Empirical techniques
  - Mathematical modeling from first principles
  - Optimization
- Process operation
- Process measurement and control

We may consider these techniques to be part of electrochemical and chemical engineering. An attempt will be made to discuss certain techniques and to give a general industrial perspective.

A. Objectives

The objectives of this chapter are to identify the features that distinguish industrial electrochemistry and its techniques from other branches of electrochemistry; categorize and illustrate the various types of electrode, electrolyte, separator, and cell systems now used; illustrate the economic basis that rules all actions and selects the techniques used; and discuss the relation of electrochemical engineering to the design and operation of industrial cells.

While the orientation of the chapter is toward industrial electrochemical processing, features common to the electrodeposition field and battery and fuel cell technology will be mentioned as they arise. There is much to be learned by crossing technological boundaries. The electrochemical technologies are all based on the same fundamental electrochemical engineering principles: thermodynamics, kinetics, mass, heat and momentum transport, current distribution, and scaling laws. Likewise, the technologies use the same applied aspects of electrochemical engineering: cell design techniques, materials selection, and economic optimization. Advances in one technology may be applied to others; for example, throwing power and current distribution calculations developed for plating are equally applicable to process cells.

B. Audience

This chapter is directed in particular toward the young electrochemist or electrochemical engineer. Industry can be quite bewildering to the fresh product of academe because the objectives of the two institutions are rather different. The university is entrusted by society to search for, transmit, and preserve knowledge. Industry, on the other hand, is intended to provide products and services at a fair return on its investment. The different goals and the separation of the two institutions has resulted in some cases in a lack of mutual understanding.

Economic forces are very powerful and all pervasive in industry. A plant or laboratory, generally treated as an economic unit, must earn a satisfactory return on investment if it is to survive. These economic forces establish procedures and shortcuts in design and operation that may appear to lack the rigor of technical analysis instilled by academe, although they have their own logic. The young scientist or engineer must develop an overall perspective if he is to make an adjustment to industry. A goal of this chapter is to outline an electrochemical engineering and economic perspective of electrochemical processing.

C. Specific Tasks

Two specific tasks were chosen for this chapter: to classify and summarize the various cell and electrode configurations used in industrial electrochemical processes, and to provide a brief perspective on the application of electrochemical engineering techniques to industrial processes with emphasis on the role of economic optimization.

Three important aspects set the techniques used in electrochemical processes apart from the laboratory techniques described in the first two volumes (1) of this series. The first aspect is that the processes considered here are many orders of magnitude larger than those involved in laboratory techniques. Instead of micrograms or grams of material in laboratory, the industrial plant may have an output of hundreds of tons per day. Instead of microamperes or amperes, the plant deals with 10^4 to nearly 10^6 A per cell.

The second aspect is that industrial processes are enormously complex compared to laboratory experiments. An objective in scientific research is to eliminate all but a few essential variables. In industrial electrochemical processes one is forced to simultaneously juggle many variables, including those of thermodynamics, kinetics, transport processes, current distribution, materials, materials handling, process control, labor relations, economics, finance, law, and so on.

The third and overriding consideration is that the process must earn a satisfactory and identifiable profit and return on investment. If a process cannot meet this criterion, a plant is not built. A plant must continue to meet this criterion by improvements or it is scrapped. The return from basic research on the other hand is more difficult to identify but in the long run it may be greater than from an existing industrial process. Society, for example, is still receiving dividends from the work of Michael Faraday.

D. Exclusions

Techniques used in the electrochemical process industries cover such a broad range that an encyclopedia would be required to describe them adequately. Therefore, it is appropriate to note areas of exclusion from this chapter.

This chapter does not give a detailed description of industrial electrochemical process. Other books (2 - 5)are addressed to this topic.

This chapter is not a course of clectrochemical plant or cell design. A need exists for a definitive book on this subject. MacMullin has made an excellent start with two pioneering papers (6, 7) relating to cell design and scaleup. Many books on chemical plant design (8-13) cover process areas other than cells in an electrochemical plant.

This chapter is not a text on electrochemical engineering, and no attempt was made to review, comprehensively the electrochemical engineering literature. Newman (14) has developed an excellent and rigorous work that covers the interrelations of thermodynamics, kinetics, mass transport, current distribution, etc., in electrochemical systems. Levich (15) has formulated a treatise, dealing in large part with electrochemical mass transport, that is invaluable to the electrochemical engineer. Periodic reviews of particular aspects of electrochemical engineering appear in a series edited by Delahay and Tobias (16). Several electrochemical texts have sections relating to industrial processes (17-20). A need remains for a quantitative book including the applied aspects of electrochemical engineering.

This chapter is not a definitive exposition on economics of electrochemical plants. Rather, it is intended to serve only as a general introduction because there are so few papers concerning industrial cells (21-29) and fuel cells (30-32) in the literature. Several works are available on chemical processes economics (33-39).

This chapter does not discuss measurement techniques in electrochemical plants. Many parameters are measured including current, potential, potential drops, concentrations, pH, temperature, pressure, pressure drops, flow rates, current efficiency, anode-cathode spacing, and so on. These measurements include standard, well-known, laboratory techniques that need no elaboration. Instruments and techniques related specifically to the electrochemical industries might also deserve documentation.

II. CELL DESIGNS

Industrial electrochemical cells deal with a wide variety of liquid and gaseous reactants and products at temperatures from below ambient to 1000°C or above. Electrolytes may be aqueous, organic, molten or solid salt, or ion-exchange membranes. The electrolyte may be very corrosive in combination with the reactants, products, or intermediates. The state of technology and materials available changes with time. It is therefore not surprising to see a large variety of shapes and sizes of electrochemical cells. MacMullin (42) has given a detailed classification of industrial cells by various functions, including type of anode and cathode, type of electolyte, and type of divider.

Many other considerations are involved in the detailed design of industrial electrochemical cells. In general, although the cells may be essential, they are only part of a total plant process that is optimized as whole. This puts certain constraints on the design and operation of cells. In turn, the plant exists in a particular, technical, economic, and political climate, which reflects back on cell design. Source and cost of raw materials and construction materials, location and size of markets, freight costs, cost and availability of electrical energy, labor cost, land cost and zoning, environmental factors, taxes, and the like also may exert an influence.

Table 1 categorizes industrial cell designs into major types. The classification was intended primarily to include cells for making industrial products; but designs used also in batteries, fuel cells, electrowinning, and electroplating are indicated. Cell design and operating techniques have grown in parallel in different industries without much cross communication. In the present discussion, attention is focused on the common features of cells used in different industries.

A classification of electrode types is given in Table 2 and of separators in Table 3.

TABLE 1

Type	Applications	Advantages	Disadvantages or limitations
I. One–Compartment A. Inert vessel, immersed electrodes 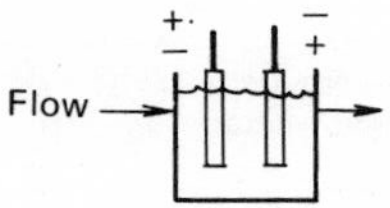 + − − + Flow	Electrorefining of Cu, Zn Electroplating, Electroforming, Anodizing, Batteries — lead–acid,	Simple Inexpensive Electrodes readily changed	Gases produced must be capable of dispersal in atmosphere Electrolyte, electrodes, reactants and products must not be affected by atmosphere
B. Vessel is one electrode 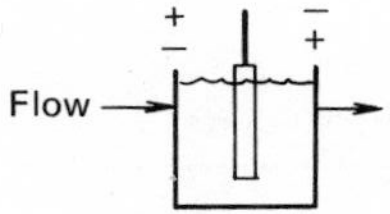 + − − + Flow	Sodium chlorate Sodium perchlorate Leclanche cell (consumable case)	Simple Inexpensive Electrodes readily changed (except for Leclanche cell)	
C. Horizontal liquid–metal cathode 1. Mercury cell 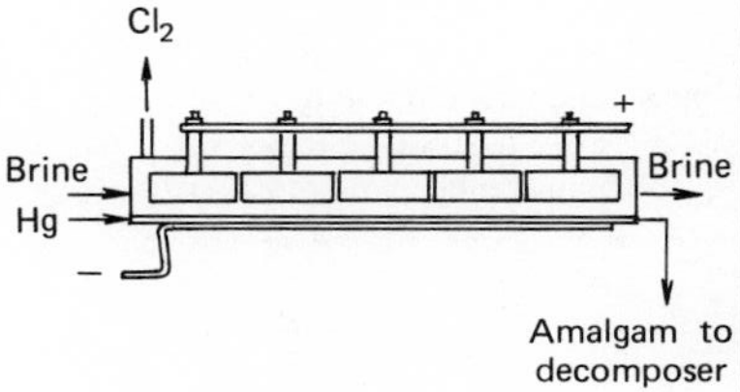 Cl_2 + Brine Hg − Brine Amalgam to decomposer	Chlorine–caustic	Produces concentrated, high–purity sodium hydroxide	Large floor area Higher energy consumption per unit of product than diaphragm cell Must avoid mercury loss to environment

TABLE 1 (Continued)

Type	Applications	Advantages	Disadvantages or limitations
C. Horizontal liquid–metal cathode (Cont'd) 2. Aluminum reduction cell 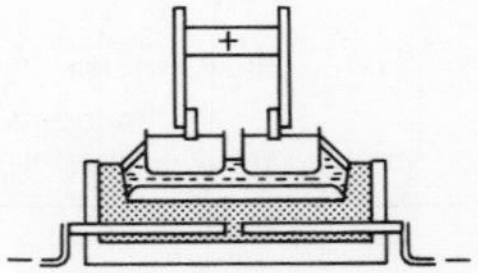	Aluminum reduction (A three–layer variation used in electrorefining aluminum)	The only cell design for producing aluminum that has reached commercial feasibility	Large floor are High capital cost
D. Tube cell 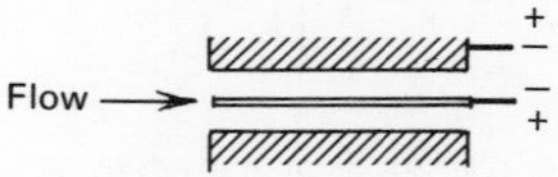 Flow	Tetraalkyllead Plating wires	Good heat transfer Good current distribution	
II. Two compartment A. Open–diaphragm cell Convection parallel to current through diaphragm 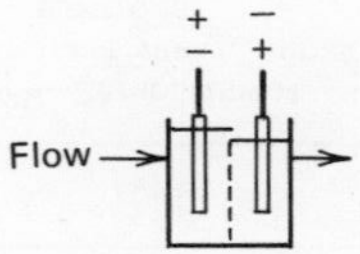 Flow	Electrolytic manganese Electrorefining Ag, Au, etc. Electroplating (bagged anodes)	Controlled concentration and pH in each compartment Each electrode is exposed to appropriate solution Impurities from one electrode are prevented from reaching other	Increased cost Plugging of diaphragms

Type	Applications	Advantages	Disadvantages or limitations
B. Closed–diaphragm cell Convection parallel to current 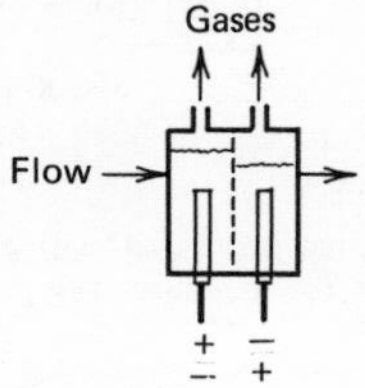 Gases Flow	Chlorine–caustic–hydrogen Hydrogen–oxygen	Same as A Separate collection of gasses	
C. Membrane cell Convection perpendicular to current 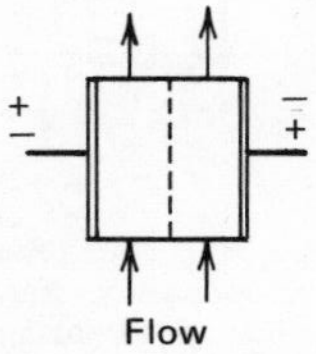 Flow	Oxidation and reduction Acid–base generator Electrodialysis Electrodimerization for adiponitrile	Electrolytes may be different and have only one common ion transported by membrane; complete separation of neutral reactants on products	

TABLE 1 (Continued)

Type	Applications	Advantages	Disadvantages or limitations
D. Cells for producing liquid metals—convection not desired Cl_2 Liquid metal − +	Magnesium Sodium Potassium Lithium	Separation of metal and chlorine products	High voltage drop Fouling of separators
II. Multicompartment Filter–press construction A. Frames Membranes or diaphragms Solid or porous electrodes Flow channels	Hydrogen–oxygen generation Hydrogen–oxygen fuel cell Other fuel cell systems Electrodialysis Adiponitrile HCl electrolysis	Compactness Precision assembly Can operate under pressure Flow channels can be built in Adaptable to a wide variety of configurations with solid and porous electrodes, membranes, diaphragms and flow channels	Limited to inert or long–lived electrodes Long–time operation between disassembly required

TABLE 2

Type	Example
I. Planar	
A. Inert	Graphite anode in chlor–alkali cell Platinized titanium or ruthanium–oxide coated titanium anodes Platinum anodes in perchlorate cell Iron cathodes in chlorate and perchlorate cells
B. Active consumable	Carbon anode in Hall–Heroult cell Copper anodes in refining cell
C. Base for deposition	Mercury cathode in chlor–alkali cell Aluminum cathode in Hall–Heroult cell Stainless steel cathodes in manganese metal cell Copper cathodes in refining cell Electroplated objects Carbon base for deposition of MnO_2
D. Phase change on surface	Anodizing aluminum Thin–film battery Solid–state coulometer
II. Mesh or screen	Steel cathode in diaphragm type Chlor–alkali cell Metal anodes in chlor–alkali cells
III. Porous	Hydrogen–oxygen fuel cell Hydrogen–oxygen generator Electrochemical fluorination
1. Gaseous reactants or products	

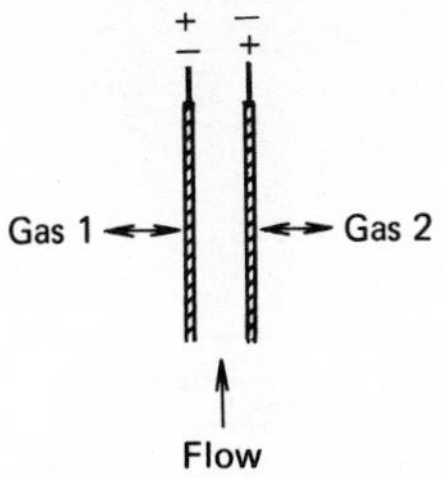

+ −
− +
Gas 1
Gas 2
Flow

TABLE 2 (Continued)

Type	Examples
2. Flow–through electrode	
(a) current and flow parallel 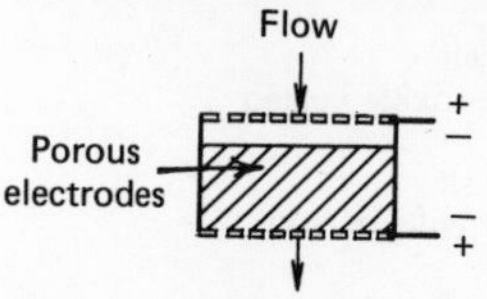 Flow Porous electrodes	Hydrazine or methanol fuel cell Oxidation or reduction of dilute constituents Concentration of dilute metal ions
(b) current and flow perpendicular 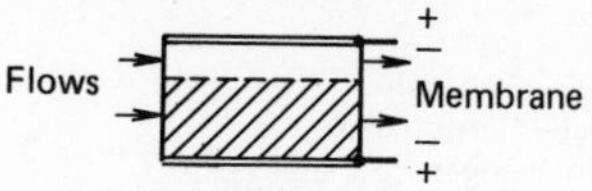 Flows Membrane	Mfg. of tetraalkyllead
B. Active phase change	Lead–acid battery plates, $Pb/PbSO_4$ and Pb/PbO_2 Other battery systems
IV. Fluidized bed	
(A) current and flow parallel 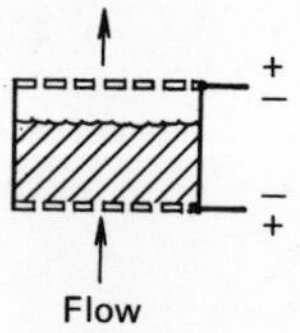 Flow	
(B) current and flow perpendicular 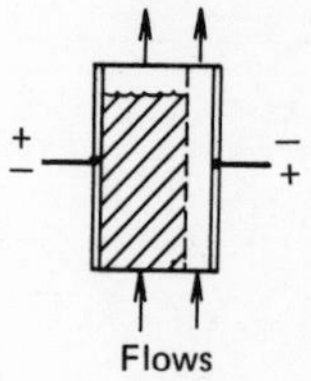 Flows	

TABLE 3

Cell Separators, Diaphragms and Membranes

Nonpermselective
Cloth
Glass
Polypropylene
Polyvinyl chloride
Fluorocarbon
Felt
Asbestos
Plastic fiber
Porous plastic
Polyethylene
Polyvinyl chloride
Fluorocarbon
Rubber
Porous ceramic
Aluminum oxide
Metal mesh
Permselective
Homogeneous (sulfonated or aminated)
Styrene-divinyl/benzene copolymer
Polyethylene
Fluorocarbon
Reinforced homogeneous
Heterogeneous (particles in binder)

A. One-Compartment Cells

Wide varieties of electrochemical reactions are carried out in Type I, simple rectangular or upright cyclindrical tanks. The inert-vessel cell is little more than a direct scaleup of a breaker cell used in the laboratory. Some common materials of construction are mild steel, coated steel, concrete, plastic-lined concrete, wood, or plastic. The advantages of the inert vessel are simplicity, low cost, and ready accessibility to change electrodes.

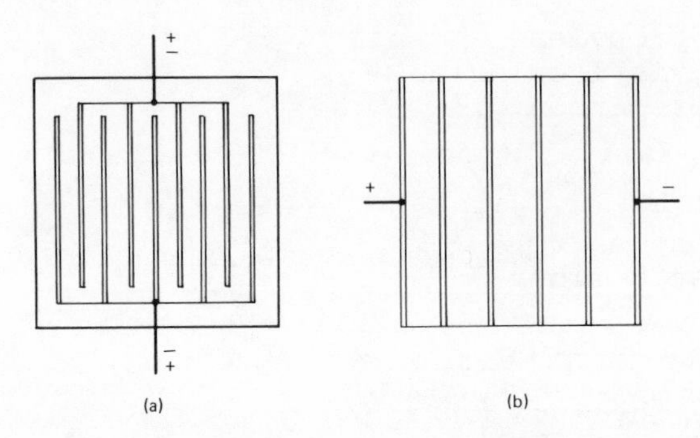

Fig. 1. Unipolar and bipolar electrode connections. (a) Unipolar (Multiple-electrode cell). (b) Bipolar (battery).

Type I cells may contain multiple electrodes that are connected in either monopolar or bipolar configurations as illustrated in Fig. 1. The unipolar or parallel connected configuration is most commonly used in Type I cells. Example processes are electrowinning, electrorefining, electroplating, electroforming and anodizing. Plates in lead-acid batteries and other battery systems are also connected in the unipolar mode.

The bipolar mode, really a battery of cells, is not commonly used in simple Type I cells because a close fit of the electrodes to the cell walls and bottom is required to avoid or minimize parasitic currents through the electrolyte between cells. An interesting exception is a system used in the manufacture of sodium chlorate in which a battery of bipolar graphite electrodes is inserted in a large concrete tank of brine. The graphite electrodes are enclosed in an insulating box open at top and bottom for circulation of electrolyte through the battery by gas lift. Parasitic current losses between cells through the electrolyte at top and bottom are compensated by simpler construction involving less bus bar with fewer ohmic-loss connections and a better match (higher voltage, lower current) to power supply impedance. These are typical of the many tradeoffs involved in cell design.

In some cases the vessel itself may serve as one electrode of the cell (Type I.B). Industrial examples are the manufacture of chlorates and perchlorates in mild steel tanks that serve as cathodes (43). A familiar example is the Leclanche flashlight cell in which the can is a consumable zinc anode. Modern practice is to enclose this cell in a steel can to avoid leakage as the zinc anode becomes perforated through discharge.

The most highly developed and extensively utilized one-compartment cells are the mercury cell (44, 45) of the chlor-alkali industry and the Hall-Heroult cell (46) of the aluminum industry. Both employ horizontal liquid metal cathodes. The sketches in Table 1 are shown in a little more detail for these cells because of their specialization and importance.

The advantage of the mercury cell process is that it directly produces a concentrated, high-purity, sodium hydroxide solution usable in the process industries. In contrast, the product of the diaphragm cell (discussed below) must be concentrated and purified. This advantage of the mercury cell is balanced against a higher energy consumption per ton of product. Again, tradeoffs are made between different cell designs.

The cathode reaction in the mercury cell is

$$Na^+ + Hg + e \rightarrow Na(Hg) \qquad (1)$$

The amalgam is sent to a decomposer in which is formed sodium hydroxide solution and denuded mercury which is returned to the cell:

$$Na(Hg) + H_2) \rightarrow NaOH + \frac{1}{2} H_2 + Hg \qquad (2)$$

The anodic reaction in the cell is

$$Cl^- \rightarrow \frac{1}{2}Cl_2 + e \qquad (3)$$

Some chemical wear of the graphite anodes occurs because of discharge of water and hypochlorite ions to form CO_2, and coincidentally, mechanical wear.

The mercury cell consists of a slightly sloping rectangular steel trough containing mercury, brine, and anodes. The bottom is machined smooth and is usually bare for making electrical contact to the flowing-film mercury cathode. The side walls are insulated with hard rubber or plastic. Anodes, which may be graphite or metallic (dimensionally stable, platinum-coated or ruthenium oxide-coated titanium), are suspended from the cover. Their positions are adjustable from the top of the cell. Graphite anodes are lowered periodically to compensate for wear. Electrical energy is a substantial part of the process cost, and the anode-cathode distance is maintained at an optimum value. An economic balance is made between energy loss by excessive anode-cathode spacing and the labor costs for adjustment. Anode adjustment has been automated in some installations. In this case, the balance is between the energy loss and the capital and operating costs of the automatic system. Modern, metal anodes do not require periodic adjustment.

Vertical electrode mercury cells have also been proposed. The mercury would flow over a mercury-wetted steel electrode. Such a concept was studied for a fuel cell to generate power from sodium amalgam and oxygen (47, 48). Vertical as well as horizontal rotating-disk mercury cells have been described (2), but they have not gained acceptance. There is presently a trend away from mercury cells because of environmental concern, illustrating that the social and political climate affects cell selection.

The configuration of the aluminum reduction cell or "pot" is largely dictated by the reactions occurring and the corrosiveness of the combination of electrolyte and aluminum. The cell occupies a large floor area for the amount of electrode area and has a very high capital cost. Nevertheless, this is the only process for manufacture of aluminum that is a commercial reality today. Although larger cells are now used, design has not changed appreciably since the process was invented by Hall and by Heroult over 80 years ago.

The overall reaction is

$$Al_2O_3 + \frac{3}{2}C \xrightarrow{(6e)} 2Al + \frac{3}{2}CO_2 \qquad (4)$$

Aluminum oxide powder is dissolved in a molten cryolite (Na_3AlF_6) "bath" at about 980°C and is electrolyzed between an aluminum cathode and a carbon anode. The molten aluminum and cryolite are contained in a carbon or graphite lining; the latter are among the few materials that have a sufficient resistance to the corrosiveness to this combination. Penetration by bath and reduction products, and slow air oxidation limit lining life to 1 1/2 to 3 years. Current is conducted out of the lining by steel collector bars. The lining is held in a steel shell with some thermal insulation in between to control heat loss. Usually some bath is frozen on the sidewall, which protects the carbon. Molten aluminum is periodically removed by siphoning into a crucible.

A relatively pure petroleum coke is used in the consumable anodes. Because aluminum is very negative in the electromotive series, most impurity metals in the materials going into the cell will find their way into it. The combination of a consumable anode requiring adjustment of position and a gaseous product formed dictates that the anode enters from the top of the cell. Anodes are either prebaked in a furnace outside of the cell or baked in place (Soderberg anode) by the temperature of the cell. Again, we see an economic tradeoff, in this case between the cost of operating a prebaking furnace and the cost of higher carbon consumption and more fuming in the potroom with Soderberg anodes. Current is carried into the anodes through steel stubs.

Numerous attempts have been made to supplant the Hall-Heroult process, but none has as yet achieved commercial success. Among these have been carbothermal reduction combined with a subhalide purification process, a process involving reduction of aluminum chloride with manganese metal (49), and electrolysis of aluminum chloride (50). There is much room for improvement because the Hall-Heroult process has a very high capital cost and a low energy efficiency (about 35 to 40%).

A rather unique variation of the type I cell (I.D) is used for manufacture of tetraaklyllead for gasoline additives (51-55). Each cell is in the form of a vertical steel tube cathode filled with a consumable lead shot anode forming a porous flow-through electrode. The shot is separated from the cathode by woven polypropylene screen or porous ceramic deposited inside of the tube. A central lead rod makes contact to the lead shot in each cell. Multiple parallel-connected cells are formed inside the vertical tubes of a sheel and tube heat exchanger. The electrolyte is Grignard reagent that flows axially through the tubes. Proposed reactions are

dissociation $$4RMgCl \rightarrow 4R^- + 4MgCl^+ \quad (5)$$

anode $$4R^- + Pb \rightarrow R_4Pb + 4e \quad (6)$$

cathode $$4MgCl^+ + 2RCl = 4e \rightarrow 2RMgCl + 2MgCl_2 \quad (7)$$

Overall $$2RMgCl + 2RCl + Pb \rightarrow R_4Pb + 2\ MgCl_2 \quad (8)$$

B. Two-Compartment Cells

A Type II, two-compartment cell with a diaphragm or membrane is used when anolyte and catholyte must be separated. Separation may be necessary because different compositions and pH are required at each electrode or in order to avoid transport of impurities or other species from one electrode to the other. A diaphragm may allow bulk flow or convection of solution from one compartment to the other as well as migration of ions and diffusion of various species. An ion-exchange membrane allows transport of predominantly one ion species or charge. Typical diaphragm and membrane materials are shows in Table 3. Simple examples of a diaphragm for avoiding contamination from one electrode to the other are bagged anodes in electroplating.

Cell Designs

A flow-through diaphragm cell is used in the electrowinning of manganese. The catholyte must be very pure and at a controlled pH to obtain high current efficiency and smooth deposition on the cathode sheets. Purified manganous sulfate, ammonium sulfate solution flows into the cathode compartments in which manganese is deposited, accompanied by hydrogen evolution:

$$Mn^{++} + 2e \rightarrow Mn \text{ (about 65 to 70\% current eff.)} \quad (9)$$

$$2H^{+} + 2e \rightarrow H_2 \quad (10)$$

The mangeanese-depleted solution flows through synthetic-cloth diaphragms into the anode compartment in which oxygen gas and hydrogen ion are generated:

$$H_2O \rightarrow \frac{1}{2}O_2 + 2H^+ + 2e \quad (11)$$

Some manganese dioxide is also deposited on the anode. The cell gases are discharged into a well-ventilated cell room in which hydrogen concentration is kept below the explosive limit. The acid anolyte is returned to the ore-leach area and to purification, an example of cells integrated with a total process.

The diaphragm cell for production of chlorine and sodium hydroxide is the most widely used flow-through cell (56). Over 70% of the chlorine and caustic produced in the United States comes from this type of cell, and the present trend in new construction is again in this direction.

The older cells have been for the most part made of concrete and have steel cathodes and graphite anodes. Plastic replaces concrete in newer cells. The present trend is toward metal (dimensionally stable) anodes whose higher initial cost is offset by longer life, cleaner cell operation, and lower cell maintenance. Diaphragms are made of asbestos, formed by pulling a vacuum on steel screens immersed in an asbestos slurry. The steel screens serve as cathodes. There is considerable interest at present in replacing asbestos diaphragms by ion-exchange membranes.

Brine is fed to the anode compartment in which chlorine is produced:

$$2Cl^- \rightarrow Cl_2 + 2e \qquad (12)$$

The brine trickles through the asbestos diaphragm to the cathode where hydroxyl ion and hydrogen are produced:

$$2H_2O \; 2e \rightarrow 2OH^- + H_2 \qquad (13)$$

About 50% of the salt is reacted per pass through the cells. The resulting catholyte cell effluent liquor containing sodium hydroxide and undecomposed salt is piped to the concentration and purification plant from which recovered solid salt is recovered and recycled. Chlorine and hydrogen are likewise piped to purification.

Membrane cells with flow-by electrolyte convection offer opportunities that have only recently been exploited commercially, for example, in the production of adiponitrile and in electrodialysis. The recent advent of suitable ion exchange membranes has made possible the utilization of the flow-by cell.

The method to make adiponitrile now employing flow-by membrane cells was discovered by Baizer (57) in 1963. It caused a revival of interest in industrial electroorganic reactions. The process consists of three basic steps:

Acrylonitrile is mixed with a suitable McKee (quaternary ammonium) salt, which increases electrical conductivity and the solubility of acrilonitrile.

This mixture is electrolyzed and the product of adiponitrile plus spent salt is continuously withdrawn.

The product mixture is extracted with a suitable solvent and pure adiponitrile is recovered; the unreacted acrylonitrile and McKee salt are recycled.

The commercial process is described by Danly (58). The principal cathode reaction on lead alloy cathodes is

$$2CH_2{=}CHCN + H_2O + 2e \rightarrow NC(CH_2)_4CN + 2OH^- \qquad (14)$$

A cation permselective membrane separates the two compartments. The anode reaction is oxygen generation from dilute sulfuric acid. Multiple, electrically cascaded cells are assembled in filter-press construction.

Another potential application for flow-by membrane cells is simultaneous generation from a neutral salt of an acid and an alkaline solution for process pH control. The process streams themselves might be put through the cell if there are no competing electrochemical reactions. The cost of electrical energy is generally small, compared to acids and bases. For example, sodium hydroxide at $7.00/100 lb and sulfuric acid at $35.00/ton give hydroxyl and hydrogen ions, respectively, at 1.23¢/g mol and 0.19¢/g mol. Electrical energy at 10 mills/kwhr for a cell at 5 V gives a cost of 0.13¢/g mol acid plus base, or less than 10% of the purchased reagent costs.

Two-compartment cells in which strong convection is not desired are used for producing liquid metals from their fused chlorides. Examples are the Downs cell for sodium and the corresponding electroysis cell for magnesium. The function of the divider is to separate the molten metal from the chlorine gas to avoid recombination. The divider may be of ceramic at the top of the cell or an electrically isolated metal screen extending down between the electrodes. There is obviously an optimization to be made here between a large spacing and minimum recombination of products with large bath voltage drop and the converse.

C. Multicompartment Cells

The Type III, filter-press construction is often used for multicompartment cells and batteries of cells. Applications include hydrogen and oxygen generation, hydrogen-oxygen and other fuel cell systems, electrolysis of HCl, manufacture of adiponitrile, and electrodialysis. Advantages include compactness, precision assembly, capability of building flow channels into the frames, adaptability to a wide variety of configurations with solid and porous electrodes, membranes or diaphragms and operation under pressure. Because assembly and disassembly are labor intensive, the filter-press design lived anodes are used and for which there is relatively long operation between disassemblies.

III. ELECTROCHEMICAL ENGINEERING

A. Characteristics of Industrial Processes and R&D

Characteristic features of successful industrial electrochemical processes as presently practiced are:

they convert electrical to chemical energy,
usually on a large scale,
with relatively high current density,
using continuous processes,
to produce products consumed by society,
and to operate at a profit and satisfactory return on investment.

knowledge, but to develop new products, improve products, increase sales, or maintain or increase profit. Only operations, experiments, or measurements that are consistent with the short- and long-range business goals of the organization are condoned.

The overriding consideration in the design of a new process or plant is making a satisfactory return on investment (ROI):

$$\text{ROI} = \frac{\text{annual profit}}{\text{investment}} \tag{15}$$

The return on investment must exceed a certain level determined by the company before proceeding with design and construction. It is outside the scope of this chapter to justify the criterion or expound on the philosophy of return on investment. Suffice it to say that investments are made with savings - somebody's savings - of which there is a finite amount and it is considered wasteful not to work capital investments as hard as possible.

Economic considerations have not penetrated the electrochemical literature to a significant extent, and many students trained in electrochemistry enter industry with a lack of understanding of these most important considerations affecting their work. It is not the purpose here to present a course in economics or plant design and operation but to put the techniques of industrial electrochemistry in perspective.

The approach to electrochemical cell and plant design has been largely empirical in the past, being really more art than science. The reason was twofold: cells embody an extremely complex set of interrelations that were not understood quantitatively, and high precision was not necessary. Many of the economic optima for cells are rather broad in respect to the design and operating parameters. Approximate empirical relationships and judgment and intuition were, therefore, adequate to satisfy optimizations with Eq. 15.

As the electrochemical industries have matured and become more competitive, and as cells and plants have become larger and more costly, a need has developed for more precise design. At the same time, more has become known about electrochemical engineering fundamentals, and computers have become available for complex computations. We are therefore in a transition period to more precise application of electrochemical engineering principles to cell and plant design. One of the purposes of this chapter is to indicate how the new techniques relate to economic considerations.

B. Electrochemical Engineering

We briefly examine the various interrelated elements of electrochemical engineering given in Table 4 as a background to the economic considerations. System behavior depends on the important phenomena controlling that behavior, which in turn provide parameters for economic calculations.

TABLE 4

Elements of Electrochemical Engineering

Fundamental aspects
Thermodynamics
Mass and energy balance
Kinetics
Transport processes
Potential and flux distribution
Scaling laws
Applied aspects
Cell and plant design
Economics, optimization
Laboratory and pilot plant experimentation
Process control

Thermodynamics tells whether a spontaneous process is possible and the maximum potential developed in a galvanic cell or the minimum potential required to drive an electrolytic process cell. A mass balance of reactants, intermediates and products is based on desired production rate, stoichiometry, current efficiency and yield. Laboratory and pilot plant experimentation may be required to establish current efficiencies and yields. Energy and thermal balances can be divided into a reversible part, predictable from thermodynamics, and an irreversible part, contributed by ohmic losses and overpotentials.

It is not always necessary to understand the detailed mechanism and kinetics of an electrochemical reaction in order to commercialize it -- witness the Hall-Heroult process for production of aluminum used industrially for over 80 years and for which detailed kinetics are still unknown. Activation overpotential is usually not a dominant part of the total potential of a cell and, furthermore, it varies only slightly with current (logarithmic) in comparison with linear ohmic losses. The slope of the linearized overpotential current density curve is important in the analysis and prediction of current distribution.

The economics of industrial processes require that high current densities be used; in many cases the current density approaches the mass-transport limit. An understanding of electrochemical mass transport may therefore be essential in plant design. Convective and diffusional transport are similar to conventional chemical engineering practice, but electrolytic migration of ions adds a new dimension to electrochemical mass transport. Heat-transport problems in electrochemical cells can usually be solved by conventional chemical engineering techniques.

Potential and flux distribution considerations are somewhat unique to electrochemistry (14). While the solution of the Laplace equation to obtain primary current distribution has its analogies in electrostatics, the interactions with mass-transport and kinetic parameters give a secondary or tertiary limiting current density that is so important to plating, anodizing, electrophoretic coating of paints, and other processes. Electrode resistance and gas bubbles in the electrolyte may play a further role in determining the distribution of current density.

Cell design is the art of combining all of the elements of electrochemical engineering and selecting appropriate materials and geometry to give a profitable unit. The cells must be integrated into the rest of the plant which, in turn, is determined by well-known chemical engineering and electrical engineering procedures.

As already mentioned, optimizations in the last analysis are made on the basis of economics. Thousands of optimizations may be made during the design of a plant. Some are routine handbook type such as selections of economic pipe diameters (although even this can be complicated by corrosion and by the need to select from a number of possible materials). An example of an optimization unique to electrochemical cells is selection of the proper anode-to-cathode distance in aluminum reduction cells. The smaller the spacing, the lower the electrolyte ohmic drop, an appreciable part of the total cell voltage. On the other hand, smaller spacing gives lower current efficiency owing to easier transport of material from one electrode to the other where it reacts.

Laboratory and pilot plant experimentation are resorted to when an element or combination of elements in Table 4 cannot be determined by calculation or be found in the literature. If many uncertainties remain or are generated during the design phase, a prototype cell may be built and operated prior to the full-scale plant, but this is usually an expensive process. A technique described by MacMullin (6) is "scale-down by dissection," in which a full-size electrode pair from a multi-electrode assembly is tested. Here, as in all other aspects of the plant, the decision is made in terms of dollars; the cost of operating the prototype cell is balanced against the cost of correcting possible malfunctions in the final plant.

C. Innovation

An idea for a new process, a process improvement or a new plant may enter a company from a variety of sources. Once in a company, the idea may follow various routes, resulting in implementation, rejection at some stage, or shelving for possible future implementation. MacMullin (6) has outlined an orderly procedure for electrochemical development shown in Table 5.

TABLE 5

Outline for Electrochemical Development (6)

The steps to be followed are:

1. Ascertain the objectives of management.
2. Assign qualified personnel and/or consultants.
3. Make a search of the literature, including patents.
4. Determine what fundamental information is lacking.

TABLE 5 (Continued)

---

5. Estimate the extent to which analogies can be made as a first approach to design.
6. Draw up a tentative process flow sheet for full-scale plant, including feed preparation, electrolysis, recovery of products, and power supply. This frequently narrows down the conditions under which electrolysis is to be carried out.
7. Make a preliminary economic evaluation of full-scale plant, based on what is known or can reasonably be guessed. This investigation involves the following:

   (a) Preliminary cell design and cost versus cell size.
   (b) Circuit design and overall circuit cost versus cell size.
   (c) Total capital cost, manufacturing cost, and return on investment, on conservative basis.

If the evaluation justifies further effort, additional steps are taken as follows:

8. Secure the missing or doubtful fundamental data in the laboratory.
9. Scale down the preliminary commercial cell to an experimental size and provide for varying the parameters and materials of construction.
10. Test the cell, modify as required, and record all meaningful data.
11. Interpret results and make complete material, voltage, and energy balances.

If the results still make sense, proceed to:

12. Extract the significant design factors for scale-up.
13. Reassess the original "commercial" cell and modify as required.
14. Scale down to the next experimental size, which will be also a scale-up of the last model tested.

TABLE 5 (Continued)

15. Repeat until objective is reached (or has to be abandoned).

New process ideas will most likely enter through a Research and Development Laboratory or through Engineering. The idea may be home grown or it may originate in the literature or be introduced by a consultant or other outside agent. A development path such as that in Table 5 may be followed entirely through one cycle within the R&D laboratory or it may branch into Engineering. Examples of new electrochemical developments are a process for adiponitrile (57, 58) and an aluminum chloride process for manufacture of aluminum (50).

Process improvement ideas may originate inside or outside a company and be carried by R&D or Engineering. If carried by Engineering, the experimental steps in Table 5 will probably be delegated to the R&D Laboratory. Examples of process improvements are metal anodes and ion-exchange membranes in chlor-alkali cells.

The impetus for a new plant such as for manufacture of aluminum, chlorine and caustic, or chlorate probably will originate in Corporate or Division management. Engineering will probably be delegated to incorporate the latest proven technology and to design the most economic plant for the existing economic conditions.

Whatever the development, its source, or route through a company, economic evaluation (step 7, Table 5), and optimization play a major role. Economic considerations sit in judgment of every technical idea proposed for implementation. The idea may be repeatedly judged as it is refined, with a decision at each round as to proceeding further.

A young electrochemist or electrochemical engineer in industry should make it a point early in his career to understand the economic facts that govern his professional life. He must sort out the real problems in order to avoid working on the wrong problem (as far as the company is concerned) just because it is scientifically interesting. He needs to develop a systems approach based on economics. He must sell his ideas to management on their economic as well as technical merits. It is probably easier for him to develop an understanding of economics and to translate a technical development into economic terms than to convey the raw technical ideas to management and let them make the economic translation.

D. Economic Aspects of Electrochemical Processes

The remainder of this chapter emphasizes economic aspects of electrochemical processes. The objective is to develop a framework for relating electrochemical science and engineering to the industrial world.

As stated earlier, making a satisfactory return on its investment is the overriding criterion of an industrial enterprise. This criterion is applied to R&D. The investment in R&D must have a reasonable probability of successful implementation and of bringing in increased profits in the further from a new process or process improvement.

As mentioned above, return on investment is defined as the ratio of annual profit to the capital investment made in a venture:

$$\mathrm{ROI} = \frac{\text{annual profit}}{\text{investment}} \tag{16}$$

Incremental return on investment is frequently applied to process improvements or to stepwise optimization of a process:

$$\Delta\,(\mathrm{ROI}) = \frac{\Delta(\text{annual profit})}{\Delta(\text{investment})} \tag{17}$$

A company usually establishes criteria for undertaking a new venture such as a minimum 25% return or incremental return before Federal taxes. A lower return may be acceptable for a process improvement undertaken to maintain a competitive position.

Once the investment has been made and the plant has been built and is in operation, the emphasis is on profit. This is the domain of the plant managers and operating divisions. Profit can be defined in terms of sales, unit price and unit costs:

$$\text{Profit} = \text{sales (unit sales price - unit costs)} \quad (18)$$

Sales are related to production by

$$\text{Production - sales = inventory change} \quad (19)$$

TABLE 6

Typical Composition of Capital Investment in a Chemical Plant

Item	Group	Cost category	Investment type	Total
Purchased equipment	Equipment	Direct costs 70-85%	Fixed capital investment 80-90%	Total capital investment
Installation	Equipment	Direct costs 70-85%	Fixed capital investment 80-90%	Total capital investment
Instrumentation and controls	Equipment	Direct costs 70-85%	Fixed capital investment 80-90%	Total capital investment
Piping, installed	Equipment	Direct costs 70-85%	Fixed capital investment 80-90%	Total capital investment
Electrical, installed	Equipment	Direct costs 70-85%	Fixed capital investment 80-90%	Total capital investment
	Buildings	Direct costs 70-85%	Fixed capital investment 80-90%	Total capital investment
	Service facilities and yard improvements	Direct costs 70-85%	Fixed capital investment 80-90%	Total capital investment
	Land	Direct costs 70-85%	Fixed capital investment 80-90%	Total capital investment
	Engineering and supervision	Indirect costs 15-30%	Fixed capital investment 80-90%	Total capital investment
	Construction expenses	Indirect costs 15-30%	Fixed capital investment 80-90%	Total capital investment
	Contingency	Indirect costs 15-30%	Fixed capital investment 80-90%	Total capital investment
Raw materials	Inventory		Working capital 10-20%	Total capital investment
Supplies	Inventory		Working capital 10-20%	Total capital investment
Material in process	Inventory		Working capital 10-20%	Total capital investment
Finished product	Inventory		Working capital 10-20%	Total capital investment
	Cash		Working capital 10-20%	Total capital investment
	Accounts receivable, payable		Working capital 10-20%	Total capital investment
	Taxes payable		Working capital 10-20%	Total capital investment

In the present discussion, the inventory change will be considered to be zero or negligible over a period of a year. Sales may then be replaced by production rate. Unit sales price is determined by the market place and will be treated as a constant in the present analysis. Competition is such in the chemical and electrochemical industries that an individual supplier can rarely control sales price.

A typical composition of total capital investment for a chemical plant is indicated in Table 6. Methods for estimating costs of all of the elements are well established (33-39) and are not discussed here. The methods are also applicable to electrochemical plants. A breakdown of total product costs for a typical chemical process plant is shown in Table 7.

TABLE 7

Typical Costs for a Chemical Process Plant (37)

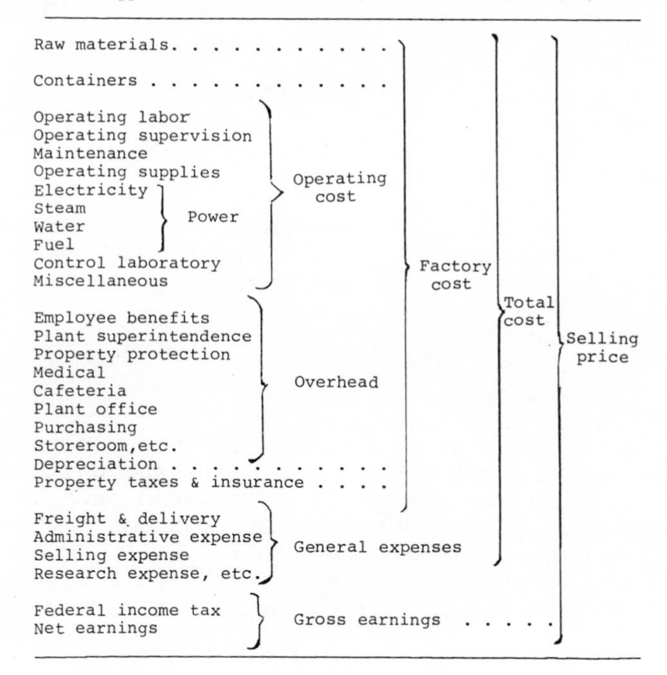

Raw materials			Factory cost	Total cost	Selling price
Containers			Factory cost	Total cost	Selling price
Operating labor		Operating cost	Factory cost	Total cost	Selling price
Operating supervision		Operating cost	Factory cost	Total cost	Selling price
Maintenance		Operating cost	Factory cost	Total cost	Selling price
Operating supplies		Operating cost	Factory cost	Total cost	Selling price
Electricity	Power	Operating cost	Factory cost	Total cost	Selling price
Steam	Power	Operating cost	Factory cost	Total cost	Selling price
Water	Power	Operating cost	Factory cost	Total cost	Selling price
Fuel	Power	Operating cost	Factory cost	Total cost	Selling price
Control laboratory		Operating cost	Factory cost	Total cost	Selling price
Miscellaneous		Operating cost	Factory cost	Total cost	Selling price
Employee benefits		Overhead	Factory cost	Total cost	Selling price
Plant superintendence		Overhead	Factory cost	Total cost	Selling price
Property protection		Overhead	Factory cost	Total cost	Selling price
Medical		Overhead	Factory cost	Total cost	Selling price
Cafeteria		Overhead	Factory cost	Total cost	Selling price
Plant office		Overhead	Factory cost	Total cost	Selling price
Purchasing		Overhead	Factory cost	Total cost	Selling price
Storeroom,etc.		Overhead	Factory cost	Total cost	Selling price
Depreciation			Factory cost	Total cost	Selling price
Property taxes & insurance			Factory cost	Total cost	Selling price
Freight & delivery		General expenses		Total cost	Selling price
Administrative expense		General expenses		Total cost	Selling price
Selling expense		General expenses		Total cost	Selling price
Research expense, etc.		General expenses		Total cost	Selling price
Federal income tax		Gross earnings			Selling price
Net earnings		Gross earnings			Selling price

It is apparent from Tables 6 and 7 that accurate assessment of the effect of plant and process parameters on profit and return on investment is complex. Such calculations may now be done by computer. For the present purposes we examine a few limiting cases in order to get a feel for first-order effects.

The key in any economic analysis is to include everything that is significant but not to include more than is necessary. The principle is analogous to a material balance around a process or part of a process. A circle is drawn around the project in question and all streams entering and leaving are measured. Similarly, in an economic balance, a circle is imagined around a plant, a process, or a single piece of equipment and all cost streams entering and leaving the circle are measured or estimated.

Unit costs shows in Table 7 can be lumped in various ways depending on their variation with the parameters under study. It is convenient to lump them in the following way for an electrochemical process:

$$\begin{aligned}\text{Unit cost} = {} & \text{raw materials} + \text{utilities (other than direct electrical energy)} + \text{direct labor} \\ & + \text{maintenance and repairs} + \text{overhead, administrative, and marketing} + \text{fixed charges} \\ & + \text{direct electrical energy} \qquad (20)\end{aligned}$$

A characteristic of inorganic electrolytic processes is that electrical energy forms a large part of the unit cost and its cost will be treated separately. (The raw materials cost for electroorganic processes may be greater so that electrical energy is not as significant in the unit cost (7).) Unit fixed costs vary inversely as production and will be also treated separately. The other unit costs shown in Eq. 20 may as a first approximation be assumed constant and lumped further for the analyses made here.*

$$cc = c_m + c_u + c_{dl} + c_{mr} + c_{oh} \tag{21}$$

The effect of electrochemical parameters on profit and return on investment will be examined. Production and unit electrical energy cost can be defined in terms of Faradays law+:

$$p = \frac{NISM\varepsilon_c f}{zF(1000)}, \text{ kg/yr} \tag{22}$$

and

$$c_e = \frac{zFVC_e}{M\varepsilon_c \varepsilon_r (3600)}, \text{ \$/kg} \tag{23}$$

Fixed cost per unit of production can be expressed by

$$c_f = \frac{\beta I_c}{P}, \text{ \$/kg} \tag{24}$$

Combining Eqs. 18 through 24, profit can then be expressed by

$$P = \frac{NISM\varepsilon_c f}{zF(1000)} \left[s - c_c - \frac{\beta I_c zF(1000)}{NISM\varepsilon_c f} - \frac{zFVC_e}{M\varepsilon_c \varepsilon_r (3600)} \right] \tag{25}$$

This equation has general applicability for electrochemical processes to examine the gross effects of current, current efficiency, and cell voltage on profit. Various limiting cases can be assumed for other unit costs in order to make calculations to get a feel for the effect of electrochemical parameters. Relations between current, current efficiency and

*All symbols are defined in the Table of Nomenclature.

+If there are chemical or physical losses, yield of product may also be included in the term f.

voltage, and other cell parameters may be empirical or in some cases determined from first principles of electrochemical science and engineering. It should be emphasized, however, that opitimizations of real industrial systems are specific to a piece of equipment or a plant, and may be much more detailed than indicated here.

The terms inside the brackets of Eq. 25 will be rearranged into dimensionless form so that we can better see the effect of the electrochemical parameters. This can be accomplished by multiplying the inside by (3600) M/zFV_oC_e) and the outside by its reciprocal, which gives

$$P = Nf\varepsilon_c \underbrace{\left[\frac{C_e I S V_o}{(3600)\ (1000)}\right]}_{\text{annual electrical energy cost per cell at the reversible potential, } \varepsilon_c = 1 \text{ and } f = 1.}$$

$$\left[\underbrace{\frac{(3600)\quad (s-c_c)M}{zFV_oC_e}}_{\text{dimensionless ratio of unit gross profit (excluding energy \& fixed cost) to unit energy cost}} - \underbrace{\frac{(3600)\ (1000)\beta I_c}{NIS\varepsilon_c fV_oC_e}}_{\text{dimensionless ratio of unit fixed cost to unit energy cost}} - \underbrace{\frac{1}{\varepsilon_c\varepsilon_v\varepsilon_r}}_{\text{reciprocal of overall plant electrolysis energy efficiency}}\right] \tag{26}$$

Inserting numerical values for the constant gives

$$P = 8.76\ Nf\varepsilon_c IV_o C_e \left[\frac{0.0373(s - c_c M}{zV_o C_e} - \frac{0.11\beta I_c}{Nf\varepsilon_c IV_o C_e} - \frac{1}{\varepsilon_c \varepsilon_v \varepsilon_r}\right] \quad (27)$$

which can be put in Eq. 16 to give

$$ROI = \frac{8.76Nf\varepsilon_c IV_o C_e}{I_c} \frac{0.0373\ (s-c_c)M}{zV_o C_e} - \frac{0.14\beta I_c}{Nf\varepsilon_c IV_o C_e} - \frac{1\ 1}{\varepsilon_c \varepsilon_v \varepsilon_r} \quad (28)$$

It is apparent that several types of optimizations can be made. In the design phase, attainment of a specified incremental return on investment is usually the criterion. Figure 2 shows a typical relation of profit to cost (related to size) of a piece of equipment being optimized. Profit is maximum at point O. This point is not the economic optimum based on return on investment, though, because the incremental return becomes zero here. The optimum equipment size is at a point of diminishing returns, A, at which the incremental return, α, is equal to some established value such as 25% before Federal taxes for a new process.

Once a plant is built and is in operation, the investment is fixed so the plant manager focuses on the numerator of Eq. 16. If he can sell all the production the plant is capable of delivering, he will try to maximize profit, Eq. 18. If sales are limited to a constant level he will try to maximize profit margin (unit sales price = unit cost) or minimize unit costs. The particular form that the optimization equations assume thus depends on the type of optimization that is made.

Optimizations may be made graphically, as indicated in Fig. 2, or analytically. The graphical method has the advantage that the sharpness of the peaks can be seen, showing the sensitivity of the optimization to the plant parameters. Analytical solutions are presented here for the purpose of getting a feel for which parameters are important in some electrochemical processes.

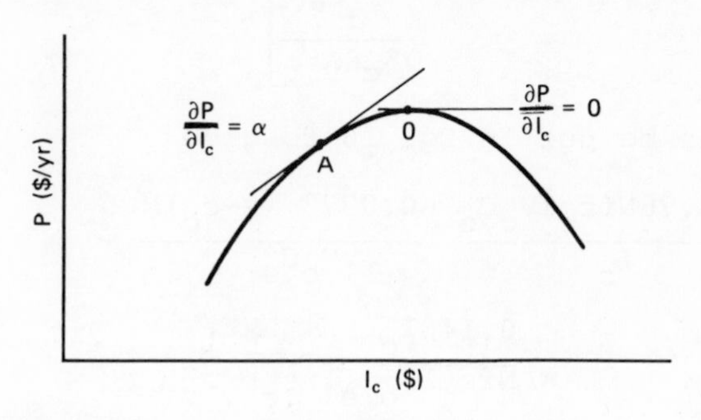

Fig. 2. Typical relationship of profit to investment in equipment.

TABLE 8

Some optimizations to be made in cell and plant design

Cell design
Cell materials
Cell size
Current and number of cells
Current density
Anode-cathode spacing
Frequency of anode adjustment
Unipolar versus bipolar
Circulation or agitation rate
Mechanization versus labor
Concentration
Temperature
Pressure
Selection of diaphragm or membrane material
Metal or graphite anodes
Inside or outside location of cells
Instrumentation

Some of the many optimizations that are made in the course of design and operation of electrochemical cells and plants are indicated in Table 8. A few limiting cases are examined in order to demonstrate the effect of some electrochemical parameters such as voltage, current, and current efficiency. It will be seen that optimum design and operation are related to available technology and to economic conditions, that is, the strength of the market and the relative cost of raw materials, energy, and labor. Periodic reoptimization of an existing plant may be necessary as technological developments become available and as economic conditions change.

The economics of different cell designs (Table 8, first item) is too complex to cover here in any detail. The cell design may be rather specifically determined by the process or there may be significantly different alternatives, for example, the competition in the chlor-alkali industry between diaphragm cells and mercury cells. The economic advantages of one over the other are close (25) because the choice of which one is installed varies from year to year and from country to country. Presnetly the hazard of mercury pollution is shifting new facilities to diaphragm cells (59). Similarly, there has been a competition between prebaked anode and Soderberg (baked in place) anode cells in the aluminum industry. New facilities are tending toward the prebake cell (60), also influenced by environmental considerations. The aluminum chloride process could completely change aluminum cell designs, because graphite anodes could be relatively permanent and horizontal aluminum electrodes may no longer be necessary.

The subject of cell materials is also too large to cover in any detail here. Tradeoffs occur between first cost and replacement of various materials. New plastics such as PVC have permitted design of more precise, lower-maintenance, and longer-lasting cells. Metal anodes are revolutionizing the chlor-alkali industry, and new diaphragm and membrane materials are on the way.

Chemical plants use a small number of very large reactors because of the economics of scale. Capital costs of equipment scale up with a 0.6 to 0.7 power dependency on size, giving lower unit fixed costs for larger units. Unit operating labor and repair costs also generally decrease with equipment size.

Electrochemical plants, in contrast, are characterized by a large number of smaller reactors (cells). Electrochemical processes are therefore capital intensive and labor intensive as compared to chemical processes. Limitations on cell size have been due to a number of technical factors.

Ohmic losses in the electrolyte and cell structure and mass transport limitations for reactants and products limit the range of practical economic current densities. Scale-up has therefore been achieved by increasing electrode area.

There are practical limits to the amount of planar electrode area that can be put into a given cell due to such factors as the amount of structure to maintain accurate anode-to-cathode spacing, ohmic drop along or through electrodes, increased ohmic drop in electrolyte due to accumulated gas bubbles, and distrotion of a liquid metal cathode by electro-magnetic effects.

A long series of cells was formerly needed to use the 500 to 1000 V available at mercury arc rectifier terminals to get maximum transformation efficiency. With silicon rectifiers, economic cell lines operating as low as 200 V are economic.

Economic forces and improving technology have resulted in significant increases in the sizes of industrial cells as shown in Table 9. Much of the earlier progress resulted from trial-and-error empiricism. In recent years the advent of silicon and mechanical rectifiers, metal anodes, inert porous membranes, ion-exchange membranes, and PVC and other plastics has provided new options in cell design. Full application of electrochemical engineering principles and procedures should bring further improvements.

For a given size cell and required production, there is an optimum number of cells and load per cell depending on economic factors. We can make a first-order approximation by making the following assumptions in Eq. 27:

TABLE 9

Historical Changes in Cell Loads
(in kiloamperes)

Year	Diaphram Chlor-Alkali	Mercury Chlor-Alkali	Hall-Heroult Aluminum
188 (56,61)	2.5		1.8
1894 (44)		0.5	
1940 (62)	0.5 - 7	0.13 - 15	8 - 40
1950 (2)			40 - 50
1960 (2)	10 - 27	15 - 170	50 - 125
1965 (63)	30 - 55	35 - 200	
1968 (63)		150 - 300	
1972 (59)	- 150		

$f, \varepsilon_c, \varepsilon_r, V_o, C_e, s, c_c,$ and β are constant

$$NI = K_4 \text{ (constant production rate)} \tag{29}$$

$$I_c = N\,C_c \text{ in which } C_c \text{ is constant} \tag{30}$$

$$\varepsilon_v = \frac{V_o}{V_1 + RI} = \frac{V_o}{V_1 + RK_4C_c/I_c} \tag{31}$$

Equation 27 can then be rewritten as

$$P = K_1 \left[K_2 - K_3I_c - \frac{(V_1 + RK_4C_c/I_c}{\varepsilon_c\varepsilon_rV_o}\right] \tag{32}$$

Differentiating and setting $\partial P/\partial I_c = \alpha$ gives the following relation after substituting back in the values of the constant, K:

$$Iopt - \left[\frac{0.114(\alpha + \beta)\varepsilon_rC_c}{RfC_e}\right] 1/2 \tag{33}$$

The optimum load depends predominantly on the ratio of fixed cell cost to electrical energy cost. Bus and connection resistances may be lumped in the effective cell resistance. Hubbard (21) gives a detailed analysis of a diaphragm chlor-alkali cell including effects of temperature changes with current. Schmidt (2) considers a number of other factors including more detailed investment calculations for electrolysis installations in general. Seelmann et al. (24) present results of computer calculations for large mercury cells that include a great many more variables.

A related analysis can be made for the optimum current density or size of cell for a given load. Assume N, f, ε_c, ε_r, I, V_o, C_e, s, c_c, and β are constant. $C_c = K_a A^n$ in which $0.6 \leq n \leq 1.0$ and $A = 1/i$ so that

$$C_c = \frac{K_4 I^n}{i^n} \tag{34}$$

$$Ic = NC_c = \frac{K_4 I^n N}{i^n} \tag{35}$$

$$\varepsilon_v = \frac{V_o}{V_1 + ri} = \frac{V_o}{V_1 + rI(K_4 N/I_c)^1/_n} \tag{36}$$

Equation 27 can then be rewritten

$$P = K_1 \left[K_2 - K_3 I_c - \frac{B_1 + rI(K\ N/I_c)^1/_n}{\varepsilon_c \varepsilon_r V_o} \right] \tag{37}$$

Differentiating and setting $\delta P/\mu I = \alpha$ gives the following relation after substituting back in the values of the constants:

$$\text{iopt} = \left[\frac{0.114 K_4 n \varepsilon_r (\alpha + \beta)}{rfC_e I^{(1-n}} \quad \frac{1}{(n + 1)} \right] \tag{38}$$

Optimum current density is seen to be determined by the ratio of fixed costs to energy costs as shown earlier by Ibl and Adam (22). Hine (26) indicates tha that $n = 0.74$ for a particular mercury cell design. Optimum current density decreases with size of cell as noted earlier by Seelmann et al (24).

An illustration of possible application of electrochemical engineering techniques to an optimization is in decreasing the specific resistance term, r, in Eq. 38. Hine (26) points out that the resistance of the electrolyte in a mercury cell with graphite anodes is about twice the calculated value based on electrolyte resistivity. This high resistance is attributed to gas bubbles containing chlorine and water vapor. Economic balances can be made between the cost and benefits of obtaining a smaller fraction of gas entrainment. Higher electrolyte flow rates produced by pumping or by internal circulation by gas lift (6) may be used to decrease gas entrainment. Porous or screen metal anodes are now being used to minimize chlorine bubbles going into the electrolyte phase.

Equation 38 indicates that graphite-anode, diaphragm, chlor-alkali cells should operate at a much lower current density than mercury cells because of the larger average anode-cathode spacing and thus the larger value of r. Operating data confirm this (63). The lower current density is compensated by a considerably larger electrode area per cell because of the vertical attitude of the electrodes.

The optimum anode-cathode spacing is of considerable interest for all types of cells. In mercury chlor-alkali cells, the spacing is as close as practically possible (about 2 to 3 mm). Electrical shorting between anode and cathode, increased electrolyte resistance by gas bubbles, and decreased current efficiency (63) by chlorine transport to the cathode set a lower limit. In diaphragm chlor-alkali cells with graphite anodes, the spacing increases as the anodes wear. Dimensionally stable metal anodes permit a constant much closer average spacing and permit higher current densities.

An approximate analysis for optimum anode-cathode distance can be made for an aluminum reduction cell. It has been known empirically for some time, although not understood quantitatively, that current efficiency increases with anode-cathode spacing. The current inefficiency appears to be mass-transport controlled, but it is not certain whether the diffusing species are dissolved monovalent aluminum or sodium from cathode to anode, or dissolved carbon dioxide from anode to cathode, or a combination (64). Although, the mechanism is not understood in a quantitative way based on first principles, an optimization may be made using empirical relationships. In the past, it has usually been less expensive to develop empirical relations than a basic understanding. This situation is changing with the establishment of a firm scientific basis for electrochemical engineering.

Two empirical relationships (65) proposed to relate current efficiency to anode-cathode distance are due to Lorenz,

$$\varepsilon_c = 1 - 0.12 \left(\frac{\ell_o}{\ell}\right) 0.8 \qquad (39)$$

in which ℓ_o = 1 cm, and to Abramov

$$\varepsilon_c = 0.5 \left(\frac{i_o \ell_o}{i \ell} \right) \qquad (40)$$

in which $i_o = 1\ A/cm^2$ and ℓ_o = 1 cm. Cell voltage is related approximately to current density and anode-cathode distance by

$$V = V_1 = \rho i \ell \qquad (41)$$

Equations 40 and 41 will be used to illustrate the optimization procedure. In order to carry out a real optimization on a plant, relationships for current efficiency and voltage are required for the particular cells in that plant.

The two limiting cases we examine are operation at constant current and at constant production ($I\varepsilon_c$ = constant). Assuming constant N, f, V_o, C_e, s, c_c, β, I_c, and ε_r in Eq. 27 gives for case I (I = constant),

$$P = K_1 \left[K_2 \varepsilon_c - K_3 - \frac{1}{\varepsilon_v \varepsilon_r} \right] \qquad (42)$$

and for case II ($I\varepsilon_c$ = constant),

$$P = K_1' \left[K_2' - K_3' - \frac{1}{\varepsilon_c \varepsilon_v \varepsilon_r} \right] \tag{42a}$$

Using the Lorenz relation, Eq. 40, and Eq. 41, and setting $\partial P/\partial \ell = 0$ gives for case I,

$$\ell opt = \left[\frac{(0.8)(0.12)(0.0374)(s-c_c)M\varepsilon_r}{z\ C_e \rho i} \right]^{\frac{1}{1.8}} \tag{43}$$

and the implicit relation for case II,

$$[\ell opt]^{1.8}\ (1.8)(0.12)\ell opt - \frac{(0.8)(0.12)V}{\rho i} \tag{44}$$

It can be seen that the case case II optimum occurs at the condition of maximum energy efficiency, that is, $\partial \varepsilon_e/\partial \ell = 0$. In general, when all of the factors that may affect the economic optimum are taken into consideration, the optimum anode-cathode spacing will be somewhat different from the point of maximum energy efficiency. Optimizations made on purely technical grounds may be, but are usually not, the same as an economic optimum.

The optimum frequency of anode adjustment is an important consideration in mercury cells with graphite anodes and in aluminum reduction cells. An analysis of the optimum frequency is straightforward and includes a balance between the cost of extra operating labor to adjust anodes and savings in electrical energy through lower cell voltage. Equation 27 may be written

$$P = K_1 \left[K_2 - K_3 c_a - \frac{\bar{V}}{V_o \varepsilon_c \varepsilon_r} \right] \tag{45}$$

in which

$$c_a = \frac{365 N N_a C_d}{N_d\ \tau_p\ P} \tag{46}$$

and

$$\bar{V} = V_2 = \frac{1}{2} \tau_p \frac{dV}{d\tau} \tag{47}$$

assuming that anode-cathode spacing changes linearly with time. Following the usual optimization procedure, the optimum adjustment period is

$$\tau p(\mathrm{opt}) = \left[\frac{83.3\ N_a \varepsilon_r C_d}{If(dV/d\tau)N_d C_e}\right]^{1/2} \tag{48}$$

The following order-of-magnitude parameter values may be used for a mercury chlor-alkali cell:

$\varepsilon_r = 0.97$

$N_a = 10^2$ anodes/cell (63)

$C_d = 10^2$ \$/day fully burdened

$I = 10^5$A (63)

$f = 1$

$(\frac{dV}{d\tau}) = 10^{-2}$ V/day (26)

$N_d = 10^3$/day, "guestimated" number of anode adjustments per man

$C_e = 10^{-2}$ \$/kwhr

The optimum anode adjustment period is 9.1 days for these parameters values compared to about one week (63) used in the industry.

A similar calculation can be made for aluminum reduction cells. For the overall reaction 4, assuming 100% current efficiency and equal anode and cathode areas, there is obtained,

$$\frac{d\ell}{d\tau} = \frac{(3600)(24)i}{zF}\left[\frac{M_c}{\rho_c} - \frac{M_a}{\rho_a}\right] \tag{49}$$

The metal pad thickens as the carbon anode recedes. The rate of change of cell voltage is then

$$\frac{dV}{d\tau} = \frac{i}{\kappa}\left(\frac{d\ell}{d\tau}\right) = \frac{(3600)(24)i^2}{\kappa zF}\left[\frac{M_c}{\rho_c} - \frac{M_a}{\rho_a}\right] \tag{50}$$

Using $i = 1 A/cm^2$ and $\kappa = 2.4\ \Omega\text{-}1\ cm\text{-}1$ (65), $z = 6$ (reaction 4), $M_C = (3/2)\ (12)$, $\rho c = 1.9$, $M_a = (2)\ (27)$, and $\rho a = 2.7$, gives $dV/d\tau \cong -0.65$ V/day. Other appropriate parameters in Eq. 42B are

N_a = 1 anode/cell (Soderberg) or ganged anodes in a prebake cell

$\varepsilon_r = 0.97$

$C_d = 10^2$ \$/day fully burdened labor cost

$I = 10^5$ A (Table 9)

$f = 1$

$N_d = 10^2\ day^{-1}$ (guestimate)

$C_e = 0.5 \times 10^{-2}$ \$/kwhr (aluminum plants in general move closer to low-cost electrical energy than chlor-alkali plants)

The optimum anode adjustment period is 0.5 days for these approximate parameter values. Common practice in the industry is to tap aluminum from the cells once a day and make anode adjustments at the same time.

In the final analysis, the question of unipolar or bipolar electrode arrangements is decided on economic grounds. Bipolar electrodes have in general a lower resistance because current flows through the electrode thickness rather than along the length, and because interelectrode connections with ohmic loss are eliminated. This is counterbalanced by parasitic currents between cells, mentioned earlier, which in effect decrease current efficiency. The parasitic currents can be minimized in filter-press cells by using small-cross-section, maximum-feasible-length, electrolyte passages between cells. This introduces another optimization between pressure drop and increased pumping costs versus saving in current efficiency, a problem arising too in design of fuel cells. The problem of parasitic currents also occurs between individual series cells in a cell line. A vertical, gravity-flow device that breaks the electrolyte into droplets is often used to break the electrolytic current between cells.

Circulation or agitation of electrolyte may be used to increase the limiting current density in mass-transport-limited processes (22,58). An increase in production or decrease in size of cells is balanced by increased capital cost for pumps and piping and operating and maintenance costs of the pumps. A number of studies have recently been directed toward increasing mass transport rates in practice (66-68).

As labor and energy rates have increased, there has been a trend toward the mechanization of cell operation to operate closer to optimum conditions. Automatic adjustment of mercury cell anodes (69) and continuous ore feeding of aluminum cells are examples. Capital and operating costs of the mechanical devices are balanced against savings in labor and energy.

Certain parameters must be measured for control of the process, for example, temperatures, pressures, concentrations, pH values, and flow rates. In an electrochemical process, current, voltage, anode-cathode spacing, and current efficiencies are very important. Selection of instruments and frequency of measurements are determined on economic optimizations.

Hundreds of other optimizations are made in every plant and cell design. Optimum concentrations, temperatures, and pressures must be determined. The advantages of new technical developments such as metal anodes and ion-exchange membranes must be balanced against their increased first cost. A recent trend in the chlor-alkali industry is to locate mercury cells out of doors to save building costs (70). A great deal of art remains in electrolytic cell and plant design. The process of making optimizations itself requires an investment of time and money and thus some judgment of what optimizations are relevant.

NOMENCLATURE

A — electrode area, cm^2
ca — unit anode adjustment cost, \$/kg
c_c — lumped constant unit cost, \$/kg
c_e — unit electrical energy cost, \$/kg
$d_{d\ell}$ — unit direct labor cost, \$/kg
c_f — unit fixed cost, \$/kg
c_m — unit raw materials cost, \$/kg
c_{mr} — unit maintenance and repairs cost, \$/kg
c_{oh} — unit overhead costs, \$/kg
c_u — unit utilities cost, \$/kg
C_d — labor cost, fully burdened, \$/day
C_e — energy cost, \$/kwhr
C_c — cost of a cell, \$
f — fraction of cells or fraction of time in operation, dimensionless (product yield may also be included in f)
F — Faraday, 96,500 C/equiv.
i — current density, A/cm^2
i_{opt} — optimum current density, A/cm^2
I — current, A
I_c — capital investment, \$
$\bar{I}_{opt}$ — optimum current, A
K_1, K_2, K_3, K_4 — constants, dimensional or dimensionless
ℓ — interelectrode spacing, cm
M — molecular weight, g/mole
n — constant exponent, dimensionless
N — number of cells in plant, dimensionless
N_a — number of anodes per cell, dimensionless
N_d — number of anode adjustments per man day
P — production, kg/yr
P — profit, \$/yr
r — cell effective area resistivity, ohm cm^2
R — cell effective resistance, ohm
ROI — return on investment, yr-1
s — unit sales price, \$/kg
S — 3600X24X365 seconds per year
V — cell potential, V
V_o — reversible potential, V
V_1 — zero-current intercept potential, V
V_2 — base cell potential at load, V
$\bar{V}$ — average cell potential, V
z — equivalents per mole

α	lower accepted limit on incremental return on investment, e.g., 0.25 for new plant
β	franction of capital investment paid per year for depreciation, property taxes, insurance, rent, and interest
ε_c	current efficiency
$\varepsilon_e=\varepsilon_c\varepsilon_v$	cell energy efficiency
ε_r	rectifier conversion efficiency
ε_v	cell voltage efficiency
κ	conductivity, ohm-1 cm-1
ρ	resistivity, ohm cm
ρ_a	density of aluminum
ρ_c	density of carbon anode
τ	time, days
τ_p	period between anode adjustments, days
τ_p(opt)	optimum adjustment period, days

REFERENCES

1. E. Yeager and A. J. Salkind, Techniques of Electrochemistry, Vol. I, Wiley-Interscience, New York, 1972; Vol. II, 1973.
2. C. L. Mantell, Electrochemical Engineering, 4th ed., McGraw-Hill, New York, 1960.
3. C. A. Hampel, Ed., The Encyclopedia of Electrochemistry, Reinhold, New York, 1964.
4. J. S. Sconce, Ed., Chlorine, Its Manufacture, Properties and Uses, Reinhold, New York, 1962.
5. A. T. Kuhn, Industrial Electrochemical Processes, Elsevier, New York, 1971.
6. R. B. MacMullin, Electrochem. Tech. 1, 5 (1963).
7. R. B. MacMullin, Electrochem. Tech. 2, 106 (1964).
8. R. H. Perry and C. H. Chilton, Chemical Engineers Handbook, 5th ed., McGraw-Hill, New York, 1973.
9. F. C. Vilbrandt and C. E. Dryden, Chemical Engineering Plant Design, 4th ed., McGraw-Hill, New York, 1959.
10. E. E. Ludwig, Applied Process Design for Chemical and Petrochemical Plants, Gulf Publishing, Houston, 1964.

11. R. Landau, Ed., The Chemical Plant, Reinhold, New York, 1966.
12. D. G. Jordan, Chemical Process Development, Interscience, New York, 1968.
13. J. R. Backhurst and J. H. Harker, Process Plant Design, American Elsevier, New York, 1973.
14. J. S. Newman, Electrochemical Systems, Prentice-Hall, Englewood Cliffs, N.J., 1973.
15. V. G. Levich, Physicochemical Hydrodynamics (Translation), Prentice-Hall, Englewood Cliffs, N. J., 1962.
16. P. Delahay and C. W. Tobias, Eds., Advances in Electrochemistry and Electrochemical Engineering, Vol. 1, Interscience, New York, 1961, et. seq.
17. E. C. Potter, Electrochemistry, MacMillan, New York, 1956.
18. G. Milazzo, Electrochemistry, American Elsevier, New York, 1963.
19. P. Gallone, Principi Des Processi Elettrochimici, Tamburini Editore, Milan, 1970.
20. J. O'M. Bockris and A. K. N. Reddy, Modern Electrochemistry, Plenum, New York, 1970.
21. D. O. Hubbard, Chem En. Prog. 46, 435 (1950).
22. N. Ibl and E. Adam, Chem. Ing. Tech. 37, 573 (1965).
23. H. Roscher, Chem Tech. 18, 298 (1966).
24. G. Seelmann, F. Glos, and D. Wagner, The Chemical Engineer, CE 46 (1968).
25. W. W. Smith, The Chemical Engineer, CE 52 (1968).
26. F. Hine, Electrochem. Tech. 6, 69 (1968); J. Electrochem. Soc. 117, 139 (1970).
27. N. Ibl, Chem. Ing. Tech. 41, 208 (1969).
28. A. Schmidt, J. Electrochem. Soc. 118, 2046 (1971).
29. N. Ibl and P. M. Robertson, Electrochim. Acta 18, 897 (1973).
30. M. Eisenberg, Advances in Electrochemistry and Electrochemical Engineering, Vol. 2, Interscience, New York, 1962.
31. J. Van Winkle and W. N. Carson, Jr., Electrochem. Tech. 1, 18 (1963).

32. C. G. Von Fredersdorff, in Fuel Cells II, G. J. Young, Ed., Reinhold, New York, 1963.
33. C. Tyler, Chemical Engineering Economics, McGraw-Hill, New York, 1948.
34. R. S. Aries and R. S. Newton, Chemical Engineering Cost Estimation, McGraw-Hill, New York, 1955.
35. J. O. Osburn and Kammermeyer, Money and the Chemical Engineer, Prentice-Hall, Englewood Cliffs, N. J., 1958.
36. H. E. Schweyer, Process Engineering Economics, McGraw-Hill, New York, 1955.
37. C. H. Chilton, Ed., Cost Engineering in the Process Industries, McGraw-Hill, New York, 1960.
38. M. S. Peters and K. D. Timmerhaus, Plant Design and Economics for Chemical Engineers, 2nd ed., McGraw-Hill, New York, 1968.
39. H. Popper, Ed., Modern Cost-Engineering Techniques, McGraw-Hill, New York, 1970.
40. J. Hengstenberg, B. Sturm, and O. Winkler, Eds., Messen and Regeln in der Chemischen Technik, Springer-Verlag, Berlin, 1957.
41. G. C. Carrol, Industrial Process Measuring Instruments, McGraw-Hill, New York, 1962.
42. R. B. MacMullin, J. Electrochem. Soc. 120, 135C (1973).
43. J. C. Schumacher, Perchlorates, Reinhold, New York, 1960.
44. R. B. MacMullin, in Chlorine, Its Manufacture, Properties and Use, J. S. Sconce, Ed., Reinhold, New York, 1962, p. 127.
45. H. A. Sommers, Electrochem. Tech., 5, 108 (1967).
46. R. A. Lewis, in The Encyclopedia of Electrochemistry, C. A. Hampel, Ed., Reinhold, New York, 1964, p. 32.
47. E. Yeager, in Fuel Cells, W. Mitchell, Ed., Academic Press, New York, 1963, p. 299.
48. A. J. LeDuc, J. G. Kourilo, and C. Lurie, J. Electrochem. Soc., 116, 546 (1969).
49. Chem. Eng. News, p. 11 (Feb. 26, 1973).
50. Wall Street Journal, January 12, 1973.
51. L. L. Bott, Hydrocarbon Proc. 44, No. 1, 115 (1965).

52. J. H. Prescott, Chem. Eng., 238 (Nov. 8, 1965).
53. D. G. Braithwaite (to Nalco Chemical), U. S. Pat. 3,391,067 (1968).
54. D. G. Braithwaite, J. S. d'Amico, P. L. Grass, and W. Hanzel (to Nalco Chemical), W. Ger. Pat. 1,221,223 (1966).
55. L. L. Bott (to Nalco Chemical), U. S. Pat. 3,479, 274 (1969).
56. M. S. Kircher, in Chlorine, Its Manufacture, Properties and Use, J. S. Sconce, Ed., Reinhold, New York, 1962, p. 127.
57. M. M. Baizer, J. Electrochem. Soc. 111, 215 (1964).
58. D. Danly, in Organic Electrochemistry, M. M. Baizer, Ed., Marcel Dekker, New York, 1973, p. 907.
59. A. T. Emery and J. Parker, Report of the Electrolytic Industries for the Year 1972, J. Electrochem. Soc. 120, 321C (1973).
60. G. Gerard and W. King, in Encylopedia of the Chemical Elements, C. A. Hampel, Ed., Reinhold, New York, 1968.
61. J. D. Edwards, F. C. Frary, and J. Jeffries, The Aluminum Industry, McGraw-Hill, New York, 1930.
62. C. L. Mantell, Industrial Electrochemistry, 2nd ed., McGraw-Hill, New York, 1940.
63. H. A. Sommers, Chem. Eng. Prog. 61, 94 (1965).
64. D. Bratland, K. Grjotheim, C. Krohn, and K. Motzfeldt, J. Metals 13 (October 1967).
65. T. G. Pearson, The Chemical Background of the Aluminum Industry, Royal Institute of Chemistry, Lectures Monographs and Reports, No. 3, London, 1955.
66. A. R. Despic, Abs. 245, Extended Abstracts 73-1, The Electrochemical Society, May 1973.
67. P. LeGoff et. al, Abs. 246, Extended Abstracts 73-1, The Electrochemical Society, May 1973.
68. V. A. Ettel et. al, Abs. 249, Extended Abstracts 73-1, The Electrochemical Society, May 1973.
69. G. J. Lewis, Chem. Engr. CE 38, No. 216, (March 1968).
70. P. J. Kleinholtz, Chem. Eng. Prog. 70, 59 (1974).

II. ELECTROCHEMICAL MACHINING

James P. Hoare

and

Mitchell A. LaBoda

Electrochemistry Department
Research Laboratories
General Motors Corporation
Warren, Michigan

I. INTRODUCTION

As pointed out by DeBarr and Oliver (1), power-driven machine tools for the most part have been developed in relatively recent times (over the past 180 years). In general, the guiding principle behind conventional machining techniques is the passing of a sharpened body of harder material (the tool) through a body of softer material (the workpiece) to remove chips or shavings from the workpiece so that it may be formed according to a specified pattern. With continued progress in the development of structural materials, it became necessary to cut materials of ever-increasing hardness. Consequently, the tool materials advanced in hardness from tempered tool steel through alloy steels containing tungsten and mangancese and the "high-speed steels" containing tungsten, chromium, and vanadium (probably in the form of carbides), to cobalt-cemented tungsten carbide and mixed carbide (tungsten, titanium, niobium, and tantalum) materials. Because of the brittle nature of the ceramic tool materials, the machining of most very hard materials must be carried out by grinding operations, which severely limits the degree of intricacy of the part to be machined. With the advent of the advanced technological industries, such as those concerned with the aerospace and nuclear energy fields, the demands for new and more efficient machining techniques have become critical.

Besides removing metal mechanically, which produces only physical changes in the materials concerned during conventional machining, there is another way of removing metal that involves chemical changes in the metal removed. This process is corrosion. If the corrosion process could be controlled, one should be able to develop a machining technique with which rapid removal of metal could be obtained independently of the hardness of the material to be worked with the production of excellent surface finished. Such a process has been developed and is known as electrochemical machining (ECM) (1a).

A. Mechanisms of Corrosion

In a simple metal dissolution process involving a metal in contact with a liquid phase, an atom leaves the metal lattice by giving up electrons to the metal surface and enters the liquid phase as an ion. This oxidation reaction (loss of electrons) would continue in the absence of other reactions until the electrostatic attraction of the excess electrons on the metal surface for the positive ions in solution balanced the driving force of the metal dissolution process. At this point the metal dissolution process would come to a halt. If, however, a sink for these electrons exists in the form of a reduction reaction (gain of electrons) taking place at sites on the metal surface different from the oxidation sites, the dissolution or corrosion of the metal could continue unitl the metal phase disappeared.

1. Local Cells

According to this viewpoint, a metal corrodes in a liquid medium by an oxidation reaction occurring at certain sites (usually high-energy sites such as peaks or corners and edges of a dislocation) and a reduction reaction at other sites (usually protected sites such as those in a valley between peaks). Since the metal surface is an equipotential surface, the oxidation (anodic) and the reduction (cathodic) processes must take place at the same potential. In most cases, the standard or the open-circuit potential of the anodic and cathodic reactions is not identical; and as a result, one or both of these reactions will be polarized, as shown in Fig. 1, to a common potential known as the mixed potential (2) lying somewhere between the two open-circuit values.

The potential of the reduction reaction on open circuit has a value of E_A, and as cathodic current is drawn, the potential falls to less noble values along curve A. Similarly, the potential of the oxidation reaction an open circuit has a value of E_B, and as current is drawn, the potential shifts to more noble potentials along curve B. Where these curves cross ithe point that determines the conditions under which the corrosion or local cell is established. The common potential is the mixed potential, E_m; at this point, the local anodic current is equal to the local cathodic current, which is called the local cell or corrosion current. A more detailed account of such mechanisms is given elsewhere (e.g., 3). In most cases, the anodic reaction is the dissolution of the metal, and the cathodic reaction is the reduction of oxygen or the

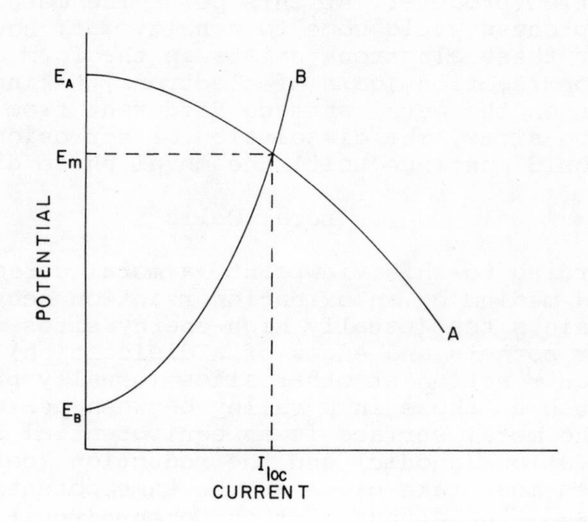

Fig. 1. A sketch of local cell polarization curves.

evolution of hydrogen. Current passes in the local cell from anodic to cathodic sites in the metal surface by electronic conduction, and the circuit is completed by the electrolytic conduction of ions through the liquid phase (electrolyte) from the cathodic to the anodic sites.

2. Selective Attack

Under certain conditions (4-8), the anodic reaction may be confined to a small number of sites, causing localized corrosion or pitting. This is a highly destructive condition to be avoided since a metal sheet may be perforated in a relatively short period of time.

Metallurgists use corrosive reagents for selective attack of crystal faces so that the grains appear in different degrees of lightness or darkness under the microscope. If the selective attack is at grain boundaries, then the boundaries can be made to show up as dark lines.

3. General Attack

With a proper choice of electrolyte, a metal sample may be made to corrode in a uniform manner over the entire surface. This process is known as general corrosion. In this case, the identification of the active sites is difficult. As explained by Evans (9), the corrosion process begins at susceptible points, such as lattice defects, corners, or edges, and when this active material is removed, the attack shifts to some other more active site. Consequently, the anodic attack wanders over the entire exposed surface, producing general corrosion.

This behavior is exploited in the process known as chemical milling, chemical solution machining, or pattern etch machining (10-16) in which the desired curring pattern is applied to the metal surface in the form of a chemically resistant film. When the workpiece is exposed to the corrosive medium, only the unprotected regions not covered by the film will be corroded. A more precise method involves photographically reproducing the master drawing on film (12). After the metal surface is covered with a light-sensitive, chemically resistant material (photo resist), the photographed mask is laid out on the metal surface and exposed to light. The unexposed areas of the photo resist are removed chemically, and the workpiece is then exposed to the corrosive reagent. Since the rate of metal removal depends only upon the kinetics of the local cell corrosion process, the depth of the cut is determined for a given temperature and concentration of the chemical etching reagent (such as $FcCl_3$ for metals and HF for galss) by the length of time the workpiece is in contact with the etching agent. Recent applications of this process in the field of microelectronics have been very successful in making wiring for printed circuits (15) and in the fabrication of integrated circuits(13).

4. Bimetallic Corrosion

The rate of a local cell corrosion process is relatively slow because it is controlled by the kinetics of either the anodic or the cathodic reaction or both. By placing another metal on which the cathodic reaction can take place more easily in contact with the corroding metal, the rate of corrosion can be increased still further. This is true if the local cell is under cathodic control (3), which is the more common case (4, p. 876). Such processes are known as bimetallic corrosion (17, 18) and are important in design and construction considerations (e.g., 19-22).

5. Anodic Corrosion

A further increase in the corrosion rate can be obtained if, instead of contacting the working metal with a second metal, an external potential is applied to the working metal such that it is made anodic with respect to a physically separate counterelectrode. In this way, only an anodic process takes place on the working metal, and the rate of corrosion may be increased or decreased by raising or lowering the applied potential. One of the principal roles played by the electrolyte, then, is to maintain the anodic and the cathodic processes physically separated. Anodic corrosion is the term applied to such processes.

In many solutions, unlimited corrosion rates with a continual increase in potential are not possible because most metals become passive eventually. If the potential that is controlled by a potentiostat is increased stepwise with current recorded at each step, the anodic polarization curve obtained is similar to the idealized curve shown in Fig. 2.

Below a certain threshold potential, A in Fig. 2, the metal is inert in the given electrolyte, and the corrosion rate is zero. Above the point A, the corrosion current increases with increasing potential; if no passive films form, the rate increases along the curve toward the point C. In the usual case, however, a passive film begins to form on the metal surface at point B and blocks the active sites with a consequent falloff in the corrosion rate. The surface becomes covered with the protective film at D where the current has fallen to a very low value and the metal is said to be passive. With increasing potential, a point is reached (E in Fig. 2) where

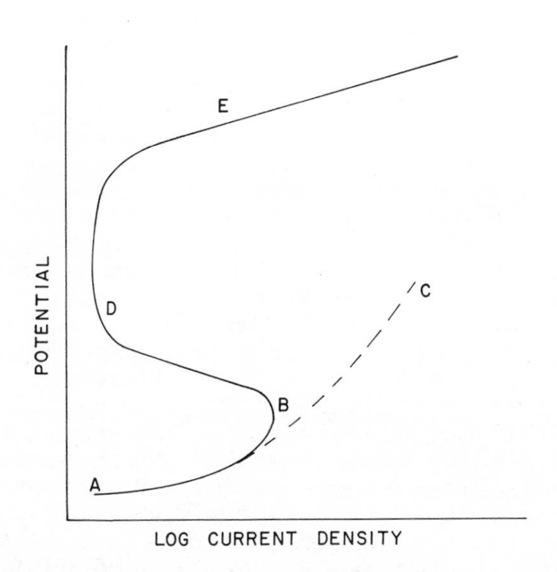

Fig. 2. A sketch of anodic polarization curve obtained potentiostatically on a given anode in a given solution.

either the film breaks down, permitting metal dissolution, or a new anodic process such as the evolution of oxygen takes place, either of which causes the current to increase again. Film breakdown can occur by the interaction of adsorbed anions from solution with the adsorbed film (23). The region of renewed increase in current is called the transpassive region. The second principal role played by the electrolyte is the determination of conditions affecting the buildup and breakdown of passivating films on metal anode surfaces for a given potential.

The new methods of metal removal, such as electrolytic grinding and electrochemical machining, make use of the anodic corrosion process. It is now possible to discuss the development of such processes.

B. Electropolishing

As early as 1929, Jacquet (24-26) disclosed that, if a metal (such as Cu, Sn, Pb, or Al) were anodized in a suitable electrolyte at a high enough current density, the metal surface would become smooth with a high degree of brilliance. He called this process electrolytic polishing; it is now known by the simpler term electropolishing. Since this process is an anodic process, the current is an important variable, but of equal importance is the composition of the electrolyte and to some lesser extent the temperature. There are some polishing baths that operate over a wide range of temperatures and current densities above a certain threshold current density. Ameskal (27) has given a rather extensive listing of electropolishing baths suitable for a number of metals and alloys, including the stainless steels. A complete history of the development of the electropolishing techniques from a laboratory curiosity to an industrial process has been traced by Jacquet (24).

1. Leveling Mechanism

Although most of the investigations of this process are concerned with electropolishing of Cu in H_3PO_4 solutions (28-30), studies of the electropolishing of Ni in H_2SO_4 (31) and HCl (32) and of stainless steels (33,34) give corroborating results. To account for the electropolishing action, any theory must take into consideration a macroscopic effect (the wearing down of peaks, the leveling process) and a microscopic effect (the prevention of etch patterns, the smoothing effect).

When copper was first placed in a water solution of H_3PO_4 at a suitable current density, the surface was etched; and after a while, a thin salt layer was formed (25). Under this layer, a brightening of the surface took place. Jacquet (25) recorded a potential-current curve for Cu in H_3PO_4 solutions. During the etching period, the current rose rapidly over a short potential range to a maximum value. With further increases in potential, the current fell to a limiting value. It is in this limiting current region that electropolishing takes place.

From these observations, Jacquet concluded that the metal surface is covered with a layer of salt which is thicker than the surface roughness as diagrammed in Fig. 3. Since the distance from a peak to the boundary of the film (A in Fig. 3) is shorter than from a valley (B in Fig. 3), the iR drop at A would be less than at B. Consequently, the rate of dissolution would be greater at the peaks, producing the leveling action.

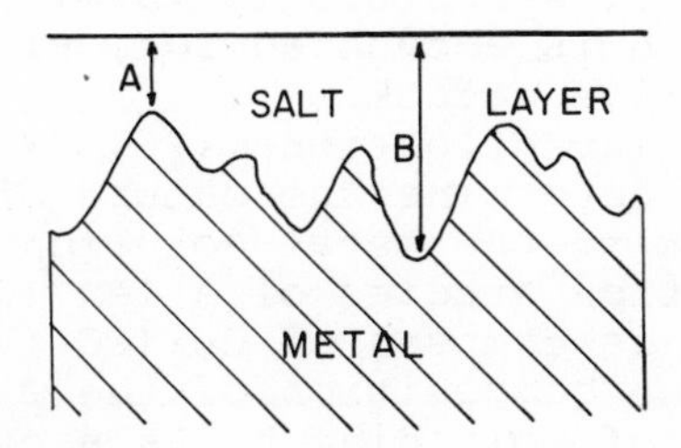

Fig. 3. A sketch of the salt film present on a metal surface that is undergoing electropolishing in a given reagent.

Elmore (35) considered the film resistance model of Jacquet too simple and believed that diffusion plays an important role. According to this viewpoint (35), etching of the metal surface continues until the salt layer is built up; then the anodic corrosion process is limited by the diffusion of Cu^{++} ions through the layer to the bulk of solution. Since the diffusion length is shorter at A in Fig. 3 than at B, the rate of reaction will be greater at the peaks, and a leveling action will take place. Elmore performed a mathematical analysis of this diffusion model.

2. Smoothing Mechanism

These theories accounted for the leveling action but not the smoothing action of the electropolishing process. Edwards (36) concluded from viscosity studies that the polishing action did not depend on reaching a critical concentration of Cu^{++} ions at the anode-solution interface. In fact, no change was observed when Cu^{++} ions were added to the solution. The process however, was found to be diffusion controlled, but the limiting step is the diffusion of PO_4^{-3} ions through the diffusion layer since it was maintained that Cu^{++} ions could not go into solution unless they were complexed with the phosphate ion. With this model, a leveling action is explained as before by the shorter diffusion path at A in Fig. 3. Because a phosphate ion diffusing through the diffusion layer complexes with the first Cu^{++} ion it meets, no effect on crystal structure is involved; thus a smoothing sction is accounted for, since etching requires selective attack.

A number of investigations (26, 37-40) were carried out to determine the nature of the salt diffusion layer at the metal-solution interface. Cathodic reduction studies of Allen (37) suggest that the layer is compsed of $Cu_3(PO_4)_2$, but determinations of the composition of the electrolyte at various stages of polishing by Balasev and Nikitin (38) suggest the presence of both $CuH(PO_4)_2$ and $Cu_3(PO_4)_2$. From electron diffraction experiments, Williams and Barrett (40) found that the composition of the salt film is $Cu_3(PO_4)_2$. However, $CuHPO_4$ is formed in the first 200 sec of polishing but is then converted to $Cu_3(PO_4)_2$.

As noted by Hoar (31, 41) and later by Higgins (32) from an alaysis of the potentiostatic polarization curves for the electropolishing of Ni in acid solutions, current did not flow below a certain potential value. Above this point, a region existed in which the current rose rapidly for small changes in the potential (see Fig. 4). After a peak value was reached, the current fell to a limiting value where electropolishing took place with increasing potential. At high potential, either O_2 or Cl_2 (depending on the acid anion) was liberated. With decreasing potential, the polarization curve was retraced except for the disappearance of the peak as noted in Fig. 4.

This initial rapid rise in the current to a peak value suggested to Hoar that a compact oxide film is formed on the metal surface and reaches a steady-state thickness because it dissolves as fast as it is formed. The polishing process is diffusion controlled by the migration of Ni^{++} ions through vacancies in

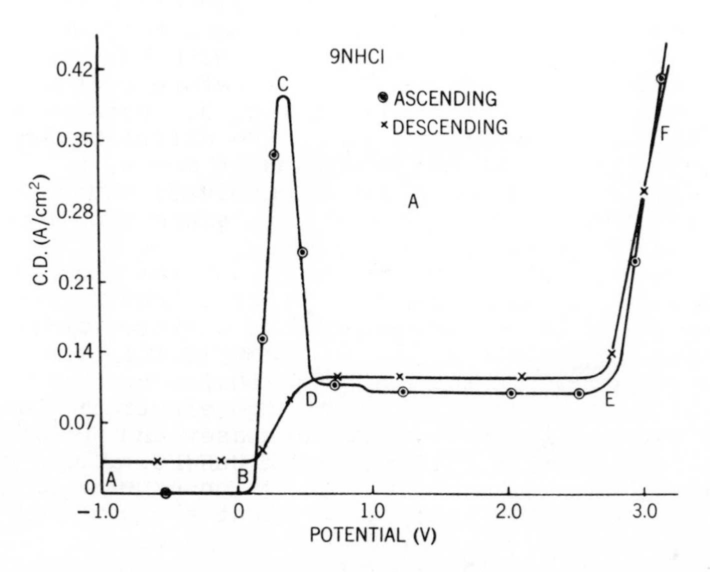

Fig. 4. Potentiostatic polarization curves obtained on Ni in 9N HCl (32). Electropolishing occurs in region DE. (By permission of The Electrochemical Society, Inc.)

the oxide film. A leveling effect is produced because the diffusion layer is thicker in the valleys than on the peaks. A smoothing effect is obtained because the position of the vacancies in the oxide film determines where the ion leave the surface, and since there is a random distribution of vacancies, there is a random dissolution of the metal. Higgins found that the limiting current depended on the rate of stirring the solution as expected for a diffusion-controlled process. He estimated a value between 0.002 and 0.003 cm for the thickness of the diffusion layer. The compact oxide film is probably of the order of about 1000 Å thick (42). Higgins agrees that an oxide film may be present beneath the salt diffusion layer, and he found similar results for the electropolishing of Cu in KCN.

To show that an oxide film is present on Cu when electropolished in H_3PO_4, Hoar and Farthing (30) arranged their electrodes so that small drops of Hg could be dropped from a microburette onto the Cu anode at any point during the electropolishing process. The Hg drops wetted the Cu surface in the region before the limiting current where etching of the surface takes place. When the potential was raised to the limiting current region, the Hg drops rolled into balls and did not wet the surface. Assuming that Hg wets only an oxide film-free surface, they concluded (30) that, in the limiting current region where polishing occurred, the metal surface was covered by an oxide film. It seems that the electropolishing model composed of a compact oxide film beneath a salt diffusion layer gives a satisfactory account of the observed experimental results. In recent studies of the electropolishing of Cu (42a) using ellipsometric techniques (42b), the existence of a thin oxide film (40 Å thick) on the metal surface beneath a thick layer of a salt film (between 2000 and 3000 Å has been confirmed).

C. Jet Etching

It was desirable to localize the anodic corrosion process in order to drill small holes or draw fine grooves in a metal sample, so Uhlir (43) directed the current flowing between the cathode and anode down a tube made of a nonconducting material, such as glass or plastic, as shown in the sketch of Fig. 5. In this system, the metal piece to be drilled was connected to the anodic side of a d. c. electric power source and immersed in a container of electrolyte. The end of the tube was drawn down to a fine tip, and a Pt wire that served as the counter-electrode (cathode) was sealed in the upper end of the tube. After the tube was filled with electrolyte, it was positioned over the workpiece to be drilled. Most of the current flows from the tube tip to the region of the metal surface directly beneath the tip, because other paths through the solution are longer, higher-resistance paths.

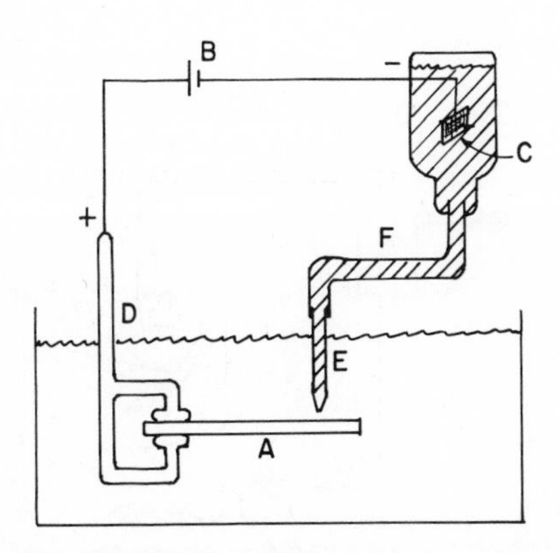

Fig. 5. Sketch of the essential components of the virtual cathode. A, the workpiece; B, d.c. power source (battery); C, negative electrode; D, clamp; E, virtual cathode; F, flexible solution-filled pipe.

Since the tube filled with electrolyte gave the same overall results as a metal cathode, Uhlir called the cathode arrangement a virtual cathode; he referred to the process as electrolytic etching. Localization of the current is limited by the fact that as the tip diameter becomes smaller, the ohmic resistance increases, the current density falls, and the drilling rate becomes slower. The rate of etching increased with increasing current density until a point was reached where the corrosion rate dropped suddenly to a small value. Uhlir attributed this behavior to boiling of the electrolyte. In these studies, Ge and Mo were drilled with 10% KOH; Fe and Cu, with 5% H_2SO_4; tungsten carbide, with 10% NaCN; and Ag, with NH_4NO_3. According to Uhlir, this process may be used to remove broken drills or taps from a workpiece in a manner similar to the anodic corrosion methods described by Bleiweis and Fusco (44).

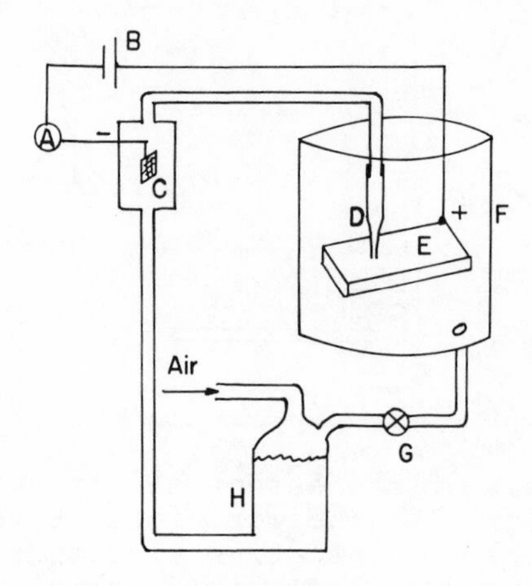

Fig. 6. Sketch of the essential components of the jet etching process. A, the ammeter; B, d.c. power source; C, negative electrode; D, jet; E, workpiece; F, spash container; G, stopcock; H, reservoir.

Tilley and Williams (45) described a process, as shown in Fig. 6, in which the electrolyte was forced through the nonconducting tube of Uhlir's virtual cathode by compressed gas and the resulting jet of solution was directed against the workpiece connected as the anode. A Pt wire welded in the wall of the tube served as the counterelectrode (cathode). Etching takes place directly under the jet because the ohmic resistance of the solution limits any other current path. The great advantage of the flowing system is the elimination of uneven etching due to the presence of gas bubbles and the accumulation of reaction products in the reaction zone which is characteristic of the static methods. Currents of the order of 10/Ain.2 at potential between 2 and 20 V have been reported (46) for the jet etching of semiconductors, and the value of the current depends on the flow rate of the electrolyte and on the potential.

Jet-etching of p-Ge can be carried out in acid or basic solutions of $NaClO_4$ and of p-Si in solutions of fluorides and HF (46). However, the anodic corrosion of n-Ge or n-Si that contains a low concentration of holes requires strong illumination of the semiconductor to produce electron-hole pairs so that positive ions can enter solution (47).

Another strong point of the jet-etching process over conventional metal removal processes is the ability to remove layers of metal without destroying or modifying the structure of the underlying layers. Kelly and Nutting (48) reported the preparation of thin foils of steel as transmission samples for electron diffraction studies. Vertical and horizontal driving cams scanned the jet of solution over the surface area to be machined. A solution of three volumes of HCl to one of $CuCl_2$ solution was used as the corrosive electrolyte. The final thinning was carried out by electropolishing in a 10-to-1 mixture of acetic and perchloric acids (from 50 to 100 μ after jet-etching to 20 μ thick after electropolishing).

II. ELECTROLYTIC GRINDING

If the anode and cathode are held stationary as in electropolishing, several undesirable effects take place. First, the distance between the electrodes widens during the anodic corrosion process until the iR drop across the gap limits the current flow to the point where metal removal stops. When the electrodes are placed close together again, however, metal removal resumes. In the second place, accumulation of corrosion products in the gap space can bring the metal removal process to a halt. Movement of the electrolyte can remedy this situation by sweeping the corrosion products out of the gap, but the electrolyte flow must be great enough to prevent vaporization of the electrolyte with the passage of large current densities. Finally, the formation of compact oxide films on the anode surface can effectively block the corrosion process.

A. The Grinding Wheel

The first attempts to develop an electrolytic machining operation that would embody the requirements mentioned above were mechanical. In the first published account of such a machining process by Keeleric (49, 50) called electrolytic grinding or electrolytically assisted grinding (51), an electrically conducting wheel mounted on a drive shaft with nonconducting bushings was used as the cathode in the machining of an anodic workpiece. A sketch of the apparatus is presented in Fig. 7.

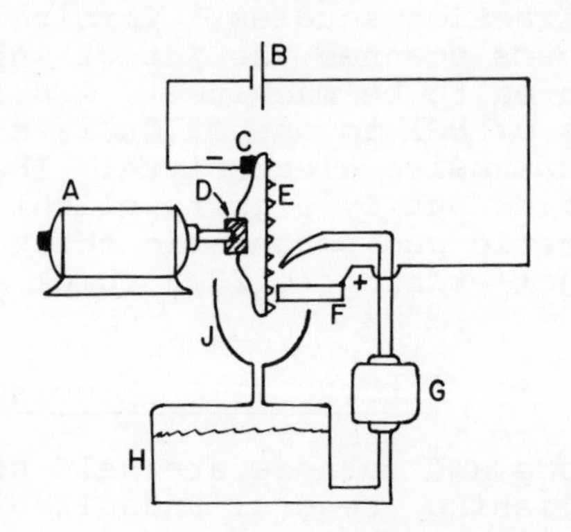

Fig. 7. Sketch of essential components of an electrolytic grinding operation. A, the motor; B, d.c. power source; C, brush; D, insulated bushing; E, diamond-studded, electrically conducting grinding wheel; F, workpiece; G, pump; H, reservoir; J, splash basin.

1. Structure of the Wheel

To maintain a constant gap between the cathode and the anode, diamond bort or powder is embedded in the edge or face of the conducting wheel made of phosphor bronze (52) or copper (53). Stroup (53) describes how diamonds are mechanically pressed in the surface of the copper wheel and bonded tightly to the surface by plating a thin film of nickel on the diamond-studded copper surface. The diamonds embedded in the phosphor bronze (72 carats/in.^{3}) were esposed by anodic etching of the wheel according to Cole (52). Chemical etching may also be used to expose the diamonds. The diamonds should project from 0.0005 to 0.0025 in. (49, 51, 52, 54) from the face of the wheel. Grit materials other than diamonds have been investigated. Alumina and silicon carbide have been used (51, 52, 55) for the machining of steels with performance equivalent to that of diamonds. The best performance is still obtained with diamonds for the machining of cemented carbides.

2. Electrolyte Flow

Current is fed to the wheel from the rectifier through slip rings or brushes. To get productive cutting of the workpiece, high current densities are needed, from 500 to 1600 A/in.^{2} (49, 51, 53), this demands a narrow gap distance between the cathode and the anode. The electrolyte is pumped into the gap, and the sheel is rotated at the ghigh speeds of mechanical grinding wheel operation (5000 to 6000 ft/min) to carry sufficiently large amounts of electrolyte through the gap. For a given applied voltage, the current density does not continually increase with increased speed of wheel rotation because a point is reached where the centrifual force prevents additional amounts of electrolyte to be transported by the wheel to the cathode-anode gap (58). A large electrolyte flow rate is required to prevent sparking and to carry away corrosion products. This is the third important role played by the electrolyte. Sparking cannot be tolerated and is to be avoided because heat generated by the spark can cause localized melting of the metal. Such a condition erodes the wheel and damages the surface of the workpiece. Graphitization of the diamonds has been reported (61) by the local heating due to sparks.

3. Maintenance of Gap

To prevent spattering of the electrolyte and the generation of noxious misting, the complete cutting assembly may be enclosed in a plastic box (such as Lucite). The electrolyte draining down the walls of the box is returned to the solution reservoir.

Williams (67) described a system in which the wheel was enclosed in a ring of bristles attached to a vacuum system. In this way the mist was sucked away.

Since one of the main roles played by the diamonds is the maintenance of the narrow gap space and not the removal of metal, only light pressure is required to hold the wheel against the workpiece. This constant applied pressure to the wheel has been obtained with spring loading (49), compressed air (51), or a servo-drive system (56). Pahlitzsch (58) pointed out that the efficiency of the diamond wheel decreases with increasing pressure applied to the wheel and increases with increasing concentration of diamonds in the surface of the wheel. If the applied pressure is too great, the diamonds cut into the metal of the workpiece, the gap is narrowed, and deleterious sparking results. Consequently, most machines employ some device for detecting sparking. One such device described by Comstock (57) utilized an electronic detector of the high-frequency noise generated by sparks to fire a thyratron, which reduced the voltage.

4. Corrosion Product Removal

The other main purpose of the diamonds is the scraping off of insulating oxide layers so that the anodic corrosion process may continue (50-53). Cole (52, 59) has determined that over 99% of the metal removal in electrolytic grinding of steels is electrolytic. This percentage falls to values of about 75% (58, 60) for the machining of cemented carbides because, as suggested by Reinhart and Grünwald (61), the metal is removed electrolytically, leaving a weakened carbide skeleton that is removed by mechanical grinding.

5. Wheel Wear

Since only a minimum pressure is applied to the diamond wheel and since nearly 100% of the metal removal takes place electrolytically, tool wear is reduced considerably (54). Schwartz (62) has estimated that a diamond wheel used for electrolytic grinding should last about 46 times longer than a similar wheel used for mechanical grinding. In addition, electrolytic grinding rates may reach values nearly twice as fast as mechanical grinding for the difficult-to-machine alloys (49, 50, 52, 57, 63). Machining rates of the order of 0.1 in.3/min are recorded; they are independent of the hardness of the material machined. As in mechanical machining, the shape of the diamond wheel determines the shape of the cut made in the workpiece.

B. The Electrolyte

At first, solutions of NaCl were used (50), but these electrolytes proved to be too corrosive for the materials from which the electrolytic grinding machines were made. Solutions of $NaSiO_4$ (water glass) have been used (54, 64), but it was found (54) that machining in this electrolyte was slow and gave off fumes. The best electrolyte from the standpoint of minimum corrosiveness and high conductivity was found (49, 51, 52, 54) to be 5% to 10% solutions of $NaNO_3$. Rust inhibitors, such as Na_2CO_3, Na_3PO_4, and $NaNO_2$, have been added to the nitrate electrolyte (52, 58, 61). To keep the corrosion products in solution, Pahlitzsch (58) reports that complexing agents have been added: N Na_3PO_4 or Rochelle salts for the machining of W and Ti, NH_4NO_3 for Co, H_3BO_4 for ferrous alloys, and $NaNO_2$ or Na_2CO_3 for tungsten carbides. As the conductivity of the electrolyte is increased, the amount of metal removed increases for a given applied voltage. Potassium salts seem to be better than sodium salts, probably because potassium salts are more soluble and the K^+ ion is more mobile than the Na^+ ion (58). As a general rule, though, potassium salts are more expensive than sodium salts.

C. Advantages

1. Lack of Induced Stress

One of the great advantages of the electrolytic grinding process is the ability to remove metal rapidly without inducing stress in the workpiece even with relatively deep cuts. In studies of the amount of stress induced in steel during mechanical grinding, Frisch and Thomsen (65) found that the stressed layers penetrated to deeper levels with increased depth of grinding. Using mild steel, Frisch and Cole (66) showed that deep cuts could be made with electrolytic grinding without inducing stresses in the metal. At the deepest cuts, only insignificant tensile or compression stresses were induced. As the metal removal rate, which is proportional to the current density and electrolyte flow rate, was increased, the process went from pure electrolytic grinding to some mechanical grinding and from occasional sparking to continuous sparking. Since the current density depends on the solution flow rate, one must have a uniform flow of electrolyte and a uniform distribution of solution over the workpiece to get uniform cutting.

2. Good Surface Finish

Mechanical grinding always leaves scratches, grooves, or tool registry on the surface of the machined metal. On the contrary, the surface of metals machined by electrolytic grinding under certain conditions may be smooth and free of scratches or grooves (49, 50). Profilometer measurements of the surface smoothness range between 10 and 40 μin. As pointed out by Stroup (53), such delicate structures as stainless steel honeycomb can be machined with electrolytic grinding because this process does not produce burring, thermal or mechanical stress in the workpiece, and requires only a minimum of applied pressure on the grinding wheel. Pahlitzsch (58) noted that as the current density increased to high values the roughness of the surface increased, an effect which he attributed to the high voltage required in such a current range. In the presence of high electric fields, particles of metal can be pulled out of the surface. Also, some sparking may be present.

3. Hardware

Pictures of typical electrolytic grinding machines may be found in the articles written by

Williams (67), Cole (52), Pahlitzsch (58), and Metzger and Keeleric (50). In fact, Metzger and Keeleric describe in some detail how to convert a mechanical grinding machine into an electrolytic grinding machine. Williams (67) presents a review of this process along with a list of applications for which electrolytic grinding is suitable. A tool designed for electrolytic honing has been described (73), but the process appears to be expensive. A short account of the electrolytic grinding process is presented by DeBarr and Oliver (1, pp. 35-40).

Although the Soviet workers (68) are known for their work in electrical discharge machining (EDM) in which metal is removed by the erosion of metal from the spark discharge between electrodes in a dielectric fluid (69), Ryabinov (70) describes considerable activity in the field of electrolytic grinding. In spite of several early references to Gusev by Ryabinov (see Ref. 70, pp. 16 and 81), the earliest published report of Gusev (71) is in 1952, coinciding with the reports of Keeleric (45, 50). It is possible that the electrolytic machining process was developed concurrently in the Soviet Union and the United States.

III. ELECTROCHEMICAL MACHINING

In the discussion of a paper on electrolytic griding by Cole (52), Van der Horst suggested that the wheel be eliminated and the cathode held at a fixed distance from the workpiece. This would eliminate the commutation problems to the wheel. It was pointed out that such a device had already been developed by the Anocut Engineering Company of Chicago, Illinois, and was announced in 1959 (74) as the first ECM machine offered commercially.

A. Gap Distance and Feed Rate

To obtain a commercially feasible metal removal rate (∿1.0 in.^{3}/min), which is identical with the current density (1000 A/in.^{2}), the cathodic tool must be maintained at a very small gap distance (between 0.002 and 0.050 in.) from the anodic workpiece; otherwise an intolerable iR loss will be encountered over the longer current paths of a wider gap space. Since metal removal during the ECM process increases this gap distance, there must be relative motion between the anode and cathode (either a movable cathode and a fixed anode or vice versa) obtained by a mechanical feed drive system.

If the feed rate is made too large for a given metal removal rate (current density), it is possible to create a short circuit between the anode and cathode, producing catastrophic damage to the tool (spark erosion, thermal melting, welding, and tearing of the metal). Protective sensing devices (usually based on detecting increasing rates of current) are used to avoid this costly event by shutting down the current within microseconds when the gap closes beyond a certain specified safe distance (101).

B. The Electrolyte

Conduction of the current across the gap between the cathode and the anode is carried out by the ions of a suitable electrolyte of high conductivity. Such an electrolyte is a water solution of NaCl (5-25% by wt) (74-81). Acids such as HCl or H_2SO_4 have been used, but in this case these solutions are more corrosive to the machine than is NaCl, and metal may be plated out on the cathode (79, 82, 83). Under such condition, the shape of the tool is changed, producing unwanted changes in the shaping of the workpiece. Consequently, only neutral salt solutions are practical. Other neutral salts, such as KCl, $CaCl_3$, and $NaNO_3$ (75, 79), have been used, but NaCl appears to be the most universally used electrolyte (84, 131). Because of film formation (ref. 89, p. 22), tungsten can be machined only in NaOH electrolytes (101).

Any parameter that changes the conductivity of the electrolyte will change the current density, and hence, the metal removal rate.

1. Temperature Effects

If the electrolyte were permitted to remain stagnant in the gap between the electrodes, corrosion products from the metal dissolution collecting in the gap would increase the resistance across the gap until the metal dissolution process would come to a halt. Also, the passage of such large quantities of electricity through the thin layer of solution would produce enough heat to vaporize the solution, causing the metal removal process to stop. To offset these effects, the electrolyte is pumped through the gap (74, 77, 85) with a high-pressure pump usually rated to deliver 25-250 gal/min under operating pressures from 100 to 400 psi. It is important that the flow of solution be uniform through the gap; otherwise, uneven metal removal will destroy the integrity of the final product (76, 86).

A number of equations have been derived (75, 87-89) to determine the change in temperature of the electrolyte across the length of the gap. Typical is one set down by Cole (75) for a cavity-sinking process:

$$\Delta T = 2.23 \times 10^9 \frac{f^2 d_m^2 \rho h A}{w^2 d_e q C_p}$$

where f is the feed rate; d_m, density of anode; ρ, specific resistance of the electrolyte; h, gap width; A, active area of tool; w, equivalent weight of anode; d_e, density of electrolyte; q, electrolyte flow rate; and C_p, specific heat of electrolyte. Kubeth and Heitmann (78) have measure this ΔT experimentally under normal ECM conditions and reported changes in the temperature of the solution entering and leaving the gap in the order of 45°C. Such a change in temperature could cause a change in ρ of about 100% and in viscosity of about 50% (75) for a 10% solution of NaCl. Therefore, the temperature change should be kept below 10° (76) to avoid uneven metal removal at the hot spots. Most units employ heat exchangers in the electrolyte flow line to keep the temperature of the intake solution controlled within ± 2°F (77, 85).

2. Pressure Effects

The temperature change along the gap can be reduced by increasing the flow rate of the electrolyte. However, a point is reached above which cavitation of the solution in the gap is initiated by the high pressure gradient along the gap (75, 79, 80, 90). Cavitation breaks up the solution flow, producing striations, grooves, and a general roughening in the mcahined surface of the metal workpiece. Kawafune and co-workers (80) have derived an equation for the distribution of pressure along the gap for a radial geometry condition (such as hole cutting). The results show that, with increasing flow rate, the pressure near the inlet at first increases, passes through a maximum and then decreases. When the pressure near the inlet drops with high flow rates to values lower than the vapor pressure of the electrolyte, cavitation is initiated in the electrolyte flow. By adding ink to the flowing solution, the flow patterns were observed and photographed. From these observations, it was found that increasing the pressure at the exit (back pressure) reduced the range of cavitation. Similar conclusions were reached by Opitz (90). With proper design, cavitation can be engineered out of the ECM system for high flow rates.

3. Corrosion Product Effects

With high flow rates the corrosion products are swept out of the gap. In the electrochemical machining of ion and ferrous alloys in NaCl solutions, the iron atoms of metal lattice of the anode are oxidized to ferrous ions (91) by the transfer of two electrons to the anode. Because a dimer of the electron has never been observed, this process probably takes place in two one-electron steps which follow one another so quickly that they cannot be studied separately. Bockris and co-workers (135) have suggested that this electron transfer is aided by the presence of adsorbed OH radicals formed either from the discharge of OH^- ions, $Fe + OH^- \rightarrow FeOH + e$, or water, $Fe + H_2O \rightarrow FeOH + H^+ + e$. The radical is discharged, $FeOH \rightarrow FeOH^+ + e$, which then decomposes to liberate Fe^{++} ions, $FeOH^+ \rightarrow Fe^{++} + OH^-$. In a neutral solution, the Fe^{++} ions are quickly precipitated as a greenish sludge of $Fe(OH)_2$, which is eventually oxidized by dissolved oxygen in the solution to red hydrated ferric oxide (red rust).

As long as the hydrated oxides remain in a uniform colloidal suspension, good ECM finishes are obtained (79, 82); but if the electrolyte becomes contaminated with particulate matter, such as bits of carbon (eroded from high-carbon steel), nonmetallic inclusions, metallic grains (from nonuniform alloys), or lumps of coagulated hydrated oxides, then solution flow rates become uneven and current distribution becomes nonuniform, producing poor surface finishes. De la Rue and Tobias (92) have shown that the presence of nonconducting material suspended in an electrolyte lowers the conductivity of the electrolyte. Suspensions of high concentrations of insoluble hydroxides increase the viscosity of the electrolyte, thus affecting metal removal rates (89, p. 25).

Consequently, it is desirable to filter out this contaminating particulate matter (77, 81, 84, 85, 89) from the ciruclating electrolyte. The most common means has been the use of a filter press (89, p. 25) usually using diatomaceous earth. Since filters eventually become clogged and need periodic cleaning, other purification methods have been investigated. Ito and Shikata (81) have used a centrifuge to remove sludge, but it is reported (84, 89) that centrifuges are expensive. Settling tanks, which take up considerable space and are only 50% efficient even at low flow rates (89), have also been considered. Frothing and flotation methods followed by skimming have been tried (89), and largely unsuccessful applications of ion-exchange membranes have been investigated.

4. Electrochemical Aspects

In contrast to electronic conduction of current by physical changes (presence of a magnetic field, heat changes, sometimes the emission of light), electrolytic conduction of current through an electrolyte by ions is always associated with chemical changes at the metal-solution interface (anodic oxidation and cathodic reduction). The species carrying the current through the body of solution are not necessarily the same as those which participate in the electrochemical reaction at the electrode interface (72, 93). In the case of NaCl, the bulk of the current is carried by the sodium and chloride ions. At the anode, metal dissolution takes place more easily (at a lower potential) than the evolution of Cl_2 from Cl^- ions or O_2 from H_2O molecules, whereas at the cathode, evolution of H_2 from H_2O molecules takes place more easily than the deposition of Na from Na^+ ions.

If another electroltye that produced an insoluble film on the anode were used, or if greasy impurities filmed the anode, it might be possible to make metal dissolution so difficult that some current would go into the evolution of oxygen from water molecules. Such side processes would lower the current efficiency for the metal removal process. Current efficiencies of virtually 100% are reported by Cole (75) for the electrochemical machining of iron in NaCl where protective anodic films are not formed (94).

Since the ECM process is a constant-current or galvanostatic process, the rate of the electrochemical reaction is fixed. If very high currents are used, a point is reached where metal removal is not fast enough to keep up with the demanded rate. As a result, the additional current goes into the evolution of Cl_2 from the Cl^- ions. This viewpoint predicts that the current efficiency for the electrochemical machining of iron in NaCl should fall with increasing current density. In mass-balance studies carried out at the General Motors Research Laboratories by Mao (95), this prediction was verified experimentally. At the cathode in neutral solutions, only the evolution of H_2 from the discharge of water molecules takes place with a current efficiency of virtually 100% (95) according to the equation $2H_2O + 2e \rightarrow H_2 + 2OH^-$. Thus, the cathodic process increases the pH of the solution, a situation that is advantageous in precipitating out the metal hydroxides to be removed by the filtering system (87).

5. Gas Bubble Effects

It has been shown by Tobias (96) that the presence of gas bubbles in the electrolyte causes a decrease in the conductivity of the solution and gives non-uniform current distribution. Such a situation in the gap between the tool and workpiece would affect the metal removal rate deleteriously and must be minimized (75, 89, 98-100). To evaluate the effect of bubbles on the metal removal rate, Hopenfeld and Cole (98) used a flow cell made of clear plastic so that the gap could be photographed during actual ECM operation. There existed at the cathode surface a two-phase region, termed the "bubble layer," which increased in thickness along the gap in the direction of solution flow. This bubble-layer thickness was observed to increase with increasing current and decreasing electrolyte flow rate. Only at very low flow rates could the individual bubbles be seen, and as the flow rate was increased the bubble size diminished (112). Using a number of simplifying assumptions, Hopenfeld and Cole (98) derived an expression relating current, solution flow rate, and volume of H_2 produced; it was concluded from the relation that the most desirable way to increase the uniformity of metal removal was to increase the solution flow rate. It was also noted that the bubble layer, which thickens along the gap and tends to lower the current density, has a compensating effect on the temperature effect, which tends to increase the current density along the gap.

Recently, Tobias and co-workers (132), using an optically instrumented ECM cell (133), made a more sophisticated determination of the effect of gas bubbles on ECM parameters in the anodic dissolution of Cu in KCl solutions. Although the bubble size and the bubble-layer thickness diminished with flow rate, the increase in resistance of electrolyte due to gas evolution idid not entirely compensate for the temperature rise as noted before (98).

There has been some concern expressed about the safety hazard of the large amounts of hydrogen evolved in an ECM process (89, p. 35). Some suggestions, such as scavenging the H_2 with air, have been made (102).

C. Metal Removal Rate

Metal removal in the ECM operation is an anodic corrosion process, and, as such, is governed by Faraday's Laws. Accordingly, the amount of metal removed (assuming 100% current efficiency) is directly proportional to the quantity of charge (product of current and time) passed and the mass of material associated with unit charge (equivalent weight - atomic weight/valence). The charge associated with an equivalent weight of a metal is a constant called Faraday's constant (F) and equals 96,484 C (116). In mathematical form, these statements become $w = ItW_{eq}/F$, where w is the amount of metal removed; W_{eq}, the equivalent weight of metal; I, current passed; and t, the time current flows. In addition, the rate at which metal is removed, dw/dt, is directly proportional to the current, $dw/dt = kI$. By differentiating the mass equation, the proportionality constant is seen to be equal to W_{eq}/F. For studies of reaction rates, one generally considers the parameters on the basis of unit area, such as current density instead of current.

1. Effect of Voltage

As pointed out before, all electrochemical processes involve conduction of current through an electrolyte with chemical changes taking place at the electrode-solution interfaces. The voltage drop or potential difference across the metal-solution interface is the driving force for the electrochemical processes of oxidation (at the anode) and reduction (at the cathode), and a threshold value of the potential or driving force must be reached before the electrochemical process can take place. The table of redox potentials (117) and the electromotive series are related to this threshold driving force. Since nearly all electrochemical reactions take place with activation energy, additional driving force or potential is needed to make the reaction proceed at a given rate or current density. This excess potential is called electrochemical polarization, or under certain conditions, over-voltage. The rate of reaction is an exponential function of the polarization and may be written as the Tafel equation (118, 119); $\eta = a + b(\log i)$, where η is the polarization, i is the current density, and a and b are constants. At the high flow rates used in electrochemical machining, contributions of mass-transfer processes (diffusion of reactants to and products from the electrode-solution interface) to the polarization are negligible, as noted in studies with rotating disks (120).

The voltage drop across the cathode-anode gap is comprised, then, of the cathodic polarization of the reduction process, the ohmic drop across the electrolyte path, and the anodic polarization of the oxidation process. The value of the applied potential must be high enough to overcome these voltage drops and supply the high currents demanded by the ECM operation across the narrow solution gap. In practice, ECM processes operate at applied potentials ranging between 5 and 20 V (74, 77, Ref. 1, p. 23, Ref. 89, p. 6).

Because simple, regulated, low-voltage d.c. power sources are required, it is not difficult to obtain rectifier units that will deliver very high currents. Many units can deliver up to 20,000 A (87), and there is discussion of the need for units greater than 20,000 A (84). Myerley (85) noted that with a 20,000-A unit, steel stock could be removed at the rate of 120 in.^{3}/hr, which is to be compared with the first announcement in 1959 (74) of 18 to 60 in.^{3}/hr. Metal is removed with current densities between 30 and 3000 A/in.^{2} (76, 84, 110) and as high as 5000 A/in.^{2} (89, p. 6) and with feed rates from 0.02 to 0.5 in./min (73, 87, 110, 111, 113). Since the metal removal rate is directly proportional to the current, the feed rate decreases with the work area (110).

2. Effect of Metal

Aaron and Wolosewicz (110) give a table of the metal removal rates and the electrolyte type and concentration for electrochemical machining on a number of materials. Good metal removal rates are obtained in NaCl electrolytes with steel; molybdenum; iron-, nickel-, and cobalt-based alloys; aluminum and aluminum alloys; copper and copper-based alloys; uranium; zirconium; and semiconducting ceramics, such as B_4C, ZrB_2, ZrC, and TiB. Titanium alloys may be machined at lower rates. Apparently, better surface finishes are obtained with copper in $NaNO_3$ electrolytes. Tungsten requires NaOH-based electrolytes for successful ECM operation.

3. Effect of Geometry

As the cathodic tool penetrates the anodic workpiece, metal should be removed only from those sites on the anode surface that are directly opposite the cutting face of the tool. Metal removal from any other point is unwanted and contributes to the deformation of the finished product. For example, in hole-cutting, the sides of the hole are not straight but tapered. The tapering is caused by the presence of stray currents, and the excess metal removed is referred to as stray or wild cutting or simply overcut.

Just opposite the cutting face of the tool, there is a low-resistance path to the anode; most of the current passes to such sites. However, longer high-resistance paths extend to the side walls, and some current can flow from the tool to the side walls of the hole, causing some metal removal and producing a tapering of the hole. The longer the cathode remains in the vicinity of a given region of the anode, the more metal will be removed from that region. Consequently, the amount of taper can be lessened by increasing the feed rate. Since there are mechanical as well as electrochemical limitations to increasing the feed rate (111), there is always a certain amount of overcut obtained with NaCl-based electrolytes.

Also contributing to the magnitude of overcut is any variation in the applied voltage; hence, a well-regulated rectifier is a necessity for ECM operation. Haggerty (113) has estimated that the voltage must be held within 2% variation to maintain a dimensional tolerance of 0.001 in. A similar control of feed rate was determined to obtain the same tolerance.

It is possible when making a plunge cut to make allowances for the overcut, but such a hit-and-miss proposition is satisfactory only for the gross removal of metal. For sizing and finishing operations, such a procedure is unacceptable. The effects of overcut can be reduced by contouring the tool and designing the size and shape of the lip on the cutting face of the tool (86, 90, 110).

4. Effect of Addition Agents

Sodium chloride is a very corrosive reagent, and $NaNO_2$, $NaNO_3$, and organics such as amines have been added to NaCl as corrosion inhibitors to protect the exposed metal parts of the ECM machine (89, p. 24). In some operations, complexing or sequestering agents have been added to the electrolyte to keep the corrosion products in solution and thereby achieve better surface finishes.

One of the extensive efforts in the use of addition agents is the search for a reagent that can be added to the economically advantageous NaCl electrolyte to reduce the stray cutting so that precision work can be performed. The addition of both organic and inorganic reagents, such as potassium fluoride (136), tripotassium phosphate (137), monoethanolamine (138), triethanolamine (138), ethylene glycol, glycerine, imidazole (139), 2-aminoethanol (138), 2-mercaptobenzothiazole (140), and sodium styrene sulfonate (141), to the NaCl electrolyte formed films on the anode surface during the ECM process. These films were adherent enough to improve the overcut characteristics of the NaCl electrolyte. Depending on the type of addition agent used, it was found that a wide range of film types could be obtained, ranging from those that were voltage sensitive and were dissolved off in the high current density range to those that were so firmly adherent that only mechanical abrasion, such as electrolytic grinding, was required to allow the electrochemical machining to continue.

Although these addition agents did improve the overcut characteristics of NaCl electrolytes, they did not eliminate all the problems completely. Of these addition agents, some produced a roughening of the surface and all drastically altered the cost of the electrolyte. The normal cost of a 10% NaCl solution may be 1 to 2¢/gal. Such a cost differential is too great for the advantage gained.

Boden and Evans (121, 122) have suggested that additions of carbonate, phosphate, chromate, and ferricyanide ions to NaCl have significantly reduced the stray cutting of nickel. It has been our experience (123) that many electrolytes containing high concentrations of Cl^- ion will produce some stray cutting of steels and nickel-based alloys in spite of the presence of passivating anions.

5. Mathematical Approaches

Since the bothersome stray cutting of Cl^- ion-based electrolytes makes it impossible to use the ECM operation in precision work, a number of workers (80, 82, 89, 98, 125, 126, 166) have attempted to characterize the ECM process mathematically. In this way, the predictable behavior of controlling parameters could be known, and during the ECM process, due account could be taken of these parameters to yield a faithfully reproducible product.

Most treatments are based on Faraday's laws, Ohm's law, and potential and current distribution functions. In cases such as plunge-cutting operations, where an equilibrium gap can be maintained, the overcut predicted by derived equations agrees well with that obtained experimentally (80, Ref. 89, p. 48).

McGeough and Rasmussen (125) have attempted to derive a nondimensionaly parameter that would include all dynamic and electrochemical features of the ECM process. In this treatment, they considered the electrolyte to be merely an electrical component of the system obeying Ohm's law. It was found by Kawafune and co-workers (80) using this approach that the measured current density was considerably lower than the expected value calculated from Ohm's law.

The deviations from experiment are to be expected since the electrochemical polarization of the oxidation and reduction processes at the electrodes has been neglected. In the derivation of nondimensional parameters, these polarization effects must be included (127). To do this, single electrode polarization data must be obtained on the metals of interest as a function of electrolyte concentration and flow rate, feed rate, and cell geometry, among other ECM parameters.

A thourough and comprehensive analysis of the ECM operation using Cl^- ion-based electrolytes was carried out by Bayer and co-workers (82). In an attempt to take into account the polarization effects of the electrode reactions, they employed a transient technique between the anode and reference electrode to estimate the values of polarization. With this method, the experimental parameters were adjusted to obtain a steady-state current, after which the circuit was opened rapidly with a pulse generator and the decay of potential was recorded on an oscilloscope. At the termination of the constant current pulse, the potential drops rapidly (within 10^{-8} sec) to a point after which a slower exponential decay of potential to the open-circuit value follows (see Fig. 30 of Ref. 82). The very rapid portion of the transient with the very short time constant is the decay of the solution iR-drop, and the exponential portion with the long time constant is the decay of the electrode polarization. The potential drop across the electrolyte is a linear function of the current (Ohm's law), whereas that across the electrode-solution interface is an exponential function of the current (Tafel relationship). Unfortunately, Bayer and co-workers used the rapid decay of potential as the estimation of the electrode polarization in their equations instead of the exponential drop. Even with a computer programming of the system, unambiguous and dependable determination of the overcut was not obtained.

It appears, then, that any ECM operation using Cl^- ion-based electrolytes cannot be applied to those cases where rigorous dimensional control at high metal removal rates is essential to the realization of the final product.

D. Surface Finish

One of the appealing features of electrochemical machining is the elimination of the finishing operation that is always required in conventional machining processes. Because the metal is removed atom by atom by an anodic corrosion mechanism, there are no sharp edges, burrs, scratches, or grooves to be removed. As reported in the literature (73, 76, 81, 87, 90, 110, 128-130), profilometer readings of metal surface finishes range from 15 to 40 μin. for carbon steels. Very reflective surfaces can be obtained on stainless steels with Cl^- ion-based electrolytes (86, 101).

The type of finish obtained is strongly dependent on any parameter that affects the current distribution. One of the most critical parameters is electrolyte flow. To obtain good surface finishes, it is imperative that a uniform flow of electrolyte be maintained through the gap. Ito and co-workers (129) found that surface roughness decreased by increasing the feed rate and decreasing the electrolyte concentration. They suggested that the increase in electrolyte resistance suppressed corrosion in the valleys and improved the surface finish.

Wtith a number of metals in KNO_3 electrolytes, Cole and Hopenfeld (128) found that polishing of the surface was obtained if the current density was high enough and concluded that the polishing is a physical instead of a chemical effect. Where the current density is low (such places as the side walls), one obtains an etched or matte surface according to Aaron and Wolosewicz (110), and the surface finish may be as bad as 400 μin. (101).

Coarse-grained materials, such as cast iron, produce very rough surfaces because the matrix material is etched away first (90, 110, 129). Consequently, a good surface is expected only with homogeneous materials.

Nitrites and NaOH have been added to Cl^- ion-based electrolytes to assess their effect on the quality of finish obtained (81, 124). No effect was found with the machining of stainless steels, but a deterioration of the surface was noted for the machining of carbon steels with increasing concentration of the addition agent.

Since the fatigue strength of a given metal is a surface phenomenon (134), it was found by Gurklis (131) that metals which had been either machined by ECM or electropolished had a lower fatigue strength than those which had been mechanically machined. If the machined metal is to be subjected to variable stresses, then caution must be taken in the use of electrochemical machining. It is possible to restore much of the fatigue strength by surface treatments such as shot peening (131).

E. Types of ECM Operations

The ECM process is a very versatile one in its applications. A number of these operations are now described.

1. External Shaping

With the advent of jet aircraft and the development of high-temperature alloys, it became increasingly difficult to mass produce turbine blades. One of the first applications of electrochemical machining was in the machining of turbine blades (103-109). In this operation, the tool is shaped to fit the desired contours of the finished workpiece, and with two cathodes, both sides of the blade can be finished simultaneously (103). The diagram in Fig. 8 shows the arrangement of the essential parts of the ECM cell. According to Faust and Snavely (103), jet engine turbine blades were machined to tolerances of 0.003 in. in 5 to 10 min with electrochemical machining; the same job required 1 to 2 hr to complete with conventional machining methods.

Since no pressure is applied to the workpiece, thin foils can be machined as easily as heavy castings without concern for warp, bow, or twist (103, 105). Because the metal removal takes place by an anodic corrosion process where the metal is removed atom by atom from the metal lattics, no mechanical stresses are induced in the metal lattice, no thermal modification of metallurgical structure is generated, no burring or sharp edges are produced, and in many cases the surface finish is acceptable without further treatment (20 to 40 μin.).

There is no physical contact of the tool with the workpiece, and the only electrochemical process taking place at the cathode surface in neutral salt solution is the evolution of hydrogen. Consequently, there is no tool wear, and parts may be reproduced with good accuracy indefinitely provided damage to the tool from sparking or short circuits is avoided. Because hydrogen is liberated only at the cathode, hydrogen embrittlement of the machined workpiece cannot take place (74, 131).

Usually the cathode is made of copper, but copper-tungsten, stainless steel, brass, and titanium cathodes have been used (110). Care must be taken in the design of the tool to provide for a uniform flow of electrolyte through the gap. Tooling costs may be high, but the replicatability of the system can easily overcome this economic problem (111).

These remarks made with respect to the external shaping operation apply equally well to all operations described below.

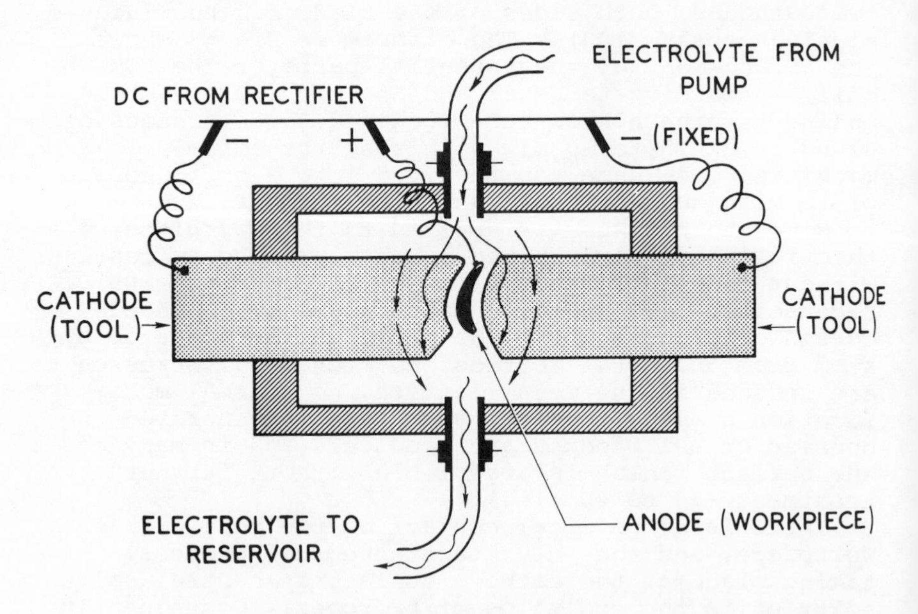

Fig. 8. (a) Diagram of the ECM machining of turbine blades (103).

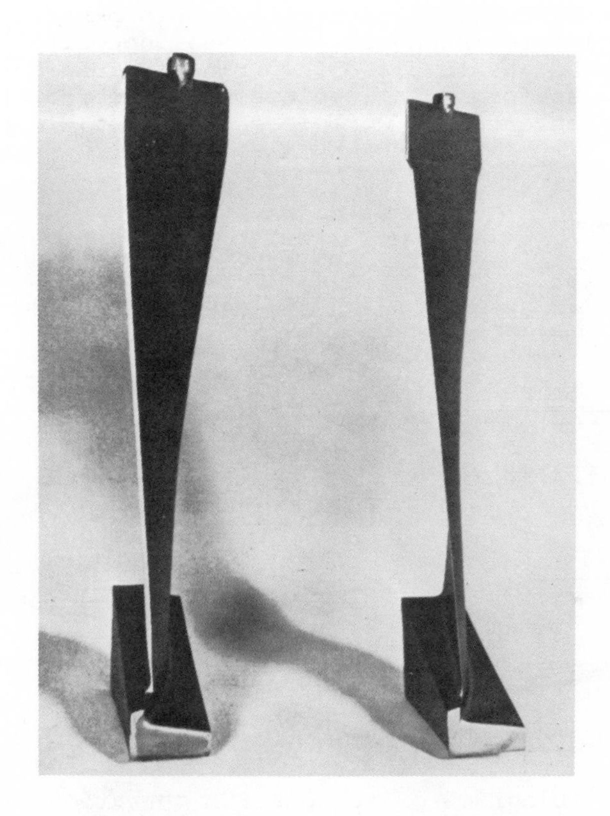

(b) Turbine blade before and after electrochemical machining (103). The blade (left) is in the as-forged condition, and the blade (right) is finished to dimensions. (By permission of Iron Age.)

2. Die Sinking

In this operation, also called cavity sinking (75, 80), the cathodic tool is fashioned in the shape of the cavity to be machined in a block of metal. Figure 9b presents a sketch of the general arrangement of the electrodes in this operation. As the face of the cathodic tool approaches the metal surface of the workpiece, metal removal takes place at the high spots first; with deeper penetration of the tool into the metal, the metal opposite the valleys begins to dissolve. Finally, the current becomes uniform at all points when the tool has completely entered the workpiece (103, Ref. 1, p. 22).

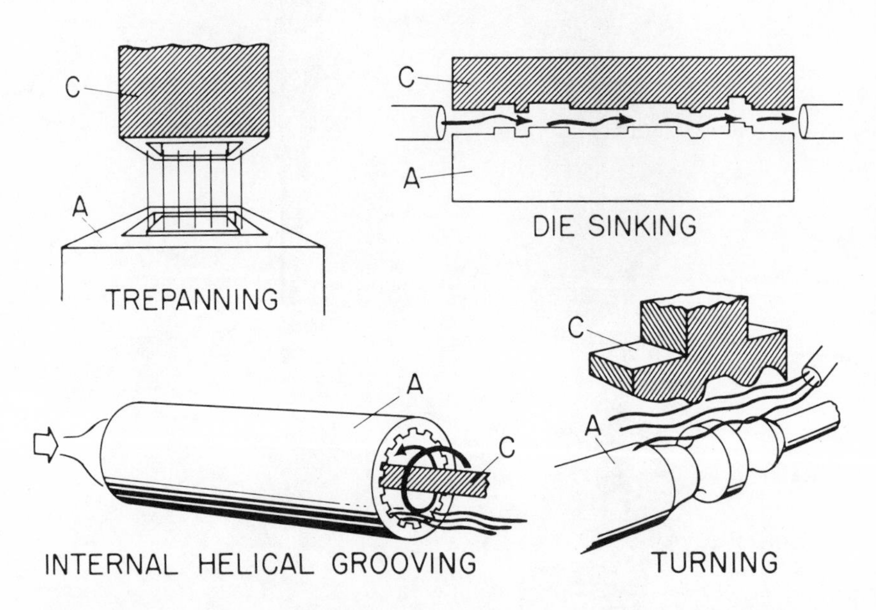

Fig. 9. Diagrams of various ECM operations (103). (By permission of Iron Age.)

In one operation, a very complex cavity, which would have required numerous operations with conventional machining methods, can be made quickly and with acceptable finishes, as demonstrated by the pipe elblow cavity in Fig. 10. Haggerty (113) states that this pipe elbow die was machined to a depth of 1-1/8 in. in 9-1/2 min with a finish of 25 μin.

Because this anodic corrosion process is independent of the mechanical properties of the metal concerned, hardened steels are no more difficult to machine than the unhardened steel. As a result, very complex shapes may be produced quickly in the hardest of alloys which resist any other method of

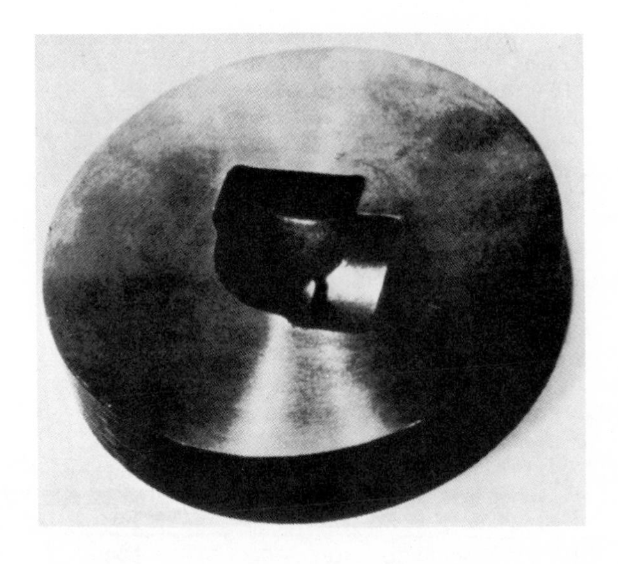

Fig. 10. Pipe elbow die cavity (113). By permission of the Society of Automotive Engineers.)

working them (82, 85, 87, 113-115) without any tool wear. As noted in Fig. 11, ECM metal removal rates become larger than conventional rates for high-strength alloys. It is in this region, then, that electrochemical machining could become the dominant metal removing process.

3. Plunge Cutting

In this operation (also known as hole cutting), a hollow tool is used through which the electrolyte is pumped as shown in Fig. 12. In some cases a reversal of the electrolyte flow is used (110). Since it is desirable for metal to be removed only from the bottom of the hole and not from the sides, the exterior walls of the tool cathode are coated with an electrically insulating material (epoxy coatings seem to give the best results).

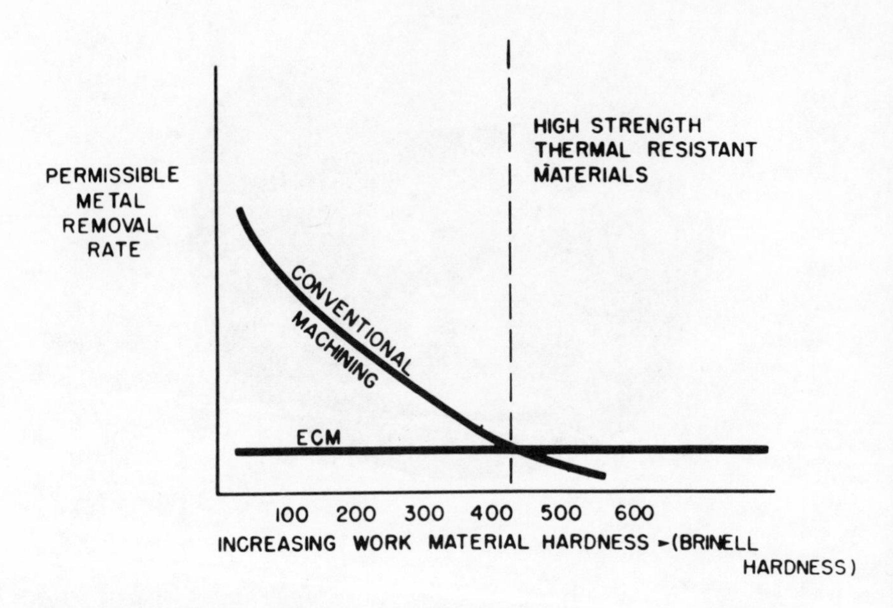

Fig. 11. Comparison of permissible metal removal rates in conventional and electrochemical machining (113). (By permission of the Society of Automotive Engineers.)

There is always a certain amount of taper along the walls of the hole (82, 89); this can be minimized somewhat by the proper design of the cathode.

By using the properly designed tool, holes of almost any conceivable complex shape (square, elliptic, hexagonal, fluted, etc.) may be machined through alloys of any hardness or as blind holes. According to Haggerty (113), a rectangular blind hole 1 x 1.5 x 1.5 in. was machined by electrochemical machining in a steel cylinder with one operation in 6 min. Conventional machining taking 43 min required three separate operations: drilling, milling (before heat treatment), and grinding (after hardening).

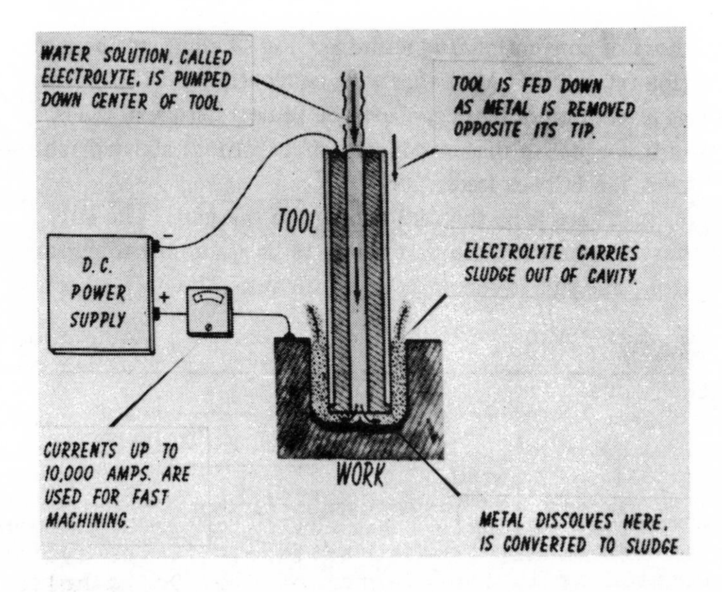

Fig. 12. Typical plunge-cutting operation (113). (By permission of the Society of Automotive Engineers.)

With through-hole cutting, there may be a small slug of metal that is trepanned out in the center of the plunge-cut. The danger of shorting occurs (89, p. 9) at the end of the cut when the slug may hinge and fall against the side of the cathode. To avoid this danger, a false or backup workpiece is attached to the backup plate with a screw. A second use of the backup plate is the prevention of a loss of solution pressure when the hole penetrates the workpiece.

In the plunge-cutting operation, metal may be removed at rates varying from 0.05 to 0.5 in./min depending on the size and complexity of the hole machined (113). The rate falls off with decreasing hole diameter and increasing complexity of shape.

4. Turning or Lathing

With a fixed cathode of proper design, symmetrical contouring of a rotating rod may be produced as demonstrated in Fig. 9d. Solution may also be passed through a slot in the cathode (89, p. 13). In this process, the machine is operated at the high ranges of current density (up to 5000 A/in.^{2}).

5. Trepanning

Complex shapes of uniform thickness with good surface finishes may be machined from a metal slab in the arrangement pictured in Fig. 9a.

6. Internal Grooving

It is possible to pass a thin or pointed cathode down the inside walls of a hollow cylinder to generate an internally grooved surface. If the cylinder is rotated as in the diagram of Fig. 9c, a helical groove similar to the rifling in a gun barrel may be produced.

7. Wire Cutting

If it is desirable to remove a large volume of metal in a roughing out operation, it is possible to pass a wire or thin metal tube through the workpiece, as shown in Fig. 13. The feed rates are very low because of the small tool cross section, which limits the current-carrying capacity of the cathode (113). Solution may be passed over the workpiece or through the thin tube and out of fine jets in the cutting face of the cathode. Cutoff operations are possible with this tool.

A pictorial review of the variety of machining operations possible with electrochemical machining may be found in the literature (106). In all these operations, the most essential component is tool design so that the proper uniform flow of electrolyte passes through the gap between the anode and the cathode (84, 101). Most cathode design is obtained by a trial-and-error process (101), and a need for the lacking mathematical descriptions of the ECM operation has been decried by Faust (97).

Fig. 13. Demonstration of wire-cutting operation (113). (By permission of the Society of Automotive Engineers.)

Of equal importance is the necessity of proper alignment of the tool and workpiece and the maintenance of this alignment throughout the ECM operation to avoid impairment of dimensional and geometric control. As a result, the components of the machine as well as the feed drive mechanism must be ruggedly designed and constructed.

Fig. 14. ECM plunge-cutting machine, circa 1959. (By permission of Anocut Engineering Company.)

F. ECM Machines

An early device built to operate in the current density range of 1000 A/ $in.^2$ with an NaCl solution pumped under a pressure of 100 to 200 psi through a narrow gap space (0.0002 to 0.05 in.) is the ECM machine pictured in Fig. 14. This model appeared in 1959. After ten years of development, versatile machines have evolved, operating routinely at 20,000 A with the capability of removing 2 in.^{3} of metal per minute. Such machines are designed to perform the ECM operations of external shaping, die sinking, plunge cutting, and trepanning.

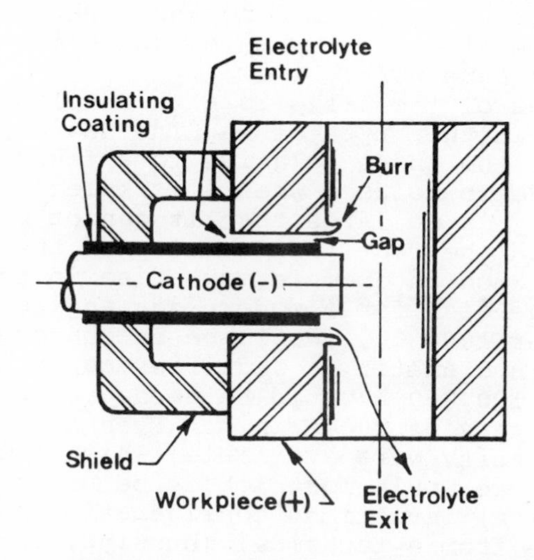

Fig. 15. Diagrammatic sketch of a typical deburring machine. (By permission of Chem-Form Division, KMS Industries, Inc.)

A second class of machines may be identified in the employment of various types of fixtures and is ideally exemplified by the electrochemical deburring machine that is finding increasing use by various groups in manufacturing. A sketch of the operating principle of this type of machine is presented in Fig. 15. It is a universal type, capable of handling a variety of electrolytes and parts with a minimum of alteration. In common usage, the fixtures are attached to any one of several repetitive stationary points so that a number of parts can be deburred at the same time.

The design and operating requirements for a deburring device are much less stringent than for a plunge-cutting ECM machine. In all cases, a wider gap space and a lower electrolyte flow are characteristic of deburring machines because the operation is dictated by the type of material (the burr) that is being removed. It is difficult to characterize the burr that is created in manufacturing because it is an unwanted, undesigned by-product and, as such, seldom has the same size and dimensions. The deburring device must be capable of removing the largest burr without damaging the workpiece by excessive metal removal when encountering a small burr in the same work.

The electrolytic lathe is a third class of ECM machines. In this type of machine design, there is movement of the anode relative to a fixed, limited-area cathode to be used when a workpiece of symmetrical geometry is so large that it cannot be accommodated within the rectifier ability of the largest machines. Such an example may arise in the machining of a turbine disk for a modern jet engine. Since an area of over 1000 in.^{2} would be encountered with a disk 3 ft in diameter, a rectifier capable of delivering over 100,000 A would be needed to provide a current density of 100 A/in.^{2}. Even if rectifiers of this capacity were available, which they are not, such a venture would most likely be unattractive from a capital investment consideration as well as undesirable from a technical (bussing, commutating) point of view. By the arrangement of the components employed in the electrolytic lathe, this entire problem is bypassed.

G. Advantages and Disadvantages

Since the ECM operation is an anodic corrosion process, the ability to machine a given metal is independent of the hardness of the material. As a result, a casting can be hardened before it is machined, thus avoiding any dimensional changes that might be brought about during the hardening operation. Very complex shapes that would be impossible to do in any other way can be machined in one operation with electrochemical machining in high-strength alloys.

There is no physical contact of the tool with the workpiece, and so there is no temperature effect due to friction forces which could cause thermal damage to an alloy. Changes in the matallurgy of a metal or alloy produced by mechanical stresses incurred in conventional machining are unknown in ECM operations. In the absence of any tool pressure, thin metal foils, expanded metal lattices, and metal foams can be machined without burring, warping, smearing, bowing, or twisting.

There is no wear to the tool cathode in neutral electrolytes since only hydrogen is evolved at the cathode. Since it is important that the tool be inflexible, copper bronzes may be used for long, thin tools instead of the usual copper metal.

The ECM metal removal rate is muc faster than that with conventional machining for the hard metals. A removal rate of 1 $in.^3$/min is common. At such a high metal removal rate, the surface roughness in very low (15 to 40 μin.), which is quite acceptable for many applications. In this way, a separate finishing operation is unnecessary.

Since low-voltage direct current is required, high-current-rated, well-regulated rectifying units are easily available, and electrical hazards are minimized.

The ECM process is expensive because the machine and the housing must be very rugged to withstand the highly pressurized electrolyte flowing from the high-speed, high-pressure pump. Also, the design and tooling of the cathode are very expensive, but these disadvantages are offset by the ability of the process to replicate parts with high reproducibility and lack of tool wear.

The ECM process is not normally used for point-tooling operations, such as marking screw threads or cutting treads in the face of a wheel. With NaCl electrolytes, sizing and finishing operations, such as the finishing of ball races, cannot practically be done because stray cutting prevents the attainment of the required dimensional control.

IV. MODERN DEVELOPMENT OF THE ECM PROCESS

To obtain useful work from and ECM machine, a certain amount of dimensional control is required, making it a necessity to hold the amount of overcut to a minimum. It is possible that large overcuts might be allowed if such overcuts were uniform and predictable. However, it appears that as the overcut becomes larger, the uniformity of the cut decreases, thus seriously limiting the amount of agreement between the predicted and the actual overcut.

In the theoretical analysis of the ECM process, most investigators (80, 82, 98, 125, 126) failed to take adequate account of the electrochemistry involved in derivations of the overcut equations. It is insufficient to consider the electrolyte as merely an ohmic element of the electrical circuit. To include all the subtleties of anodic and cathodic polarization, film formation, complex ion formation, as well as the effects of temperature, pressure, and electrolyte flow rate on these properties in the theoretical analysis is a formidable task indeed. Recently, McGeough and co-workers (159) have derived a mathematical theory for anodic smoothing, shaping, and cavity forming, taking into account activation overvoltage effects (which only modify the boundary conditions). As expected, the equations are complex. The success of predicted overcut or taper with actual ECM operation was not discussed. It comes, then, as no surprise that the actual use of mathematical models has virtually been useless in the design of tooling for the ECM production of complex shapes. As a consequence, the process of tool development becomes a tedious, empirical, expensive task when the ECM unit must generate its own geometry.

A. The $NaClO_3$ Electrolyte

Neither theoretical analyses of the process parameters nor the use of addition agents in the Cl^--based electrolytes resulted in perfect overcut control. As a result, General Motors Research Laboratories sought a completely new type of ECM electrolyte that would provide its own inherent protection of sites where metal removal was unwanted. An intensive investigation of the ECM properties of possible electrolytes was carried out, and of several hundred electrolyte combinations investigated, it was reported (142) that solutions of $NaClO_3$ provided a superior ECM electrolyte from the standpoint of excellent surface finish obtained at high metal removal rates with the desired good dimensional control. In these studies, results were obtained with small laboratory ECM test rigs including an electrolytic grinder, a through-hole-machining cell, and an outside-diameter-machining cell.

As an example of the type of dimensional control obtained, the notches machined in fully hardened 5160H steel bars with the electrolytic grinder are shown in Fig. 16 using $NaClO_3$ and NaCl electrolytes. The recordings reproduced in Fig. 17 were obtained with a Proficorder from the furfaces of a Carpenter low-alloy, nickel-chromium steel hardened to Rockwell 60C before and after a hole was sized in the through-hole machine using a $NaClO_3$ electrolyte. Values of the surface finish ranged between 2 and 5 μin., and the tolerance amounted to $\pm$ 0.0002 in. According to LaBoda and McMillan (142), both the taper and out-of-roundness were below 100 millionths of an inch.

The metal removal rate in $NaClO_3$ was nearly equal to that in NaCl at all concentrations until the solubility limit of NaCl was reached. Above this limit, the greater solubility of $NaClO_3$ permits higher metal removal rates to be reached until a leveling off of the rate occurs at about 600 g/liter, at which point the rate is twice the maximum rate attained with NaCl. The mirror-like finish is still obtained at these high metal removal rates.

1. Steady-State Polarization Studies

Since excellent dimensional control was obtained with $NaClO_3$ solutions, it is likely that stray cutting at points remote from the cathodic tool cutting face is inhibited by the presence of a film of oxide fromed on the metal surface. To investigate the film-forming process, potentiostatic polarization studies were carried out on soft iron microelectrodes in O_2-stirred electrolytes (120, 143) so that the complications of electrolyte flow, feed rate, and other mechanical parameters of the ECM process could be eliminated. The results of this work were then correlated with the actual ECM measurements.

Fig. 16. Notches cut (142) in 5160H steel with electrolytic grinder: $NaClO_3$ (top) and NaCl (bottom). (By permission of the Electrochemical Society, Inc.)

The polarization studies were carried out in solutions of $NaClO_3$ (350 g/liter), $NaNO_3$ (350 g/liter), NaCl (275 g/liter), and $Na_2Cr_2O_7$(100 g/liter); Fig. 18 shows the results of this work. Consider the curve for $NaClO_3$ first (circles in Fig. 18). At low potentials (single electrode potentials with respect to a saturated calomel reference electrode; see Ref. 3, p. 4), iron is active and goes into solution as Fe^{++} ions that can then be oxidized to Fe^{+++} ions. In this region, the current increase with increasing potential until a point is reached where an anodic oxide film begins to form on the metal surface. As

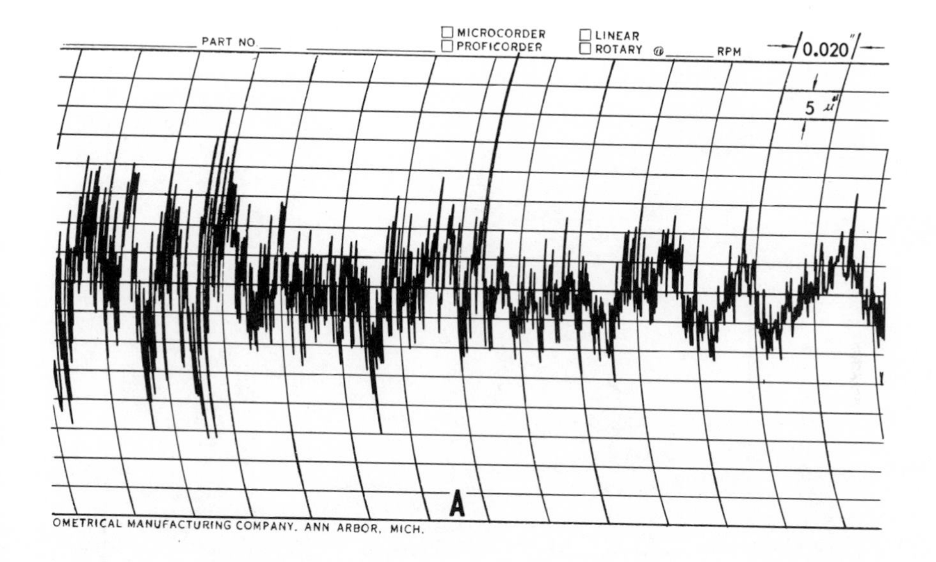

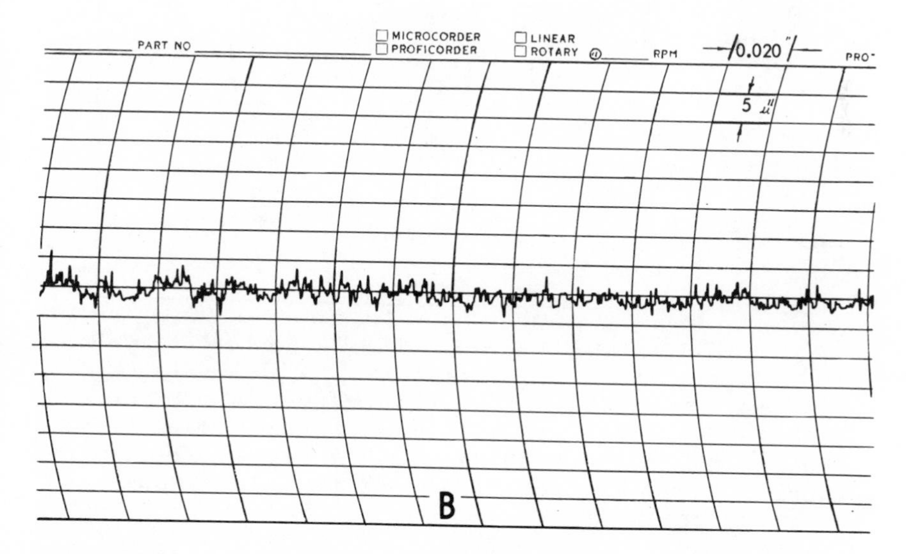

Fig. 17. Proficorder charts of typical steel samples before (A) and after (B) electrochemical machining with $NaClO_3$ (142). (By permission of The Electrochemical Society, Inc.)

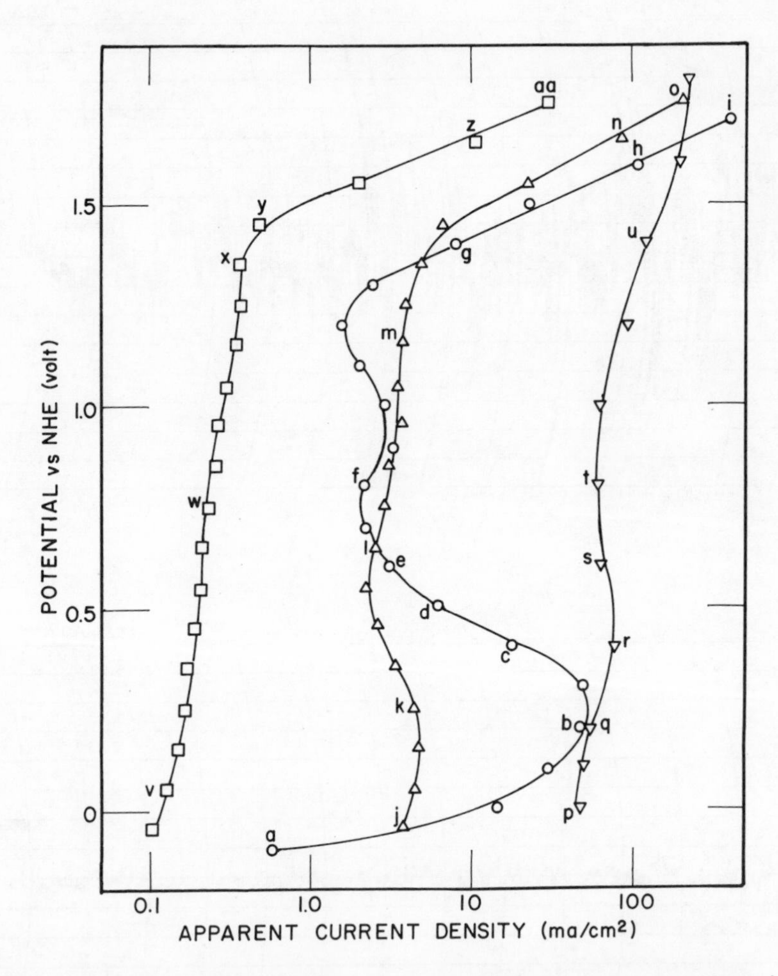

Fig. 18. Polarization curves of iron electrodes in electrolytes of $NaClO_3$, open circles; $NaNO_3$, triangles; NaCl, inverted triangles; and $Na_2Cr_2O_7$, squares (94). (By permission of the Electrochemical Society, Inc.)

this film grows with increasing anodic potentials, more of the surface is covered by the protective film, metal dissolution is hindered, the current falls to smaller values, and the metal surface is said to be passive. At still higher potentials, the anodic film begins to dissolve, possibly by interaction with anions (144, 145), and metal dissolution can take place once more in this so-called transpassive region where again the current increases with increasing potential. It is in this transpassive region that it is believed (94, 120, 143, 146) that electrochemical machining takes place.

At the high values of current density (200 to 1000 $A/in.^2$) encountered in the actual ECM operation, the iR drop across the electrolyte gap between the tool cathode and the workpiece anode may assume very large values. Since the double layer impedance (147, 148) at the anode-solution interface is in series with the resistance across the electrolyte gap, the potential appearing across the double layer that controls the rate of the anodic corrosion process will be much greater at points directly opposite the cutting face of the tool than at those remote from the tool face, such as the sides of the cut where the solution path is long. At the bottom of the cut in a plunge-cutting process, the rate-controlling potential corresponds to potentials on the polarization curve in the transpassive region where rapid metal removal takes place. At the sides of the cut, the solution iR is so large that the rate-controlling potential corresponds to points on the polarization curve in the passive region where metal dissolution cannot take place because of a protective film.

It was concluded from these studies (120, 143) that the potential range over which the transition from the passive (no cutting) to the transpassive (cutting) region occurs must be narrow to obtain good dimensional control. Such a sharp transition can be detected on the curve for $NaClO_3$ in Fig. 18.

The curve obtained in NaCl electrolyte (inverted triangles of Fig. 18) does not exhibit a passivation region, and large currents can be drawn even at low potentials. Consequently, metal removal can take place over relatively large solution paths, producing the wild cutting at the walls of the plunge cut.

If an electrolyte is used that produces a very protective and adherent film such as that produced in $Na_2Cr_2O_7$ solutions, a polarization curve similar to the squares in Fig. 18 is obtained. Although a sharp transition between the passive and the transpassive state exists, this transition occurs at such high potentials that sparking occurs before the ECM process can begin. It was observed that electrochemical machining could not take place in $Na_2Cr_2O_7$ solutions (149).

In $NaNO_3$ solutions, a curve (triangles in Fig. 18) is obtained similar to that found in $NaClO_3$ solutions but with a much less sharp transition from the passive to the transpassive state. This behavior agrees with the observation that dimensional control with $NaNO_3$ electrolytes is poorer than with $NaClO_3$ but better than with NaCl. Since the transpassive region for $NaNO_3$ lies above that for $NaClO_3$, the current density is less in $NaNO_3$ for a given potential, in agreement with the data from actual ECM observations that metal removal is slower in $NaNO_3$ than in $NaClO_3$.

2. Cathodic Film-Stripping Studies

To detect the presence of films on the electrode surface, cathodic stripping curves were obtained on the iron electrodes. After a steady-state current was reached at each potential setting of the potentiostat, the circuit was opened mechanically, and a constant current pulse developed by a pulsing circuit (150) employing a mercury-wetted relay was applied to the iron electrode. The resulting transient was displayed on an oscilloscope and recorded photographically. By assuming that all the current goes into the reduction of the oxide film and from the length of the transition time obtained from the photographed potential trace and the known value of the constant current, it is possible to determine the amount of charge, Q, associated with the adsorbed film (94). Plots of Q as a function of potential for anodized iron microelectrodes in solutions of $NaClO_3$, $NaNO_3$, NaCl, and $Na_2Cr_2O_7$ are presented in Fig. 19.

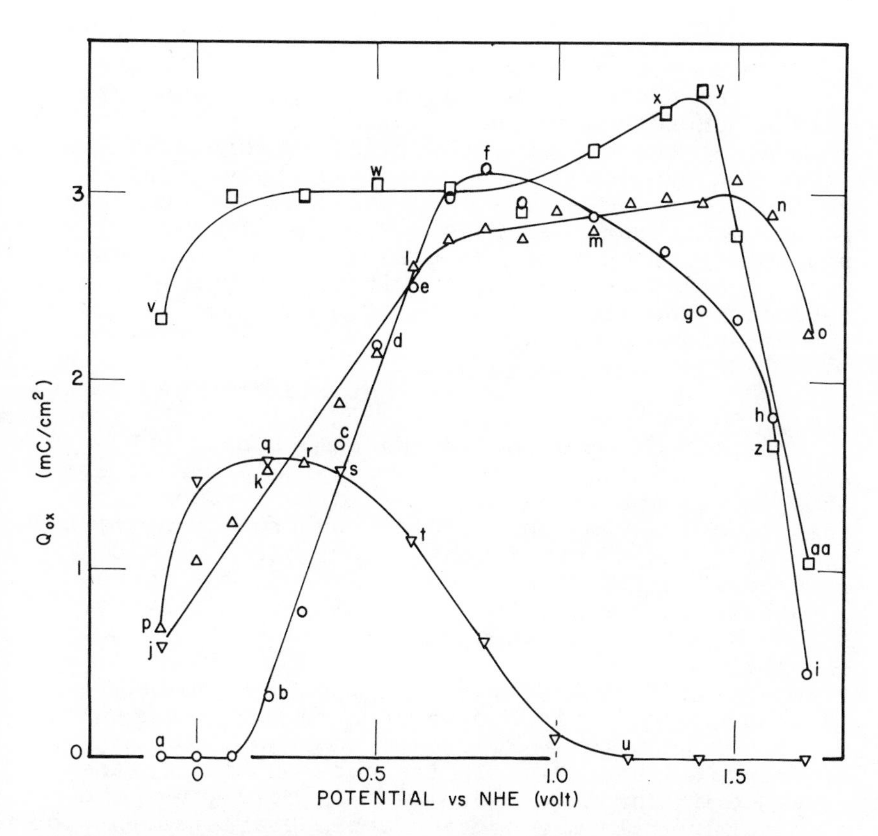

Fig. 19. A plot of the amount of charge, Q_{ox}, formed on the iron surface as a function of the potential in solutions of $NaClO_3$, circles; $NaNO_3$, triangles, NaCl, inverted triangles; and $Na_2Cr_2O_7$, squares (94). (By permission of The Electrochemical Society, Inc.)

The data in Fig. 19 indicate that the oxide film does not build up on the iron surface in $NaClO_3$ until a threshold potential is reached, after which the film grows rapidly until maximum thickness is reached in the passive region. In the transpassive region, the oxide layer thins quickly with increasing potential. A more adherent film is formed on iron in $NaNO_3$ since it is more difficult to remove in the transpassive region. A very protective film of oxide is formed on iron in $Na_2Cr_2O_7$ over virtually the entire range of potentials investigated, whereas only a nonprotective salt film, which is completely removed at the higher potentials, is formed on iron in NaCl solutions. These oxide-stripping studies support the conclusions obtained from the analysis of the polarization curves. Because it is not possible to determine the true area of the exposed iron electrode surface (94), it is not possible to determine the true thickness of the oxide film from the data in Fig. 19.

3. Nature of the Passive Film

The protecting agent in the ECM process with $NaClO_3$ solutions is a passivating film, the nature of which is not an unambiguous matter. From the results of vacuum fusion, electron probe, and electron diffraction experiments (120, 143), it was concluded that a film of α- or ∂-Fe_2O_3 was formed on steel samples machined in $NaClO_3$ electrolytes, but no film was detected on steel machined in NaCl electrolytes.

Evans (151) has referred to the passivating film on iron as "cubic iron oxide" which may be composed of either Fe_3O_4 or ∂-Fe_2O_3, or both, or some mixture of iron oxides (152). In the case of iron in the presence of chromate-containing electrolytes, the passivating film may contain considerable amounts of chromic oxide along with the cubic oxide (153). Some indication that the passive film may be composed of a layer of Fe_3O_4 below a layer of ∂-Fe_2O_3 as reported by Nagayama and Cohen (152) was observed in some cathodic stripping curves (94) that exhibited two arrests. According to the literature, the corrosion of iron is inhibited by a passive film of ∂-Fe_2O_3 in solutions of chromates (153, 154), carbonates (155), and phosphates (156). In the interest of simplicity, the passivating film on iron will be considered to be one of ∂-Fe_2O_3 in any further discussions.

The mechanism of the breakdown of passivating iron oxide films may be related to the ability of the anion to penetrate and dissolve the oxide layer (144, 145). The Cl^- ion is a very efficient agent for dissolving iron oxide films; and as noted in the data of Fig. 18 and 19, protective films are absent on iron surfaces in NaCl solutions. In contrast, the chromate ions are poor dissolvers of iron oxide films so that in such electrolytes the surface is so passivated that electrochemical machining cannot take place. We have observed that ECM metal removal is virtually nil in carbonate and phosphate electrolytes. Most likely, polarization curves in these solutions would be similar to the one for $Na_2Cr_2O_7$ in Fig. 18. Nitrate and chlorate ions behave in an intermediate manner, which accounts for the good dimensional control obtained with $NaClO_3$.

Experiments have been carried out on steel rotating disks (157 in solutions of NaCl and $NaClO_3$ (142, 158). Since little or no effect of rotational speed on the polarization values was detected, it was concluded that the metal dissolution process was under activation rather than mass transfer control. Theoretical calculations made by Fitz-Gerald, McGeough, and Marsh (159) have shown that the contribution to the overall polarization from concentration polarization is negligible, whereas that from activation polarization is significant. From an analysis of potentiostatic transients obtained on rotating steel disks, (Chin (158) found that a porous nonprotecting film was formed on the metal surface in NaCl solutions. The protective film on steel in $NaClO_3$ solutions was estimated to be about 1000 Å thick, and its rate of growth is controlled by the rate of iron dissolution.

4. Electrolyte Concentration Studies

To shed light on the effect of electrolyte concentration on ECM metal removal rates, steady-state potentiostatic polarization curves and galvanostatic stripping curves were obtained on mild steel microelectrodes in solutions of NaCl, $NaClO_3$, and $Na_2Cr_2O_7$ as a function of concentration (50 and 250 g/liter). The results (shown in Fig. 20) were compared with the electrolytic grinding of mild steel tube ends on a laboratory test rig (Fig. 21) in the same solutions as a function of the applied potential. Photomicrographs of the tube ends at 10 V are reproduced in Fig. 22 and at 20 V (30 V for $Na_2Cr_2O_7$) in Fig. 23.

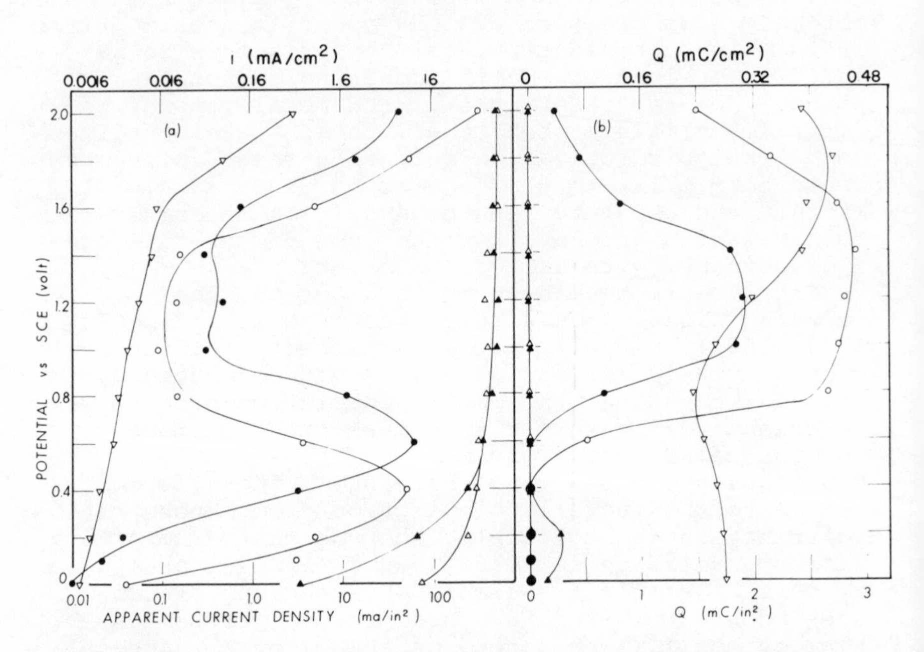

Fig. 20. (a) Steady-state potentiostatic polarization curves on mild steel, wire microelectrodes in O_2- saturated solutions of NaCl (triangles), $NaClO_3$ (circles), and $Na_2Cr_2O_7$ (inverted triangles). Filled symbols refer to solutions at 50 g/liter; open symbols, at 250 g/liter; inverted triangles, at 100 g/liter. (b) A plot of the theickness of the oxide films formed on mild steel microelectrodes in terms of m coulombs of charge, Q, per square inch determined from constant-current stripping pulses as a function of potential. Symbolic notation is the same as (a) (149). (By permission of The Electrochemical Society, Inc.)

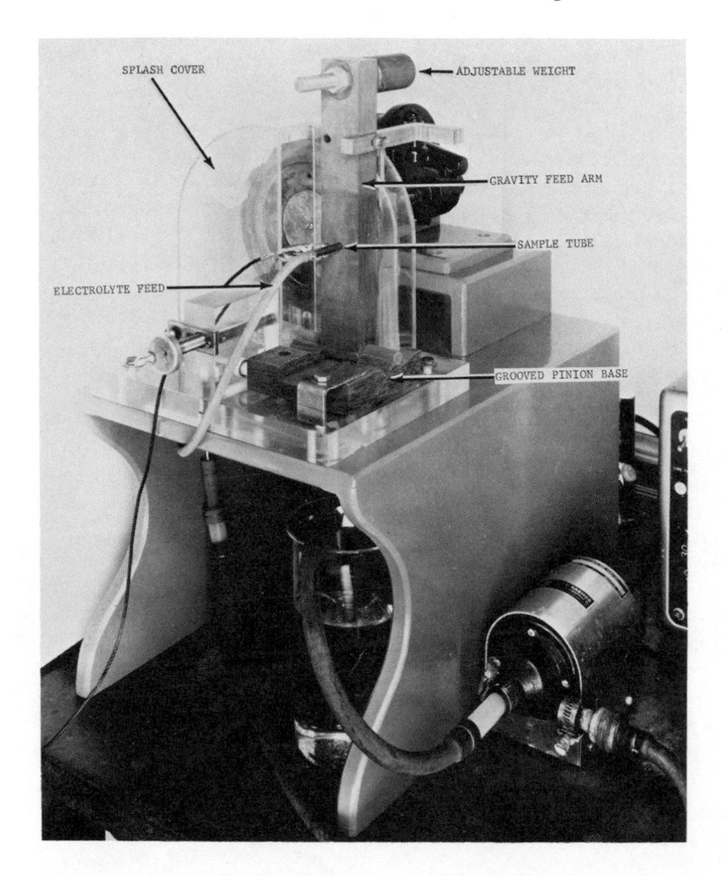

Fig. 21. The electrochemical grinding test rig (149). (By permission of The Electrochemical Society, Inc.)

From the tube end tests, it was observed that the metal removal rate was independent of the concentration of NaCl (0.079 in./min in 50 g/liter; 0.080 in./min in 250 g/liter at 20 V), which agrees well with the polarization data for NaCl (triangles) in Fig. 20 where the data for the two concentrations fall on the same curve. Because we are not interested in metal removal by stray cutting, metal removal rates are reported as the difference in lengths of the tubes before and after electrochemical machining where useful metal removal occurs only on the cross-sectional face of the tube. In addition, the rate of metal removal in NaCl was found (149) to be virtually independent of potential: 0.074 in./min at 10 V, 0.075 in./min at 15 V, 0.079 in./min at 20 V for 50 g/liter. The data of Fig. 20 for NaCl show the current density to be invariant over a wide potential range. By increasing the NaCl concentration, one merely increases the amount of wild cutting without increasing the shortening of the tube. As shown in the photomicrographs of Fig. 22 and 23, the cross-sectional area of the samples machined in NaCl had been reduced from the original value, giving evidence of cutting on both the inside and outside walls as well as the cross-sectional face of the tubes. The stray cutting increased with both potential and concentration.

When stray cutting is absent, metal removal occurs only on the cross-sectional face of the tubes as seen in Fig. 22 and 23 for tube ends machined in $NaClO_3$. The presence of a passivating film and a sharp transition from the passive to the transpassive state are observed in the data for $NaClO_3$ (circles) in Fig. 20. As the concentration of $NaClO_3$ is increased from 50 to 250 g/liter, the transpassive region is shifted to lower potentials. For a given potential, the current density (metal removal rate) is greater at the higher concentration. Metal removal at an applied voltage of 20 V was 0.048 in./min at 50 g/liter of $NaClO_3$ and was increased to 0.097 in./min at 250 g/liter (149). The agreement between basic studies and actual electrochemical machining is excellent. Further correlation is noted in the dependence on potential. The polarization curves in the transpassive region for $NaClO_3$ show that the current density increases with potential, and the data from the tube ends (149) show that the metal removal rate increased with applied potential: 0.070 in./min at 10 V, 0.097 in./min at 20 V, 0.106 in./min at 25 V for 250 g/liter $NaClO_3$.

The results of this investigation (149) indicate that ECM operation on steel in $NaClO_3$ is more advantageous at the higher salt concentrations, whereas the best results are obtained in NaCl at lower concentrations. These facts limit the operation in NaCl to conditions of lower metal removal rates since most attempts to operate at high NaCl concentrations produce such an increase in wild cutting that geometrical and dimensional control are lost.

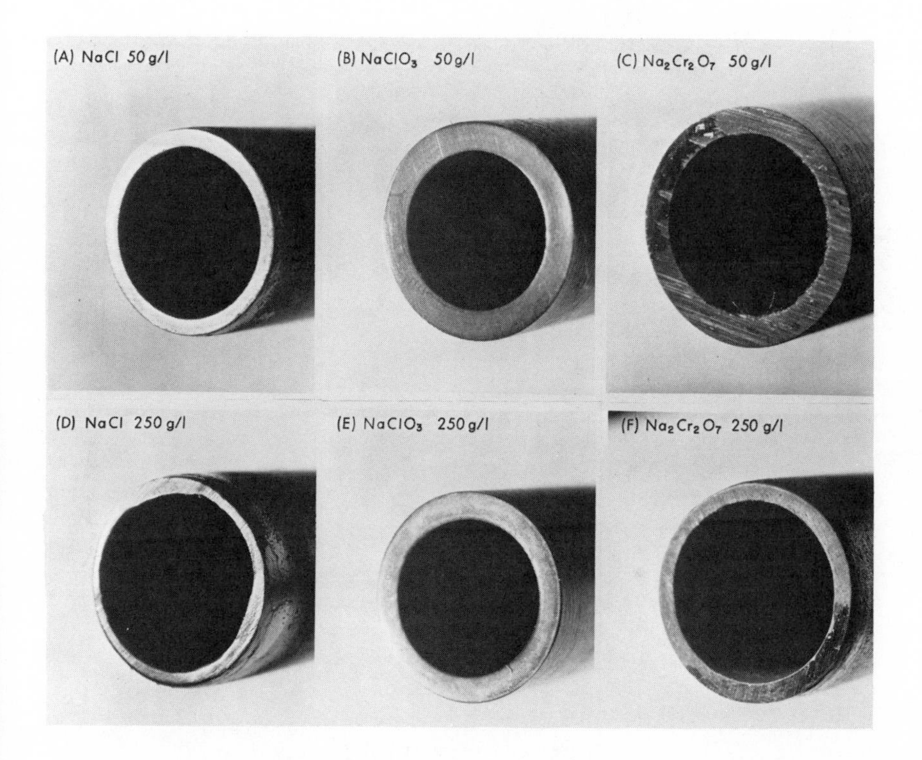

Fig. 22. Photomicrographs of tube ends after 1 min of electrochemical machining with a solution flow rate of 0.022 gal/min for two concentrations, 50 g/liter (A, B, C) and 250 g/liter (D, E, F) of NaCl (A, D), $NaClO_3$ (B, E), and $Na_2Cr_2O_7$ (C, F) at an applied voltage of 10 V (149). (By permission of The Electrochemical Society, Inc.)

As noted in Figs. 22c and 22d, there was no electrochemical machining of steel tube ends in $Na_2Cr_2O_7$ at the low applied voltage. When the voltage was raised to a point (30 V) where metal removal took place, sparking and electrical discharge machining (149) produced a totally unacceptable surface finish as seen in Figs. 23c and 23d.

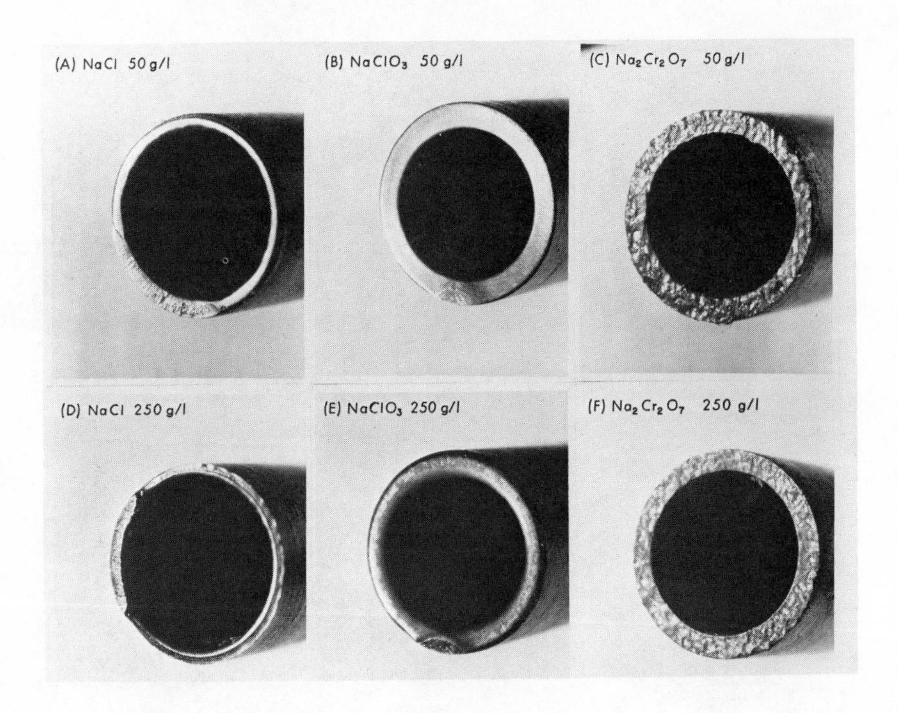

Fig. 23. Photomicrographs of tube ends under same conditions as in Fig. 22 at an applied potential of 20 V. ($Na_2Cr_2O_7$ tests were run at 30 V.) (149) (By permission of the Electrochemical Society, Inc.)

With extended periods of electrochemical machining in $NaClO_3$ solutions, the chloride ion content of the electrolyte increased until a point was reached where the solution had to be discarded because the presence of the Cl^- ion began to cause intolerable stray cutting. The sludge from electrochemical machining in NaCl is dark green, whereas the sludge from electrochemical machining in $NaClO_3$ is reddish brown. These observations indicate that the Fe^{++} ion is oxidized rapidly to the Fe^{+++} ion by the ClO_3^- ion, possibly by a reaction similar to $6Fe^{++} + ClO_3^- + 3H_2O \rightarrow 6Fe^{+++} + Cl^- + 6OH^-$.

To assess the detrimental effect of the presence of Cl^- ion in the $NaClO_3$ electrolyte, a series of experiments was carried out (120) with an experimental flow cell in which the electrolyte was pumped through a copper tube cathode and directed against an anodic test panel held at a fixed gap distance. Metal was removed from the panel in the form of a ring. When stray cutting was absent, a sharp ring-cut was observed. The electrochemical machining of steel panels was investigated with solutions of $NaClO_3$ to which increasing amounts of NaCl were added.

It was found that additions of NaCl below 50 g/liter to the $NaClO_3$ solution (350 g/liter) did not affect the nature of the ring-cut. Above 50 g/liter of NaCl, the amount of metal removed increased with the NaCl content of the electrolyte, and the area outside the ring-cut where metal was removed by stray currents also increased. This increased metal removal was independent of the $NaClO_3$ concentration in solutions of high NaCl content.

Since the perchlorate ion is known to be more stable than the chlorate ion, it is possible that $NaClO_4$ solutions may give a more stable ECM electrolyte, so measurements were made (160) on mild steel microelectrodes in solutions of $NaClO_4$ (250 g/liter) to predict the ECM behavior of the $NaClO_4$ electrolyte. The results of the polarization and film stripping curves, Fig. 24, predict that good dimensional control should be obtained with $NaClO_4$ since the polarization curve exhibits a sharp transition from the passive to the transpassive state. In the passive region, the film growth in $NaClO_4$ appears to be slower than in $NaClO_3$ because current decreases much more slowly after the film begins to form; with $NaClO_4$ solutions, the oxide film is completely removed in the transpassive region according to the stripping curve data of Fig. 24b. Such behavior predicts that a poorer surface finish will be obtained with $NaClO_4$ than with $NaClO_3$ electrolytes according to the viewpoint outlined below.

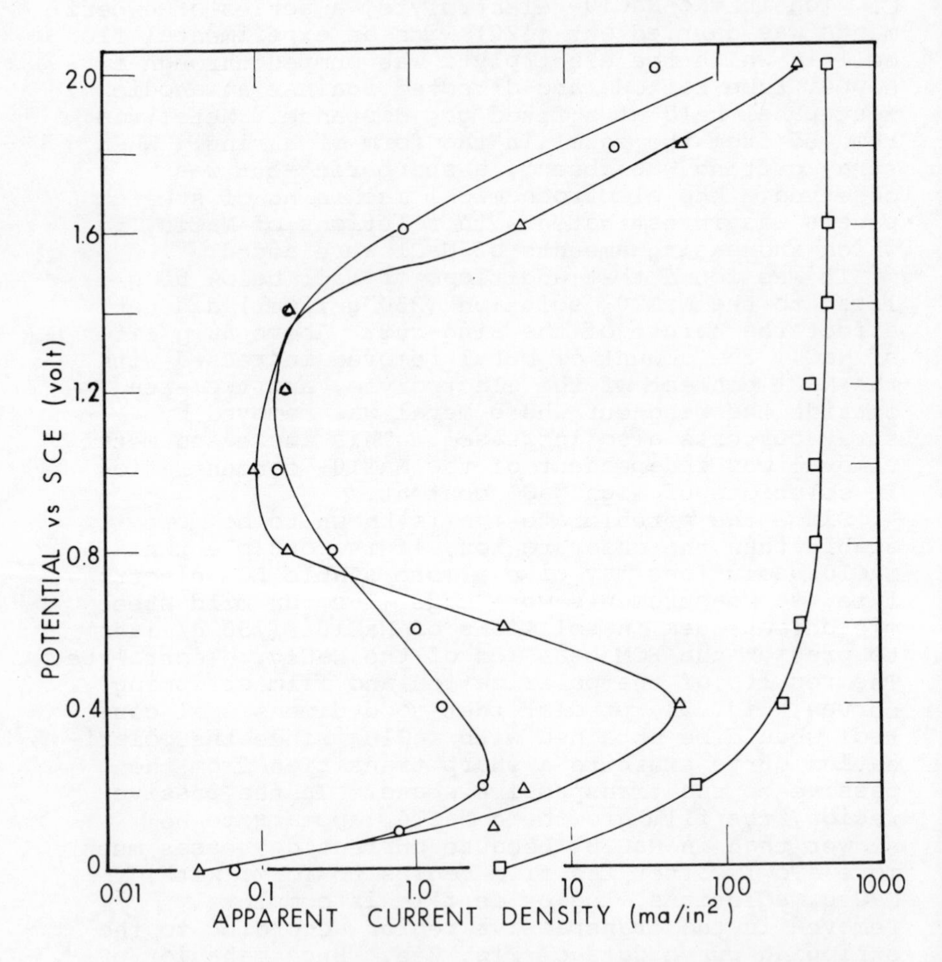

Fig. 24 (a) Steady-state polarization curves obtained on mild steel microelectrodes on O_2-stirred solutions of $NaClO_4$ (250 g/liter; circles), $NaClO_3$ (250 g/liter; triangles), and NaCl (50 g/liter; squares).

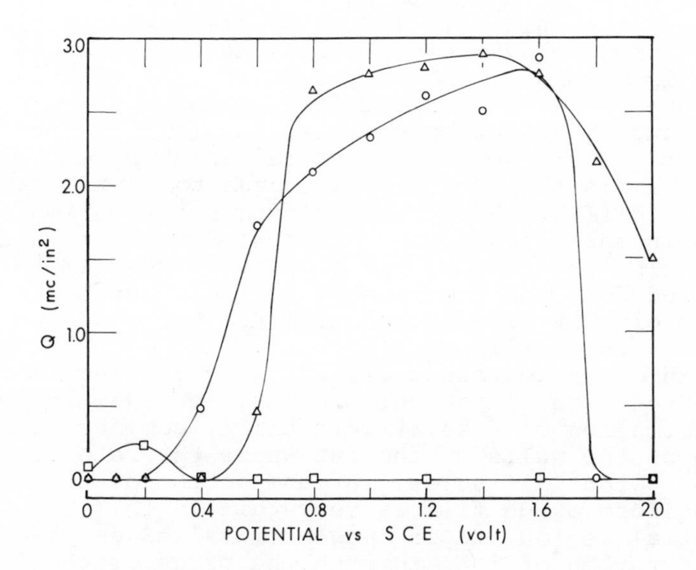

(b) Plot of the amount of charge, Q, associated with the oxide formed on mild steel anodes in O_2-stirred solutions of $NaClO_4$ (circles), $NaClO_3$ (triangles), and NaCl (squares) as a function of potential (160). (By permission of the National Association of Corrosion Engineers.)

5. Role of the Oxide Film in Surface Finishing

The mirror-like finish of steel samples machined in $NaClO_3$ electrolytes is the result of an electropolishing process that takes place by the two-layer film mechanism proposed by Hoar and Mowat (31, 41). At the high metal dissolution rates obtained with electrochemical machining, there is a layer next to the anode surface in which the concentration of the metal ions is very high and which may be called a salt film. The diffusion effects of ions in this layer produce a leveling or smoothing of the metal surface. Since this salt layer most likely exists with all electrolytes used in electrochemical machining, including NaCl (94, 158), it accounts for the fact that relatively good surface finishes

(10 to 40 μin.) on steels and high-temperature alloys are obtained with machining in all ECM electrolytes.

Below the salt film in NaClO electrolytes lies a thin, compact layer of ∂-Fe_2O_3, the thickness of which is dependent on the potential. At relatively low potentials, this oxide film is thick and protective, but at higher potentials, the film becomes thinner. Consequently, at certain high potentials this film is thin and porous enough to permit ferrous ions to diffuse through the random distribution of pores in the film to give a brightening of the surface. The presence of the compact oxide film is required for good dimensional control, but it is the nature of this film that determines how good the surface finish will be.

With ClO_3^- ion electrolytes, wild cutting is prevented in a plunge-cut process, for example, by the formation of a relatively thick, compact film of oxide on the walls of the cut where the rate-determining potential is low. At the bottom of the cut, the uniform oxide film is very thin in this high-potential region, rapid metal removal takes place by the migration of ions through the pores of the film, and electropolishing of the surface takes place under the thin uniform film. This thin film is dissolved as quickly as it is formed (31, 41) so that the film follows the receding metal surface of the anode as metal removal proceeds. Since films of ∂-Fe_2O_3 on iron surfaces are not stable in Cl^- ion-based electrolytes, mirror-bright surfaces (∿2 μin.) may not be obtained with the electrochemical machining of steel in NaCl solutions.

The effect of electrolyte flow on the electropolishing process has been investigated by Hoar and Rothwell (161). The threshold potential above which electropolishing begins rises with increasing flow rate. Although flow rates are high in electrochemical machining, the operating potential is high enough to produce the thin oxide films necessary for the electropolishing of steel in $NaClO_3$ electrolytes.

6. Agreement Between Basic Studies and Actual Electrochemical Machining

From the data in Fig. 24, it is seen that an oxide film is formed on steel in $NaClO_4$ (circles) solutions, which should give good dimensional control, but the oxide is completely removed in the transpassive region in contrast to the behavior in $NaClO_3$ (triangles) solutions where a thin oxide film can still be detected in the transpassive state. For this reason, it is predicted that a relatively poor surface finish will be obtained with $NaClO_4$ as compared to $NaClO_3$ in the low current density region.

These predictions were borne out (160) by the actual ECM machining in $NaClO_4$ solutions of steel panels with the flow cell fixture (120) and of steel ends in an ECM cell consisting of a movable anode through which electrolyte was pumped and a fixed cathode (160). The results of these studies are summarized in the photomicrographs of the tube ends in Fig. 25 and in the profile sketches of the panels in Fig. 26. It is seen that the stray cutting of the steel tube in $NaClO_4$, Fib. 25b, is much less than that for NaCl, Fig. 25a, and is similar to that for $NaClO_3$, Fig. 25c. Because the oxide film on steel in $NaClO_4$ solutions is likely to be discontinuous, if present at all, pitting and etching of the metal surface result, Fig. 26, in the low current density region. Pitting of the surface and stray cutting are absent on the panel machined in $NaClO_3$, Fig. 26.

From the excellent agreement between the basic studies of passivation phenomena and the actual ECM experiments, it is concluded that the presence of a protective, compact film of oxide is necessary for dimensional control of the ECM operation. How good the dimensional control will be depends on the sharpness of the transition from the passive to the transpassive state. The quality of the surface finish depends on the nature of the oxide film formed. The film thickness is dependent on potential so that a protective, compact, uniform film is formed on sites remote from the cathode, whereas a thin, uniform, porous oxide film is formed at sites directly opposite from the cutting face of the cathode, producing electropolishing of the metal surface. As fast as the film is formed, it is dissolved so that a steady-state thickness is reached and the film follows the machining of the workpiece.

Fig. 25. Photomicrographs of mild steel tube ends machined by electrochemical machining at 170 A/in.^{2} (a, b, c) and 240 A/in.^{2} (d, e. f) in 4.5M solutions of NaCl, $NaClO_4$, and $NaClO_3$ (160). By permission of the National Association of Corrosion Engineers.)

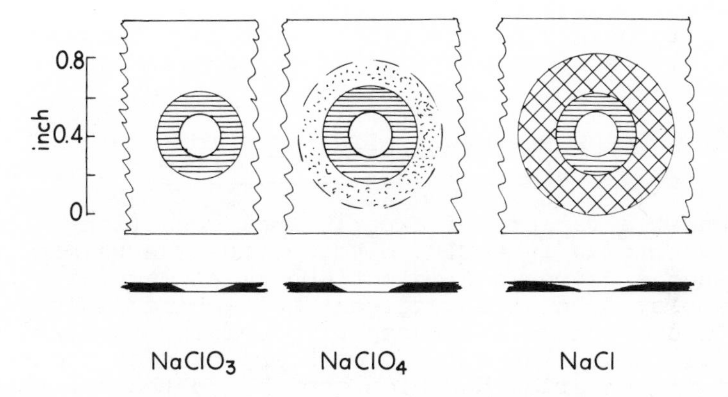

Fig. 26. Sketches drawn to scale of the corrosion patterns obtained on mild steel panels through which a tapered hole had been machined by electrochemical machining (160). The bull's-eye represents the hole; the lined area represents the tapered (high current density) region with small circle as the ID of the taper and the larger circle as the OD of the taper. In the low current density region outside of the taper area, there is a pitted area of slight metal removal in $NaClO_4$ (dotted), a region of heavy metal removal--wild cutting--in NaCl (cross-hatched), and no metal removal in $NaClO_3$. Below each panel is a side view showing the profile of the taper and the amount of wild cutting. (By permission of the National Association of Corrosion Engineers.)

7. Throwing Power

Recently, Brook and Iqbal (162) have attempted to use the electroplating concept of "throwing power" (163), which is a measure of the ability of an electrolyte. They (162) used a modified Haring-Blum cell, which consists of two cathodes spaced at unequal distances from a common anode. The throwing power is expressed as a function of the ratio of the distances to the two cathodes. Since this ratio may be obtained from more than one empirical formula (164), a consistent value may not be obtained. The use of a logarithmic function of the ratio of lengths has been shown by Chin and Wallace (165) to be a possible method for classifying ECM electrolytes. For good dimensional control, an electrolyte should have a low throwing power.

B. New Applications for Electrochemical Machining

Under a variety of experimental conditions, the stray cutting in $NaClO_3$ electrolytes can be controlled to such a degree that electrochemical machining can be employed on areas adjacent to machine-finished surfaces with an exactness of machining that was previously unavailable. The work appeared as though a stop-off material had been applied to the entire nonworked area, but no coating of any kind had been used in spite of the fact that the entire workpiece may have been wetted with the electrolyte. The use of the $NaClO_3$ electrolyte makes possible an ECM operation with such precise control and such fine surface finish that entirely new areas of ECM applications suggest themselves. Some of these applications are described in the following sections.

1. Static Fixture Finishing and Sizing

One of the new concepts in the expanding development of the ECM process is the use of ECM fixtures instead of ECM machines for finishing surfaces, and as an accurate sizing device. An ECM fixture is, basically, the electrochemical cell that consists of the anode and cathode, the cell housing, and any structure designed to provide the proper flow of electrolyte through the cell. With the same ECM machine consisting of the pumping system and controls, the rectifier, the heat exchanger, and filter, a wide variety of operations may be carried out by merely replacing one fixture with another.

A fixture similar to the one described before (142) was used for the ECM finishing and sizing of the bushings shown in cross section in Fig. 27. The wall area of the hole to be machined was 1.3 $in.^2$, and the bushing material was hardened Carpenter R.D. S. steel (C, 0.070%; Mn, 0.35%; Si, cell fixture is presented in Fig. 28.

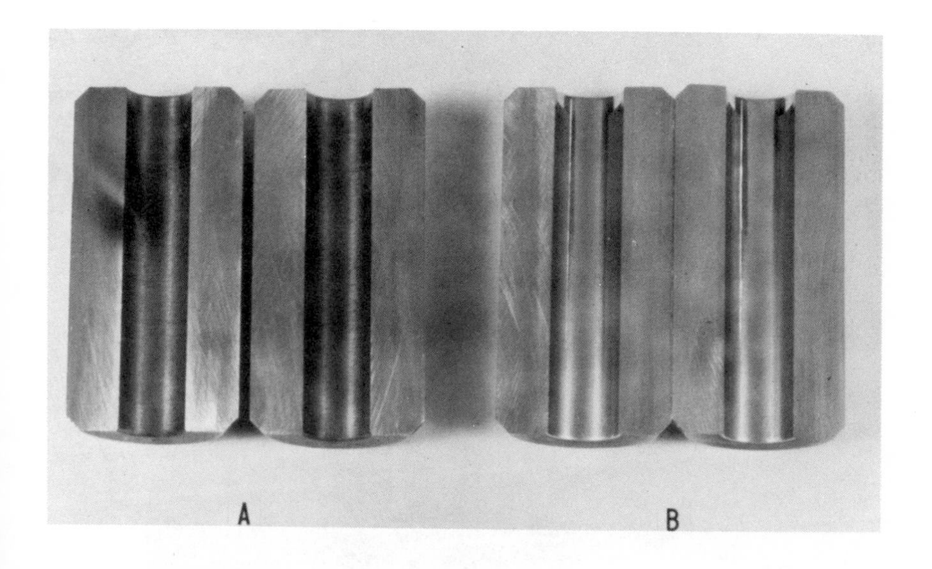

Fig. 27. Screw-machined, heat-treated bushings before (A) and after (B) electrochemical machining in $NaClO_3$ (142). (By permission of the Electrochemical Society, Inc.)

After the bushing was inserted in the cell fixture and the cell sealed, the electrolyte was pumped through the cell at pressures from 35 to 100 psig and at flow rates from 2 to 4 gal/min. The electrical circuit to the rectifier was closed manually but opened by an automatic timing device after a preset, required time. In the electrochemical machining of the average bushing, it required 300 A at 20 to 25 V (open-circuit reading) to remove 0.0033 to 0.0040 in. in 7 sec. Table 1 contains the results of a sample run. With this method, the bushings were finished with an overall average taper along the 1.5-in. hole of 20 to 90 millionths of an inch and an average out-of-roundness of 40 to 80 millionths of an inch. From an analysis of present-day machining operations, it is indicated that no machine tool exists that can remove as much metal per unit time with such precise control and excellent surface

Fig. 28. Through-hole ECM fixture for machining bushing (142). (By permission of The Electrochemical Society, Inc.)

finish from the inside of a hole in fully hardened steel as the ECM finishing does with the described fixture in $NaClO_3$ solution.

A similar cell fixture is employed to size and finish the M-16 rifle, and the machine designed for this operation is given in Fig. 29. A close-up view of the fixture showing the cathode tool and anode rifle chamber in cross section is pictured in Fig. 30. The $NaClO_3$ electrolyte is pumped between the anode and cath-

TABLE 1

Finishing and Sizing of Bushings
With Electrochemical Machining in $NaClO_3$ Electrolytes

Sample No.	I.D. Measurements (in.) Top End	Center	Bottom End	Maximum Taper (millionths)	Out-of-Round (millionths) Top	Bottom
1	0.30015	0.30013	0.30014	20	28	30
2	0.30012	0.30016	0.30020	80	45	45
3	0.29983	0.29985	0.29988	50	80	110
4	0.30087	0.30084	0.30096	90	60	65
5	0.30068	0.30066	0.30075	90	45	70
6	0.29942	0.29941	0.29943	20	55	85
7	0.29982	0.29984	0.29999	170	175	185
8	0.30000	0.29997	0.30006	90	35	50

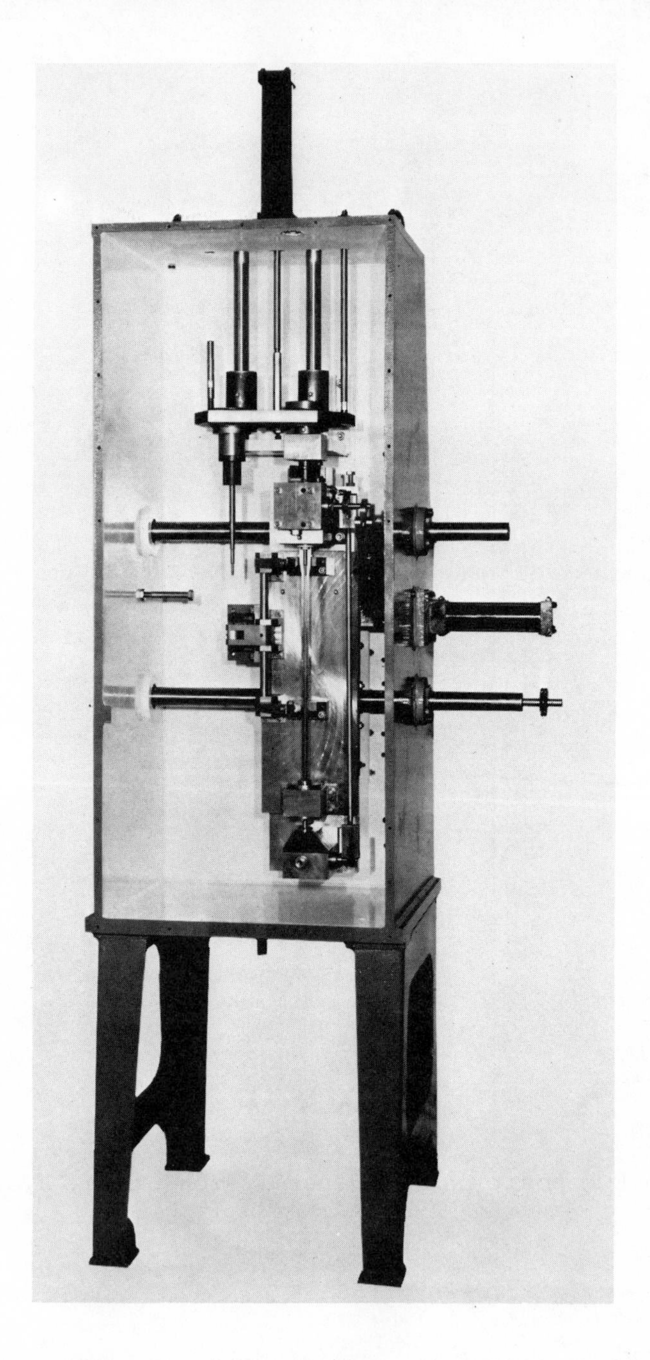

Fig. 29. The ECM rifle chamber finishing machine. (By permission of the Hydra-matic Division, General Motors Corporation.)

Fig. 30. Cross section of rifle chamber showing cathode in ECM position. (By permission of the Hydra-matic Division, General Motors Corporation.)

ode (designed to generate proper flow characteristics) so that a total metal removal of 0.03 to 0.05 in. is obtained in 5 sec of electrochemical machining. Metal removal takes place only in a fixed area of the chamber with a surface finish of 5 to 8 μin.

2. ECM Embossing

The unique ECM characteristics of the $NaClO_3$ electrolyte lend themselves to another broad class of machining operations. Because of the low overcut nature of $NaClO_3$ solutions, it is possible to design

Fig. 31. Embossing fixture for hydrostatic bearing pads showing cathodes, clamp lid, and clamps.

an arrangement of fixturing such that embossed-type patterns both relatively fine and somewhat gross may be machined by electrochemical machining in finely finished steel surfaces. A fixture for performing this type of an ECM operation is shown in Fig. 31.

In this application, the cathode is so constructed that only the front face is "visible" to the anode by cutting the cathode to the desired imprint geometry, potting it in epoxy, and grinding the face flat after the epoxy had cured. With the fixture of Fig. 31, the anode and cathode were initially spaced from 0.005 to 0.010 in. from one another, and $NaClO_3$ electrolyte (3 lb/gal) was pumped between the anode and cathode, preferably from the center outward. The duration of electrochemical machining is controlled by a timing device attached to the rectifier.

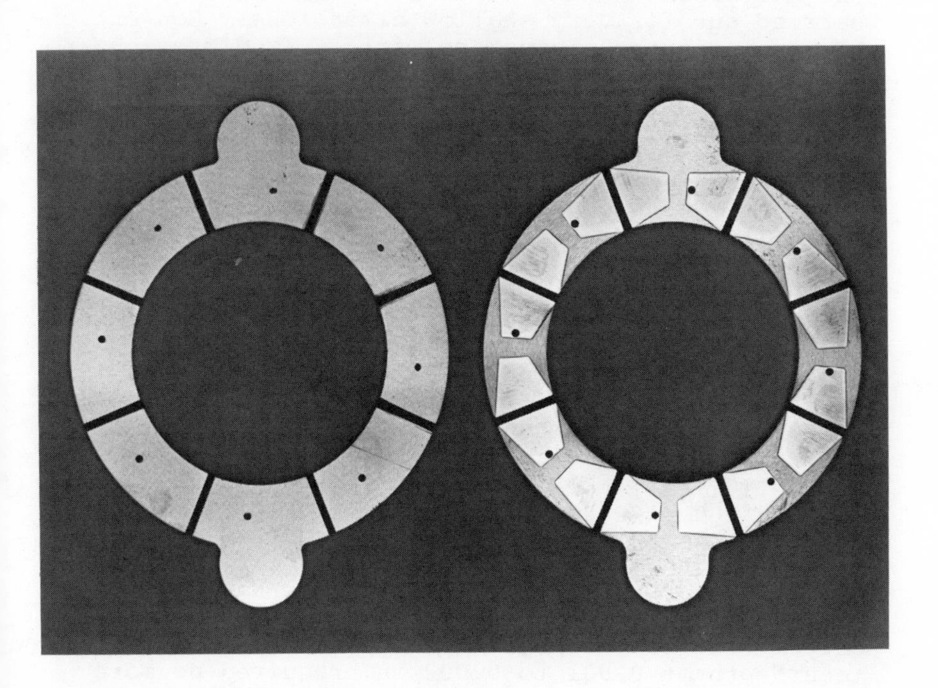

Fig. 32. Hydrostatic bearing pads (right) cut in the fixture shown in Fig. 31 from blank (13ft).

Oil pade from 0.00005 to 0.005 in. have been machined in bearings with excellent lineage, geometry, and depth control as exemplified by the hydrostatic bearing pictured in Fig. 32. In this case, the ECM time was 1 sec to machine the pads to a depth of 0.000025 in. It is to be noted that if greater depths of cutting are required in the pads, it is more difficult to control the flatness variation. Only experimentation can provide any significant control of this problem.

The value of such an ECM process becomes apparent in the knowledge that the processing of such configurations as oil pads in a steel surface is a constly, time-consuming operation with conventional machining equipment. In fact, with such methods, the shallower the depth required, the more difficult the job becomes until it is no longer possible to be carried out for very shallow dimensions. Conventional machining techniques are limited to dimensions no shallower than 0.0005 in. With ECM machining time of only 1 to 2 sec for the production of the hydrostatic bearing pads, a method for industrial production becomes entirely feasible in an area where previously none existed.

Figure 33 shows dimple indentations in steel that were finished and sized using a cell fixture similar to the one used for the bearing pads. Justification of the use of electrochemical machining on this part is made by the face that the ECM process can finish the dimples to 5 μin. in seconds, whereas hours are required by conventional means.

In an extension of the embossing-type ECM technology, a lead screw that is cut in a shaft is presented in Fig. 34 along with a blank shaft. Contained in Fig. 35 is a picture of the cell fixture used in the electrochemical machining of the screw with the shaft in machining position. As in all embossing operations, the cathode is carefully constructed and spaced in the cell. The $NaClO_3$ solution pumped between the anode and the cathode is the only moving component of the system. To cut the screw to a depth of 0.001 to 0.002 in. required no more than 2 to 3 sec with electrochemical machining. Since the line width, spacing, and depth of cut may be varied, any desired combination of parameters

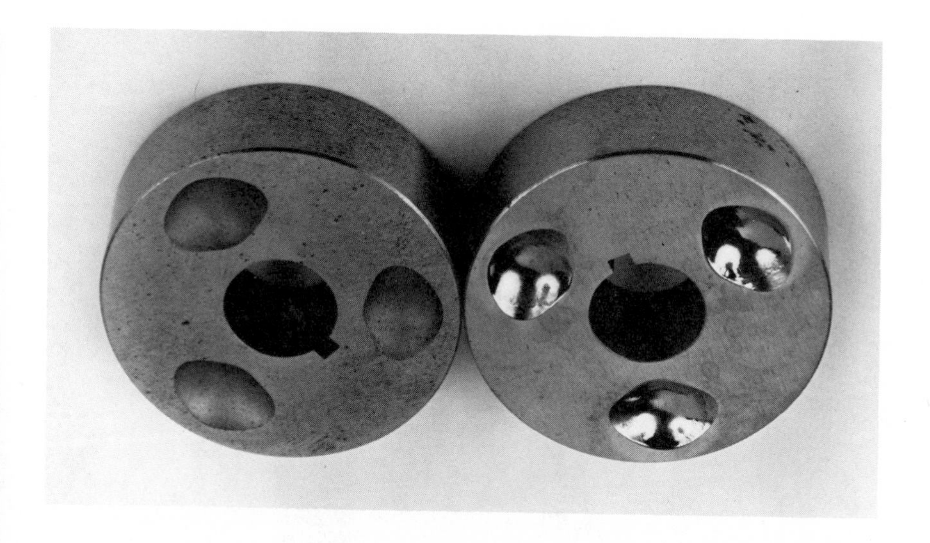

Fig. 33. ECM finishing of roller dimples to 5 μin.; blank (left) and finished product (right). (By permission of the New Departure-Hyatt Bearings Division, General Motors Corporation.)

may be obtained through trial-and-error experimentation. The reversal of the solution flow during two increments of cutting may be useful in improving the geometry and flatness variation in the bottom of screw cut.

A critical part of this operation is the correct centering of the shaft in the cell fixture so that the cathode is equidistant from the anode. If the alignment is not precise, an eccentricity of the screw depth and the shaft centerline will result. Such a situation usually is detrimental to the usage of the part.

For this application, the ECM processing possesses a staggering economic advantage over other techniques. Once the desired pattern and spacing are determined, thousands upon thousands of identical parts can be produced in one day. It is virtually impossible to produce this part by conventional means.

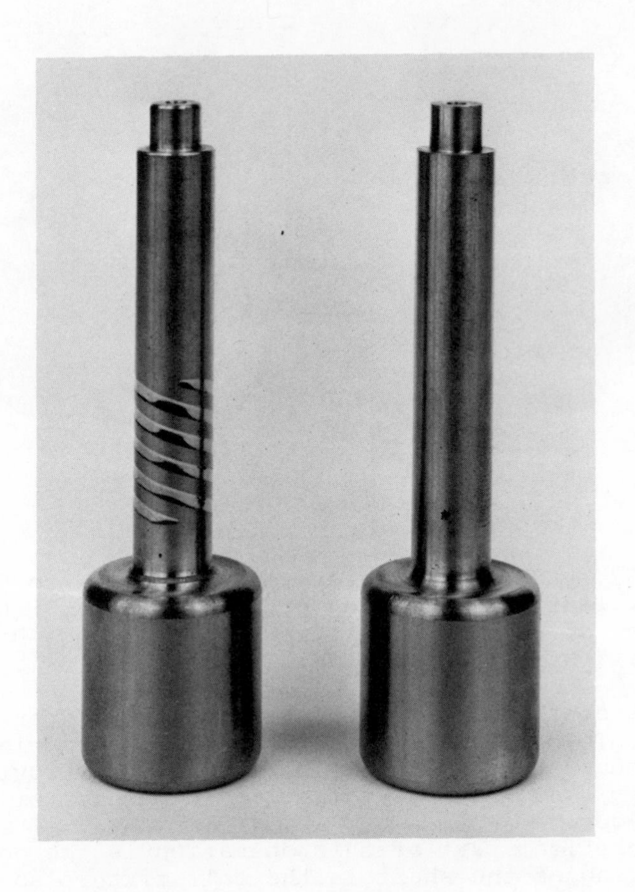

Fig. 34. Electrochemical machining of triple lead screw (right) in shaft blank (left).

Fig. 35. Fixture for cutting of lead screw shown in Fig. 34 with blank in position.

3. ECM Broaching

One of the severe limitations to the ECM process is the impracticality of the system for machining large areas. At an average current density of 250 A/in.^{2}, the usual ECM machine (10,000 A) is capable of machining about 40 in.^{2} at one time. The area limitations can be overcome in many cases by using a moving limited-area cathode that removes metal as it travels past the anode. If the geometry of the anodic workpiece is regular, it is possible to produce the part with an electrolytic grinder, and electrolytic lathe, or an electrolytic broaching machine.

One type of electrolytic broaching machine is shown in Fig. 36, and the required cell fixture is pictured in Fig. 37. The principle of the ECM broaching operation is one in which the inside diameter of an anodic workpiece is ECM machined by a cathode passing through the anode. By rearranging the components of the fixture so that the anode moves through a fixed cathode of toroidal shape, the outside diameter of the workpiece may be machined. Although most interest is directed toward machining of the inside diameter or outside diameter of workpieces having a right circular cylinder geometry, the ECM broaching of other right geometries (oval, square, rectangle, etc.) can be carried out with the proper fixturing and cathode design.

The broaching machine shown in Fig. 36 can accept a part for which the length of required stroke may range from a fraction of an inch to 20 in. and has the capability of directing the flow of the electrolyte from either the top or the bottom of the fixture. By using a Vari-Drive unit, the broach has a travel speed ranging from 0.05 to 1.0 in./sec; thus a cathodic "window" may be moved through an anodic area 20 in. long in 20 sec. At such machining rates, the process is obviously commercially interesting. With the electrolytic broach, the numerous parameters that must be controlled in the ECM process are increased by two more, namely, the stroke velocity and the accurate guidance of probe travel.

The application of ECM broaching with $NaClO_3$ electrolytes is in its infancy. With the electrolyte previously used, this type of operation would only be

Fig. 36. Electrolytic broach machine.

Fig. 37. Typical fixture for electrolytic broach machine.

used for metal removal; but with the use of $NaClO_3$ solutions, one is provided with metal removal as well as finishing, taper control, and sizing. A number of applications, such as the finishing and sizing of long rods, the skinning of machined porous surfaces, and the finishing of inside diameters instead of honing, among others, are now possible.

ACKNOWLEDGMENTS

The authors are indebted to the management of the General Motors Corporation, in particular to Dr. R.F. Thomson, Executive Director, Dr. S. E. Beacom, Technical Director, Basic and Applied Sciences, and Dr. J. L. Hartman, Head of the Electrochemistry Department, General Motors Research Laboratories, for permission to write the chapter on electrochemical machining.

Our sincere gratitude is offered to Mr. A. J. Chartrand of the Electrochemistry Department and to Mr. T. L. Davis of the Processing Department for their expert technical assistance in acquiring new data and to the personnel of the Processing, Mechanical Research, and Vehicle Research Departments of General Motors Research Laboratories for their invaluable aid in the construction of cells and fixtures and in the measurement of processing parameters. To the Hydra-matic Division, particularly Mr. W. Stone, and to the New Departure-Hyatt Bearings Division, particularly Mr. L. Davey and Mr. D. King, we express thanks for their valuable comments; also, to Mr. L. Williams and Mr. T. Tomezak of Anocut Engineering Company and to Mr. L. Nelson of Chem-Form Division, KMS Industries, Inc., for their gracious cooperation in supplying data on their equipment.

Heartfelt thanks are due to Dr. K-W. Mao, Dr. D. T. Chin, Mr. A. J. Wallace, Jr., Mr. M.E. Wheatley, and Mr. W. R. Vincent of the Electrochemistry Department of General Motors Research Laboratories for their helpful comments and suggestions. Mr. J. D. Thomas and Dr. J. L. Griffin, also of the Electrochemistry Department, read the manuscript critically and authoritatively, for which we are duly grateful.

Acknowledgements

Special thanks are given to Miss G. Sobieska of the Executive Department for her expert editing and preparation of the manuscript; to Mr. G. H. Tucker, Mr. K. O. Wetter, and their associates of the Publications Section, under Mr. A. P. Bohn, and to Mr. J. C. Seifert and Mr. W. L. Serveny of the Photographic Section of the Technical Information Department for their outstanding skill in preparing the manuscript in its final form. As always, the Library staff of General Motors Research Laboratories, under Mr. R. W. Gibson, was most cooperative and helpful, for which we are sincerely thankful.

REFERENCES

1. A. E. DeBarr and D. A. Oliver, Electrochemical Machining, Elsevier, New York, 1968, p. 11.

1a. J. P. Hoare and M. A. LaBoda, Sci. Am. 230, 30, No. 1 (1974).

2. C. Wagner and W. Traud, Z. Electrochem. 44, 391 (1938).

3. J. P. Hoare, Electrochemistry of Oxygen, Interscience, New York, 1968, pp. 357 et seq.

4. U. R. Evans, The Corrosion and Oxidation of Metals, St. Martin's Press, New York, 1960, p. 119.

5. P. M. Aziz, Corrosion 9, 85 (1953); J. Electrochem. Soc. 101, 120 (1954).

6. R. H. Brown and R. B. Mears, Trans. Electrochem. Soc. 74, 495 (1938).

7. C. Edeleanu and U. R. Evans, Trans. Faraday Soc. 47, 1121 (1951).

8. U. R. Evans and D. E. Davies, J. Chem. Soc. 1951, 2607.

9. U. R. Evans, Corrosion 7, 238 (1951).

10. O. D. Block and L. H. Cutler, Ind. Eng. Chem. 50, 153 (1958).

11. M. Byer, Materials and Methods 43, 134, No. 6 (1956).

12. M. C. Cook, Product Eng. 27, 194, No. 7 (1956).

13. W. C. Hittinger and M. Sparks, Sci. Amer. 213, 57, No. 5 (1965).

14. L. H. Sharpe and P. D. Garn, Ind. Eng. Chem. 51, 293 (1959).

15. R. L. Swiggert, Mod. Plastics 31, No. 8, 94 (1954).

16. R. C. Benton and G. D. Woodring, ASTME Paper No. 685 (1965).
17. U. R. Evans, L. C. Bannister, and S. C. Britton, Proc. Roy. Soc A131, 355 (1931).
18. M. J. Pryor and D. S. Keir, J. Electrochem. Soc. 104, 269 (1957).
19. S. F. Dorey, J. Inst. Metals 82, 497 (1953).
20. L. Kenworth, J. Inst. Metals 69, 67 (1943).
21. E. C. J. March, Electroplating 7, 88 (1954).
22. N. H. Simpson, Corrosion 13, 151t (1957).
23. T. P. Hoar, Corrosion Sci. 7, 341 (1967); J. Electrochem. Soc. 117, 17C (1970).
24. P. A. Jacquet, Sheet Metal Ind. 24, 2015 (1947).
25. P. A. Jacquet, Compt. Rend. 201, 1473 (1935); Trans. Electrochem. Soc. 69, 629 (1936).
26. P. A. Jacquet and M. Jean, Compt. Rend. 230, 1862 (1950); Rev. Met. 48, 537 (1951).
27. O. Zmeskal, Metal Prog. 47, 729 (1945).
28. J. Edwards, J. Electrochem. Soc. 100, 189C, 223C (1953).
29. W. C. Elmore, J. Appl. Phys. 10, 724 (1939); 11, 797 (1940).
30. T. P. Hoar and T. W. Farthing, Nature 169, 324 (1952).
31. T. P. Hoar and J. A. S. Mowat, Nature 165, 64 (1950).
32. J. K. Higgins, J. Electrochem. Soc. 106, 999 (1959).
33. A. Krusenstjern and H. Schlegel, Metalloberflasche 10B, 148 (1955).
34. H. Pray and C. L. Faust, Iron Age 145, No. 15, 33 (1940).
35. W. C. Elmore, J. Appl. Phys. 10, 724 (1939); 11, 797 (1940).
36. J. Edwards, J. Electrochem. Soc. 100, 189C, 223C (1953).
37. J. A. Allen, Trans. Faraday Soc. 48, 273 (1952).
38. K. P. Balasev and E. N. Nikitin, Ah. Prikl. Khim. 23, 263 (1950).
39. H. F. Walton, J. Electrochem. Soc. 97, 219 (1950).
40. E. C. Williams and M. A. Barrett, J. Electrochem. Soc. 103, 364 (1956).
41. T. P. Hoar and J. A. S. Mowat, J. Electrodepositors Tech. Soc. 26, 7 (1950).

42. J. L. Weininger and M. W. Breiter, J. Electrochem. Soc. 110, 484 (1963).
42a. M. Novak, A. K. N. Reddy, and H. Wroblowa, J. Electrochem. Soc. 117, 733 (1970).
42b. J. Kruger, Corrosion 22, 88 (1966); A. K. N. Reddy, M. L. B. Rao, and J. O'M. Bockris, J. Chem. Phys. 42, 2246 (1965).
43. A. Uhlir, Rev. Sci. Instr. 26, 965 (1955).
44. J. L. Bleiweis and A. J. Fusco, Metals and Alloys 18, 1075 (1943).
45. J. W. Tilley and R. A. Williams, Proc. IRE 41, 1706 (1953).
46. P. F. Schmidt and M. Blomgren, J. Electrochem. Soc. 106, 694 (1959).
47. P. F. Schmidt and D. A. Keeper, J. Electrochem. Soc. 106, 593 (1959).
48. P. M. Kelly and J. Nutting, J. Iron Steel Inst. 192, 246 (1959).
49. G. Keeleric, Steel 130, No. 11, 84 (1952).
50. L. H. Metzger and G. Keeleric, Am. Machinist 96, No. 23, 154 (1952).
51. J. A. Müller, Am. Machinist 97, No. 20, 122 (1953).
52. R. R. Cole, Trans. ASME 83B, 194 (1961).
53. C. R. Stroup. Am. Machinist 102, No. 1, 106 (1958).
54. F. Pearlstein, Am. Machinist 102, No. 1, 110 (1958).
55. J. S. Spizig, Werkstattstechnik Maschinenbau 1958, 641.
56. N. D. G. Mountford, Trans. Inst. Metal Finishing 40, 171 (1963).
57. G. Comstock, Aircraft Prod. 16, 488 (1954).
58. G. Pahlitzsch, International Research Production Engineering, ASME, 1963, p. 242.
59. R. R. Cole, J. Eng. Ind. 1961, 194.
60. D. Fislock, Metalworking Prod. 105, 73 (Aug. 1961).
61. H. Reinhart and W. Grünwald, Werkstatt Betrieb 1962, 212.
62. K. E. Schwartz, Industrie-Anzeiger 43, 1357 (1960).
63. H. Reinhart, Werkstatt Betrieb 1961, 529.
64. G. Koscholke, Werkstattstechnik Maschinenbau 1955, 562.

65. J. Frisch and E. G. Thomsen, Trans. ASME 73, 337 (1951).
66. J. Frisch and R. R. Cole, Trans. ASME 84B, 483 (1962).
67. L. A. Williams, Aircraft Prod. 22, 389 (1960).
68. A. L. Livshits and V. Ya. Rassakhim, Stanki Instr. 25, 12 (1954); 26, 8 (1955); Eng. Dig. 16, 429 (1955).
69. Anon., Machinery 62, No. 9, 139 (1956); No. 10, 169 (1956).
70. A. G. Ryabinov, The Electrochemical Processing of Metals and Alloys, Lenizdat, 1965.
71. V. N. Gusev, Anodic-Mechanical Machining of Metals, Mashgiz, 1952.
72. J. P. Hoare and R. Thacker, Plating 54, 553 (1967).
73. Anon., Iron Age 195, No. 15, 106 (1965).
74. Anon., Am. Machinist 103, No. 23, 99 (1959).
75. R. R. Cole, Intl. J. Prod. Res. 4, 75 (1965).
76. C. L. Faust, Trans. Inst. Metal Finishing 41, 1 (1964).
77. W. B. Kleiner, Tech. Proc. Am. Electroplaters Soc. 50, 147 (1963).
78. H. Kubeth and H. Heitmann, Industrie Anzeiger 46, 975 (1963).
79. N. D. G. Mountford, Trans. Inst. Metal Finishing 40, 171 (1963).
80. K. Kawafune, T. Mekoshiba, K. Noto, and K. Hirata, Ann. CIRP 15, 443 (1967).
81. S. Ito and N. Shikata, J. Mach. Lab. Japan 12, 50 (1966).
82. J. Bayer, M. A. Cummings, and A. U. Jollis, DDC Rept. AD 450199, Sept. 1964.
83. W. B. Kleiner, SAE Paper 618D, Jan. 14, 1963.
84. Anon., Chem. Eng. News 44, No. 34, 32 (1966).
85. V. D. Myerley, SAE Paper 670818, Oct. 2, 1967.
86. A. Haggerty and J. G. Goss, Am. Machinist 106, No. 12, 108 (1962).
87. C. A. Snavely and J. A. Cross, ASTME Creative Manufacturing Seminar, Paper SP62-39 (1962).
88. M. E. Merchant, Proceedings of the 3rd International Conference on Machine Tool Design and Research, Pergamon, Birmingham, 1962.
89. A. I. W. Moore, PERA Res. Rept. 145, p. 39, Leicestershire, 1965.

90. H. Opitz, International Research Production Engineering, ASME, 1963, p. 225.
91. W. Feitkneckt and G. Keller, Z. Anorg. Chem. 262, 61 (1950).
92. R. De la Rue and C. W. Tobias, J. Electrochem. Soc. 106, 827 (1959).
93. A. H. Meleka, Science J. 3, No. 1, 51 (1967).
94. J. P. Hoare, J. Electrochem. Soc. 117, 142 (1970).
95. K-W. Mao, private communication.
96. C. W. Tobias, J. Electrochem. Soc. 106, 833 (1959).
97. C. L. Faust, 6th International Metal Finishing Conference, London, May 1964.
98. J. Hopenfeld and R. R. Cole, Trans. ASME 88B, 455 (1966); 91B, 755 (1969).
99. S. Ito, K. Chikamori, and F. Sakurai, J. Mech. Lab. Japan 12, 37 (1966).
100. K. Chikamori, S. Ito, and F. Sakurai, Bull. Japan Soc. Prec. Eng. 2, 318 (1968).
101. Anon., Am. Machinist 111, No. 22, 149 (1967).
102. Anon., Aircraft Prod. 24, 410 (1962).
103. C. L. Faust and C. A. Snavely, Iron Age 186, No. 18, 77 (1960).
104. F. Bergsma, Ann. CIRP 16, 93 (1968).
105. Anon., Aircraft Prod. 23, 68 (1961).
106. Anon., Aircraft Prod. 23, 88 (1961).
107. R. H. Eshelman, Iron Age 190, No. 2, 109 (1962).
108. C. L. Kobrin, Iron Age 191, No. 3, 95 (1963).
109. T. W. Block, Tool Manuf. Eng. 50, 87 (1963).
110. T. E. Aaron and R. Wolosewicz, Machine Design 41, No. 28, 160 (1969).
111. L. A. Williams, ASTME Creative Manufacturing Seminar, Paper SP62-56 (1962).
112. S. G. Bankoff, Trans. ASME 82C, 265 (1960).
113. W. A. Haggerty, SAE Paper 680C, April 1963.
114. Anon., Aircraft Prod. 25, 2 (1963).
115. E. J. Krabacker, W. A. Haggerty, C. R. Allison, and M. F. Davis, International Research Production Engineering, ASME, 1963, p. 232.
116. W. M. Latimer, Oxidation Potentials, 2nd ed., Prentice Hall, Englewood Cliffs, N.J., 1952, p. 2.
117. A. J. DeBethune and N. A. S. Loud, Standard Aqueous Electrode Potentials and Temperature Coefficients, C. A. Hampel, Skokie, Ill., 1964.

118. G. Milazzo, Electrochemistry, English translation by P. J. Mill, Elsevier, Amsterdam, 1963, p. 192.
119. G. Kortüm and J. O'M. Bockris, Textbook of Electrochemistry, Elsevier, Amsterdam, 1951, p. 425.
120. J. P. Hoare, M. A. LaBoda, M. L. McMillan, and A. J. Wallace, J. Electrochem. Soc. 116, 199 (1969).
121. P. J. Boden and J. M. Evans, Nature 222, 377 (1969).
122. P. J. Boden and J. M. Evans, J. Electrochem. Soc. 116, 1715 (1969).
123. J. P. Hoare, M. A. LaBoda, and A. J. Wallace, J. Electrochem. Soc. 116, 1715 (1969).
124. S. Ito and N. Shikata, J. Mech. Lab. Japan 12, 50 (1966).
125. J. A. McGeough and H. Rasmussen, Trans. ASME 92B, 400 (1970).
126. W. Konig, D. Pahl, and H. Degenhardt, SME Paper MR70-205 (1970).
127. J. P. Hoare, Trans. ASME 92B, 402 (1970).
128. R. R. Cole and Y. Hopenfeld, Trans. ASME 85B, 395 (1963).
129. S. Ito, K. Honda, and F. Sakurai, J. Mech. Lab. Japan 11, 67 (1965).
130. H. Kurafuji and K. Suda, Ann. CIRP 14, 435 (1967).
131. J. A. Gurklis, DDC Rept. AD 613261, Jan. 1965.
132. D. Landolt, R. Acosta, R. H. Muller, and C. W. Tobias, J. Electrochem. Soc. 117, 839 (1970).
133. D. Landolt, R. H. Muller, and C. W. Tobias, J. Electrochem. Soc. 116, 1384 (1969).
134. B. Cina, Metallurgia 55, 11 (1957).
135. J. O'M. Bockris, D. Drazic, and A. R. Despic, Electrochim. Acta 4, 325 (1961); 7, 293 (1962).
136. M. A. LaBoda, U. S. Pat. 3,429,791 (1969).
137. M. A. LaBoda, U. S. Pat. 3,401,104 (1968).
138. M. A. LaBoda, U. S. Pat. 3,421,987 (1969).
139. M. A. LaBoda, U. S. Pat., 3,389,067 (1968).
140. M. A. LaBoda and W. R. Doty, U. S. Pat. 3,389,068 (1968).
141. W. R. Doty and M. A. LaBoda, U. S. Pat. 3,404,077 (1968).
142. M. A. LaBoda and M. L. McMillan, Electrochem. Tech. 5, 340, 346 (1967).

143. J. P. Hoare, Nature 219, 1034 (1968).
144. T. P. Hoar, Corrosion Sci. 7, 341 (1967).
145. T. P. Hoar, J. Electrochem. Soc. 117, 17C (1970).
146. K. Chikamori and S. Ito, Denki Kagaku Oyobi Kogyo Butsuri Kagaku 37, 603 (1969).
147. D. C. Grahame, Chem. Rev. 41, 441 (1947).
148. P. Delahay, Double Layer and Electrode Kinetic, Interscience, New York, 1965, pp. 17 et seq.
149. M. A. LaBoda and J. P. Hoare, J. Electrochem. Soc. 119, 419 (1972); 120, 1071 (1973); Coll. Czech. Chem. Comm. 36, 680 (1970).
150. J. P. Hoare, Electrochim. Acta 9, 599 (1964).
151. U. R. Evans, The Corrosion and Oxidation of Metals, St. Martin's Press, New York, 1960, p. 136.
152. M. Nagayama and M. Cohen, J. Electrochem. Soc. 109, 781 (1962); 110, 670 (1963).
153. M. Cohen and A. F. Beck, Z. Electrochem. Soc. 62, 696 (1958).
154. T. P. Hoar and U. R. Evans, J. Chem. Soc. 134, 2476 (1932).
155. M. J. Pryor and M. Cohen, J. Electrochem. Soc. 100, 203 (1953).
156. H. H. Uhlig, D. N. Triadis, and M. Stern, J. Electrochem. Soc. 102, 59 (1955).
157. A. C. Riddiford, Advances in Electrochemistry and Electrochemical Engineering, C. Tobias, Ed., Vol. 4, Interscience, New York, 1969, p. 47.
158. D. T. Chin, J. Electrochem. Soc. 119, 1181 (1972).
159. J. M. Fitz-Gerald, J. A. McGeough, and L. McL. Marsh, J. Inst. Math. Appl. 5, 387, 409 (1969); 6, 102 (1970).
160. J. P. Hoare, K-W. Mao, and A. J. Wallace, Corrosion 27, 211 (1971); 29, 143 (1973); J. Electrochem. Soc. 120, 1452 (1973); Corrosion Sci. 12, 571 (1972); 13, 799 (1973); 15, 435 (1975).
161. T. P. Hoar and G. P. Rothwell, Electrochim. Acta 9, 135 (1964); 10, 403 (1965).
162. P. A. Brook and Q. Iqbal, J. Electrochem. Soc. 116, 1458 (1969).

163. H. E. Haring and W. Blum, Trans. Am. Electrochem. Soc. 44, 313 (1923).
164. R. V. Jelinek and H. F. David, J. Electrochem. Soc. 104, 279 (1957).
165. D. T. Chin and A. J. Wallace, J. Electrochem. Soc. 118, 831 (1971).
166. J. F. Thorpe and R. D. Zerkle, Int. J. Mach. Tool Des. Res. 9, 131 (1969).

III. ELECTROCHEMICAL TECHNIQUES APPLIED TO SEMICONDUCTORS

D. R. Turner
Bell Laboratories
Murray Hill, New Jersey

and

J. L. Pankove
RCA Laboratories
Princeton, New Jersey

I. INTRODUCTION

An intelligent application of electrochemical techniques to semiconductors requires some knowledge about the solid-state physics of semiconductor materials (1). Semiconductors differ from metals and insulators in that a moderate energy gap exists between the highest electron energy levels in the valence band and the lowest energy levels in the conduction band. Semiconductors have large populations of electrons in the valence band and empty levels in the conduction band. Electrochemical reactions at semiconductor electrodes require energy level overlapping with solution levels. In metals the two energy level bands have some overlap and a large population of free electrons and vacant levels is always available for electronic conduction or an electrochemical electron transfer process at an electrode. The energy gap in an insulator is relatively large (generally more than 2 ev) so relatively few electrons from the valence band have sufficient thermal energy to bridge the energy gap to give rise to electronic conductivity. The electronic conductivity of an insulator can be improved greatly by doping the single crystal with a suitable impurity that either donates electrons to the conduction band or accepts electrons from the valence band.

Electronic conduction in semiconductors is the result of two processes: (1) conduction band electron motion and (2) electron-hole migration in the valence band. We can think of the electron hole or simply hole (the term normally used) as a missing electron from a covalent band in the semiconductor. When an electron moves from a nearby band to fill a hole, a new hole is created and in effect the hole has moved and positive charge is transferred. Conduction-band electrons are not associated with any particular semiconductor atom and they are free to move about in the crystal unless captured by a hole site (hole-electron pair recombination) or if they take part in an electron transfer process across an ohmic contact, p-n junction, or semiconductor-electrolyte interface. In addition to the recombination process which consumes equal numbers of holes and electrons, hole-electron pairs are also generated by thermal energy in the semiconductor or by external electromagnetic radiation such as visible light. The maximum radiation wavelength (minimum energy) that will produce hole-electron pairs is determined by the semiconductor band-gap energy. Larger gap energies require higher minimum light energies.

The equilibrium of holes and electrons in semiconductors has been compared with the equilibrium between hydrogen ions and hydroxyl ions in aqueous solutions. In both cases, the product of the concentration of the two charge carrier types is equal to a constant at a given temperature. A pure undoped semiconductor has an equal number of + and - charge carriers (holes and electrons). This is called intrinsic material. Doping the semiconductor with a trivalent material such as boron produces electron vacancies in the valence band. This raises the hole density and reduces the electron density. The semiconductor is called p-type since positive charges are responsible for the major part of its conductivity. A pentavalent dopant material such as antimony produces additional conduction band electrons and an n-type semiconductor is formed. The n signifies that negative charges are the principle current carrier.

Electrochemical effects occur at the surface of the semiconductor. Ions in solution are oxidized or reduced or atoms are removed or deposited at the surface. Anodic processes consume holes and/or inject electrons, whereas cathodic processes consume electrons and/or inject holes. It is helpful to consider holes as broken chemical bonds which, on arrival at the surface of the semiconductor, permit surface atoms to enter the solution as ions.

In practical applications where an electrochemical process at a semiconductor electrode is limited by the supply of holes or electrons at the interface, such as in anodic polishing, strong illumination is very effective in creating sufficient numbers of holes that are required to carry out the dissolution reaction (2, 3).

II. ELECTRODE DESIGN AND PREPARATION

Successful electrochemical processes carried out on semiconductor materials may depend on how the material is used as an electrode and the techniques used in preparing the electrode (4, 5). For electronic device applications, the semiconductor material is usually grown in single crystal rod form with a specific crystallographic orientation. The trend is toward larger-diameter ingots so that slices cut from the ingot will yield many more devices per sequence of operations. More often than not, in devices the electrochemist has no control over the geometry of the semiconductor piece he must treat electrochemically. A thin wafer is the most common shape, and electrical contact may be possible only around the edge on one side. The resistance in the semiconductor from the ring contact to the surface sites where an electrochemical reaction must take place can result in an appreciable iR drop so that the electrode potential in the center may be considerably different from that near the contact and the desired electrochemical reaction may take place at nonuniform rates across the surface of the wafer. The nonuniformity can be minimized by using a small total current and/or by modifying the electrolyte so the electrochemical polarization effect is large.

After the semiconductor wafers are cut with a mechanical saw (a thin diamond-embedded blade is usually used), the damaged surface layer is removed by an electrochemical polishing process. The details of these polishing processes are discussed later. The finished polished slice of semiconductor may be subjected to a number of gas-phase reaction and diffusion processes to produce p-n junctions, resistors, capacitors, ohmic contacts, and protective oxide or nitride layers.

Before any electrolytic treatment or precision electrical study of a semiconductor can be made, an ohmic contact is required. The resistance of the ohmic contact should be low to avoid excessive heating when high current densities are used, such as in rapid etching or electropolishing. Oxide films are readily formed on the surface of most semiconductor materials, and these can result in high contact resistance and poor contact adhesion. Techniques have been developed to make adherent low-resistance ohmic contact to practially all semiconductors. Care must be taken to ensure that the process for making the ohmic contact does not alter the semiconductor in an unpredictable way at the semiconductor-electrolyte interface.

Basic studies on electrochemical properties of semiconductor materials are usually done on massive specimens for ease of handling. Contacts are made far from the semiconductor-electrolyte interface, so relatively simple techniques are required. A good ohmic contact can be made to germanium by simply abrading the surface and soldering with an ordinary lead-tin solder, using an acid zinc chloride flux. The alloying tachniques used on polished germanium involves doping the surface layer so that no rectifying junction is formed. Most metals electroplated on p-type germanium form an ohmic contact (6). Antimony is the only known electroplated metal contact that is ohmic on n-type germanium.

Nickel deposited by chemical reduction (electroless process) provides a satisfactory ohmic contact on silicon (7). An abraded surface catalyzes electroless nickel deposition more readily than a polished surface. A typical sequence of operations is as follows: (1) lap surface on glass with a water slurry of 600-mesh silicon carbide powder, (2) wash in deionized water, and (3) immerse in a hot (∿95°C) alkaline electroless nickel bath for 3 min. This produces a deposit thickness of about 0.001 in. A typical alkaline electroless nickel bath consists of:

Nickel Chloride ($NiCl_2 . 6H_2O$)	30 g/liter
Sodium hypophophite ($NaH_2PO_2 . H_2O$)	10 g/liter
Ammonium citrate [$(NH_4)_2HC_6H_5O_7$]	65 g/liter
Ammonium chloride (NH_4Cl)	50 g/liter

Ammonium hydroxide (NH_4OH) added until solution turns from green to blue (pH 8-10).

Hydrogen gas evolution is an indication that the process is working. If no action is observed, the silicon surface can be activated by a dip in hot wt KOH solution until the surface is attacked uniformily. The silicon is quickly transferred to the electroless nickel plating bath without rinsing. The electroless nickel coating can be soldered to with an ordinary lead-tin solder and usually without a cleaning flux.

Whether or not a true ohmic contact to the semiconductor is formed by one of the methods just described is easily determined by preparing the same contact at two ends or on two sides of the semiconductor and measuring the current-voltage characteristic with current flowing in each direction. The I-V plot should be linear and unchanged by reversing current flow direction. After establishing that both contacts are ohmic, one can be removed if desirable. It is usually desirable to mask the ohmic contact area with a stable adherent insulating coating to minimize corrosion and contamination problems. A detailed account of the various types of ohmic contact and the techniques for applying them to germanium and silicon has been given by Sullivan and Warner (7).

III. ELECTROLYTIC CELL DESIGN

Experiments that seek qualitative information about electrochemical process can often be done in simple cells - a beaker with two electrodes. For quantitative electrochemical experiments, a carefully designed electrolytic cell is generally required. The design will vary depending on the kind of experiment, data required, nature of electrolyte, and degree of precision and reproducibility required in the measurements.

A well-designed electrolytic cell provides for uniform current distribution over the surface of the electrode under study and a means of accurately measuring the electrode potential. Uniform current distribution is achieved if the spacing between the anode and cathode of the cell is uniform and the solution is confined to the volume directly between the two electrodes. A number of well-designed cells have been described in the literature (8-11). A constant anode-cathode spacing is not required for uniform current distribution if the polarization voltage of the electrode being studied is large as compared to the iR voltage drop between the anode and cathode. The potential of the electrode under study is generally measured against a third electrode, that is, a standard reference electrode. These two electrodes are often connected together by means of a syphon having a capillary tip at one end which is positioned close to the electrode being studied. The design and positioning of the syphon capillary tip have been the subjects of numerous investigations (12, 13). Barnartt (12) has shown that the most satisfactory arrangement is to place the capillary tip of radium r a distance of at least 4r in front of the electrode and make a correction for the iR drop if necessary. This arrangement does not distort the primary current distribution at the electrode.

Other considerations in electrolytic cell design are (1) provisions for sweeping out an undesired gas with an inert gas, (2) stirring the electrolyte when required, (3) temperature control of the system, and (4) a means of excluding light, which is often important with semiconductor electrodes. Solutions containing hydrofluoric acid or hot caustic materials exclude the use of glass apparatus. Fortunately, some of the new polymer materials such as polyethylene and Teflon are commercially available in many forms that can be adapted for cell use.

It is generally necessary to mask part of the electrode with some insulating material. The only completely successful method involves the application of some compression of the insulator to the electrode. As for instance, tape wound in tension on a cylinder.

IV. ELECTROLYTE CONSIDERATIONS

Electrochemical reactions at an electrolyte-semiconductor interface are influenced by the nature of the electrolyte and dissolved impurities. The basic electrolyte requirement for anodic dissolution of an semiconductor is that it form solvent soluble compounds or complex ions. Details of these processes are discussed later. Solution impurities generally play no role in anodic dissolution processes, with the exception of oxidizing agents, which can cause chemical dissolution and hole injection (14). Anodic oxidation and cathodic reduction of ions in solution at semiconductor electrodes depend on the redox potential of the system and the potential where parallel processes may occur, such as anodic dissolution or cathodic discharge of hydrogen ions. On germanium, for example, the range is about 0.7 V (between +0.3 and -0.4 V in 1N H_2SO_4 and -0.5 to -1.2 V in .1N NaOH on the hydrogen scale). A compiled list of electrochemical reactions that have been studied using germanium electrodes is given in Table 1 (4, 14). The redox systems are arranged in their orders of decreasing oxidizing power.

Cathode processes on semiconductor electrodes can be influenced markedly by small amounts of impurities, particularly those that deposit out of solution and confuse one's understanding of the electrode surface. Careful purification of chemicals, solvents, and gases may be necessary. Since purification is time consuming and may complicate the experimental apparatus unnecessarily, it is desirable first to establish the level of purification required.

An effective way of removing metallic impurities from solutions used in experiments on germanium electrodes is to grind up small pieces of high-purity germanium in situ by rapid sitrring with a magnetically operated Teflon-covered stirring bar (15). Over the course of several hours, collisions between the Ge pieces produce a dense cloud of fine particles that react with impurities, extracting them from solution. Generally, the germanium powder is left in the solution throughout the experiment. Preelectrolysis may also be used to electroplate out metallic impurities on a germanium cathode. This requires several days and usually does not reduce the impurity level below that achieved by simply reacting with crushed germanium. Furthermore, powdered germanium should be effective in removing organic contaminants by adsorption. A similar technique should be useful in working with other semiconductor electrodes such as silicon.

V. REFERENCE ELECTRODES

A practical reference electrode may be defined as an electrode whose electrochemical reaction is reversible and whose potential remains constant even though small currents may be drawn from the electrode during a potential measurement. Actually, any reversible electrode with a well-defined potential can function as a reference electrode provided that the current drawn does not exceed about one-tenth the

Table 1. Electrochemical Behaviour of Redox Reactions on Germanium Electrodes

Redox System	Electrolyte	Standard Potential E° in V H_2	Potential Range with Slight Dissolution or H_2 Evolution ($i \pm 1$ mA/cm^2)	Reaction Observed	Preferred Electron Transfer Mechanism	Reference
H_2O/H_2O	1 N H_2SO_4	+1.77	+0.4 to -0.5	Reduction	?	21
MnO_2/MnO_4^-	1 N H_2SO_4	+1.66	+0.4 to -0.5	Reduction	Valence band	14,21,27
Ce+++/Ce++++	1 N H_2SO_4	+1.43	+0.4 to -0.5	Reduction	Valence band	14
$Cr^{+++}/Cr_2O_7^-$	0.1 N H_2SO_4	+1.35	+0.35 to -0.55	No reaction(?)	?	14,21
HNO_2/HNO_3	0.1-16 N H NO_3	-	- -	Reduction	Valence band	25
OH^-/H_2O_2	0.1 N NaOH	+0.99	-0.3 to -1.2	Reduction	Valence band	14
Hydroquinone/Quinone	1 N H_2SO_2	+0.70	+0.4 to -0.5	Reduction	Valence band	14
Fe++/Fe+++	1 N $HClO_4$	+0.69	+0.4 to -0.5	Reduction	Valence band	21
Fe++/Fe+++	1 N H Cl	+0.69	+0.4 to -0.5	Reduction	Valence band	14
Fe++/Fe+++	1 N H_2SO_4	+0.67	+0.4 to -0.5	Reduction	Valence band	14,17
$[Fe(CN)_6]^{4-}$/ $[Fe(CN)_6]^{3-}$	0.1 N H_2SO_4	+0.56	+0.35 to -0.55	Reduction and oxidation	Valence band	14
I_3^-/I_2	0.1 N HCl or 1 N KI	+0.54	+0.35 to -0.55	Reduction and oxidation	Valence and Conduction band	21,32
OH^-/O_2	0.1 N NaOH	+0.45	-0.3 to -1.2	Reduction	Valence	14,25
$[Fe(CN)_6]^{4-}$/ $[Fe(CN)_6]^{3-}$	0.1 N NaOH	+0.41	-0.3 to -1.2	Reduction	Valence	14,21,27
V+++/VO++	1 N H SO or 1 N HCL	+0.30	+0.4 to -0.5	No reaction	?	14
Ti+++/TiO++	1 N HCl	+0.18	+0.4 to -0.5	No reaction	?	14
V++/V+++	0.1 N H SO	-0.31	+0.35 to -0.55	Oxidation	Conduction band	14,22
Cr++/Cr+++	0.1 N HCl	-0.42	+0.35 to -0.55	No reaction(?)	?	14
$C_2O_4^-/CO_2$	0.1 N HCl	-0.49	+0.35 to -0.55	Oxidation	Conduction	32

[a]The more positive boundary for n-type Ge approximately corresponds to the potential at which the limiting current starts.

exchange current for the electrode reaction. Modern instrumentation with high input impedance ($\sim 10^{12}$ ohms) electrometers and operational amplifiers has made possible routine potential measurements where less than 1 PA of current is drawn from the reference electrodes have become practical. Ideally, the reference electrode should be as much like the electrode to be measured as possible to minimize contamination. The liquid junction potential between the reference and the electrode under study should be considered and a correction made in the measured potential if necessary. Typical reference electrodes that have been used for many years and are generally available commercially are listed in Table 2 with some appropriate data. A detailed discussion of these and other reference electrodes for use in aqueous, organic, and fused-salt media is available (16). For the sake of simplicity and elimination of contamination, it is often possible to use the same electrode for the reference electrode as electrode under study. In copper plating from copper sulfate solutions, for example, a rod of pure copper is an excellent reference electrode.

The reference electrode is usually placed in a separate container and is connected to the cell by means of a syphon and capillary tip. This minimizes contamination by reference electrode chemicals and it also makes possible potential measurements in highly corrosive electrolytes such as HF solutions.

VI. ELECTRODE POTENTIAL MEASUREMENTS

The electrode potential of a semiconductor in contact with an electrolyte is a significant measurement often made to determine basic properties about the semiconductor, semiconductor-electrolyte interface, or the mechanism of electrochemical reactions at the electrode (17). Potential measurements are usually made under a variety of conditions of electrolyte concentration, temperature, current density, polarity, presence of absence of additives, light intensity, stirring, and so on. These data are often necessary for a complete interpretation of what measured electrode potentials mean (18).

Table 2. Reference electrodes

	Reaction System	Potential vs. Hydrogen Standard at 25°C		
Hydrogen	H_2/H^+ H_2/H_2O	0 1.23	1M acid 1M alkali	Primary standard potential-defined as zero potential Soln must be O_2 free and sat. with H_2 Platinized platinum electrode normally used
Calomel	Hg/Hg_2Cl_2	+0.244 +0.280	Sat. KCl 1N KCl (neutral) 0.1N KCl	Low junction potential, best all purpose electrode
Silver-Silver Halide	Ag/AgCl	+0.222	1N HCl (acid)	Also useful in non-aqueous systems
Mercury-mercuric oxide	Hg/HgO	+0.926	1N KOH (alk.)	Most practical for strongly alkaline solutions
Mercury-mercurous sulfate	Hg/Hg_2SO_4	+0.615	1M H_2SO_4	Best for strong H_2SO_4 solutions
Lead-lead sulfate	$Pb/PbSO_4$	-0.356	1M H_2SO_4	Negative electrode of lead acid cell
Lead oxide-lead sulfate	$PbO_2/PbSO$	+01.685	1M H_2SO_4	Positive electrode of lead acid cell

Experimentally, the first considerations is a proper cell, as discussed earlier. This should include a reference electrode that is reliable and compatible with the electrolyte and ambient temperature.

A. Open-Circuit Potentials

Semiconductor electrodes that are no corroded rapidly in electrolyte solutions usually exhibit a photovoltaic effect (2). Light creates hole-electron pairs in the space-charge region of the semiconductor, thereby decreasing the potential gradient in the space-charge region. The polarity and magnitude of the open-circuit potential change when the electrode is illuminated is a function of the semiconductor material, conductivity type, minority current carrier concentration, redox potential of the electrolyte, and intensity of the light.

VII. ROTATING-DISK AND DISK-RING ELECTRODES

The rotating-disk electrode technique has become a very important method in electrochemistry for the study of mass-transport limiting electrode reactions (19, 20). Levich developed the theory for convective diffusion on a rotating-disk electrode long before it was accepted as a useful tool by electrochemists. He showed that the current density at a given potential on a rotating-disk electrode is proportional to the square root of the angular velocity of the disk, or more precisely,

$$i = kD^{2/3}\nu^{1/6}\omega^{1/2}C_O$$

where k is a constant, D is the diffusion constant, ν is the kinematic viscosity, ω is the angular velocity, and C_O is the bulk concentration of electrolyte. A plot of the current density versus $\omega^{1/2}$ extrapolated to $\omega = 0$, at any potential, should give a straight line passing through the origin if the reaction at the rotating disk is limited by convective diffusion only. Pleskov (21) has also shown that the rotating-disk electrode technique can be used to determine the current-potential relation for each of two reactions occurring simultaneously at an electrode. The separation is possible when the rate of one of the processes is limited by diffusion while the other reaction is independent of stirring. Extrapolation of these data to $\omega = 0$ results in nonzero current intercepts.

A plot of the current at $\omega = 0$ for various potentials gives the relation for the nondiffusion limiting process. The diffusion limiting characteristic is obtained by difference. Pleskov and Kabanov (22) studied the anodic oxidation of vanadium V^{++} to V^{+++} on a rotating-disk germanium electrode while the electrode was dissolving anodically at the same time.

This experiment has been useful in determining whether conduction-band or valence-band electrons were involved in electrochemical reactions at germanium electrodes. For example, rotating-disk electrodes of n- and p-type germanium were used by Pleskov (21) to distinguish between the limiting current, i_{lim}, in the reduction of KI_3 caused by the slow mass transfer of KI_3 and that caused by electron depletion in p-type germanium (Fig. 1). When the limiting current is controlled only by the mass transfer of KI_3 to the germanium, i_{lim} is directly proportional to the square root of the anular velocity as shown for n-type. Electron depletion is evident at p-type germanium. With moderate electrode illumination, the maximum i_{lim} increases, as it should if the cathode reaction is limited by the supply of electrons. Strong illumination or a hydrogen-saturated p-type electrode raises the maximum i_{lim} still higher.

A. Rotating Disk-Ring Electrodes

Modifications of the rotating-disk electrode include one or more ring electrodes outside the disk (23). These rings may be whole or split. The purpose of rings in one or more segments is to detect soluble intermediates formed at the disk that are swept past the rings in their laminar flow outward. A potential bias is applied to the ring or rings so that an oxidation or reduction current is detected only if the intermediate is formed. This has proved

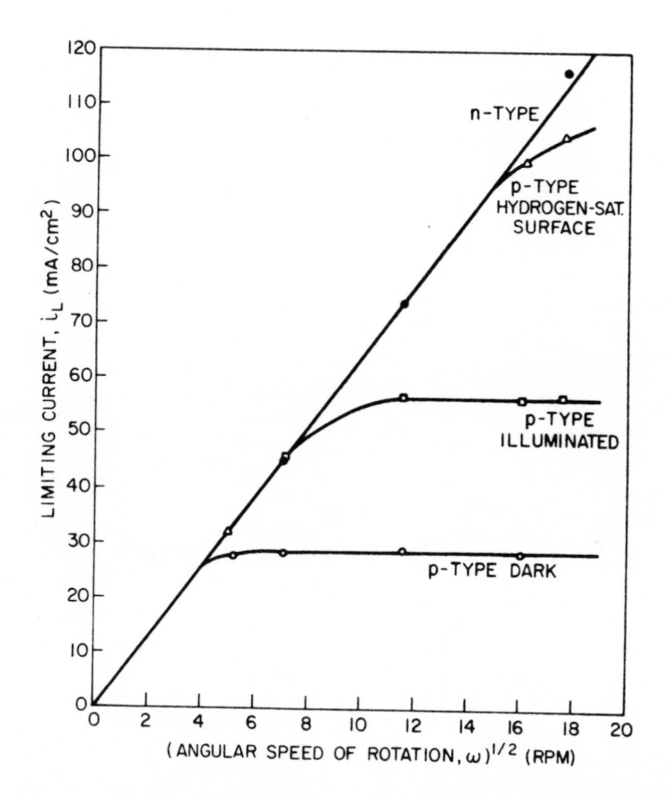

Fig. 1. Limiting current density for the reduction of KI_3 (0.096N) on 1.8 ohm-cm n-type and 2.8 ohm-cm p-type germanium rotating disk electrodes as a function of $\omega^{1/2}$ Pleskov (21).

to be a very useful tool in mechanism studies involving intermediate species, especially when the lifetime of the species is too short for detection by other convenient methods (24). Another configuration that may be useful in dealing with short-lifetime intermediates is the double ring. Intermediates formed at the inner ring need move a short distance across the electrode to be detected.

VIII. p-n JUNCTION INDICATOR ELECTRODE

The potential drop across a p-n junction in the absence of current through it, the floating potential E_f, is a measure of the deviation of the concentration of minority carriers from the equilibrium value (2). The concentration of holes in the n-type region close to the junction is given by

$$p = p_0 e^{\beta E_f} \tag{1}$$

where p_0 is the equilibrium concentration of holes on the inside and β is e/kT (=25 mV at room temperature). If the thickness of the n-type region is less than the diffusion length of holes, then p measured at the p-n junction is approximately the same as that observed at a semiconductor-electrolyte interface.

Brattain and Garrett (2) used this phenomenon as a means of detecting the sign of the carriers involved in charge-transfer processes at a germanium-electrolyte interface. A schematic diagram of the arrangement is shown in Fig. 2. The p-n junction is planar, equal in area and close to the semiconductor-electrolyte interface. The distance between the germanium-electrolyte interface and the p-n junction must be small in comparison with the bulk diffusion length for holes in the n-type Ge region. It is also assumed that the hole recomination velocities at the p-n junction and at the germanium surface exposed to the electrolyte are low. Several experimental arrangements are possible with the p-n junction indicator electrode. These are listed in Table 3. Brattain and Garrett performed two kinds of experiments: (1) they made n-type germanium anode or cathode and observed the floating potential of the p-n junction and (2) they forward-biased the p-n junction and injected holes into the n-type germanium bulk while the germanium electrolyte interface was biased anodically to the point of current saturation (i.e., limiting hole current). The latter experiment is a direct means of measuring the current multiplication factor for the anodic dissolution reaction.

The hole injection current from solution into germanium is determined from floating potential data by employing equations derived by Brattain and Garrett (2). The nonequilibrium holes diffuse across

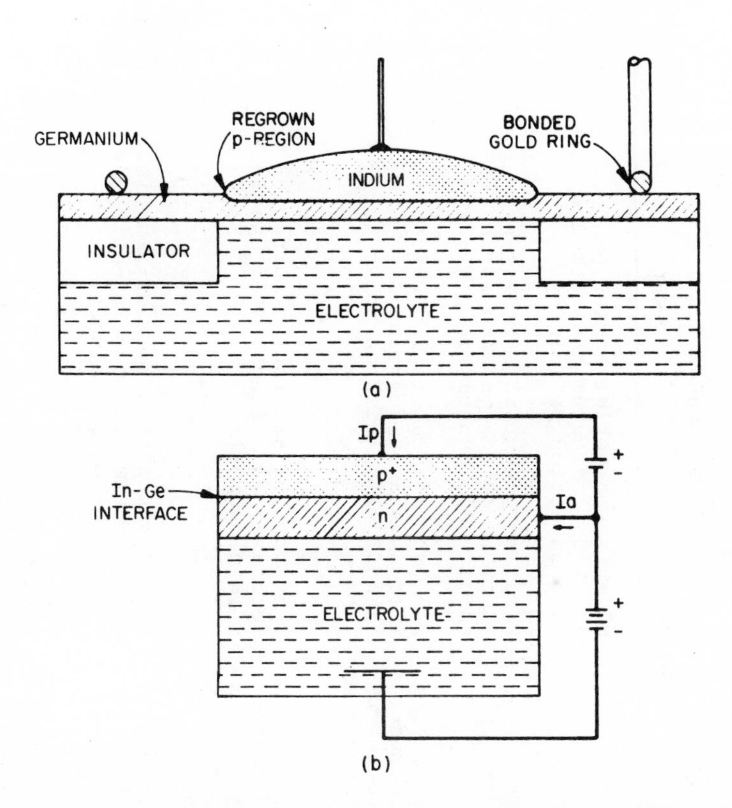

Fig. 2. Schematic diagram and experimental details of the p-n junction technique of Brattain and Garrett (2).

the n-region and their presence is detected by the floating potential across the p-n junction. To correct for the local hole generation near the p-n junction, Eq. 1 is modified to include a minimum value of the floating potential E_f^{min}:

$$\frac{p}{p_0} = \frac{e^{\beta E_f} - e^{\beta E_f^{min}}}{1 - e\beta E_f^{min}} \tag{2}$$

Table 3. Various bias arrangements and results of germanium p-n junction indicator electrode experiment

	Electrolytic Cell Ge Baised	p-n Junction	Experimental Results	Ref.
1	Cathodic current varied	Floating	Floating potential across p-n junction indicates the cathodic hole injection efficiency	2, 25
2	Anodic current varied	Floating	Floating potential across p-n junction decreases with increased anodic hole consumption	25
3	Cathodic current varied	Reverse biased at a constant voltage in current saturation region	p-n junction saturation current increases as a function of cathodically injected holes	21
4	Anodic current varied	Reverse biased at a constant voltage in current saturation region	p-n junction saturation current decreases as a function of anodic hole consumption	
5	Anodic at a constant voltage in current saturation region	Forward biased current varied	Anodic saturation cell current increases as a function of hole injection by forward biased p-n junction	2
6	Anodic at a constant voltage in current saturation region	Reverse biased current varied	Anodic saturation cell current decreases as a function of holes consumed by reverse biased p-n junction	26

The injected hole current, i_p, with cathodic current is given by

$$- i_p = i_s \left(\frac{p}{p_0} - 1 \right) \tag{3}$$

where i_s is the saturation current across the p-n junction in the absence of injection. Combining Eqs. 2 and 3 and rearranging, we have

$$i_p = i_s \left(\frac{e^{\beta E_f} - 1}{e^{\beta E_f^{min}} - 1} \right) \tag{4}$$

Atypical plot of the calculated hole injection current into n-type germanium versus measured floating potential is shown in Fig. 3. The p-n junction saturation characteristic and hole-generation properties on the n-side affect the result. Experimental results of floating potential versus anode and cathode current densities with nitric acid at various concentrations are illustrated in Fig. 4 (25). When the germanium is anode, the reaction is germanium dissolution. At the knee of the curve, the surface region is depleted of holes and the floating potential is E_f^{min}. The saturation current density of the p-n junction is calculated from the anodic current at the knee divided by the multiplication factor for Ge dissolution.

The floating potentials measured when the n-Ge is made cathode in nitric acid solutions give an indication of how much of the reaction involves valence-band electrons, that is, hole injection. In concentrated nitric acid, for example, the hole injection is very high and in fact the hole injection efficiency is greater than 100%, which means that there is also chemical etching of germanium that injects holes (18).

Various modifications of the p-n junction indicator have been used to determine whether the electron transfer process in oxidation-reduction (redox) reactions is with electrons in the valence or conduction band of the semiconductor (21, 25, 26). Table 3 gives a summary of the various bias arrangements and qualitative results of experiments.

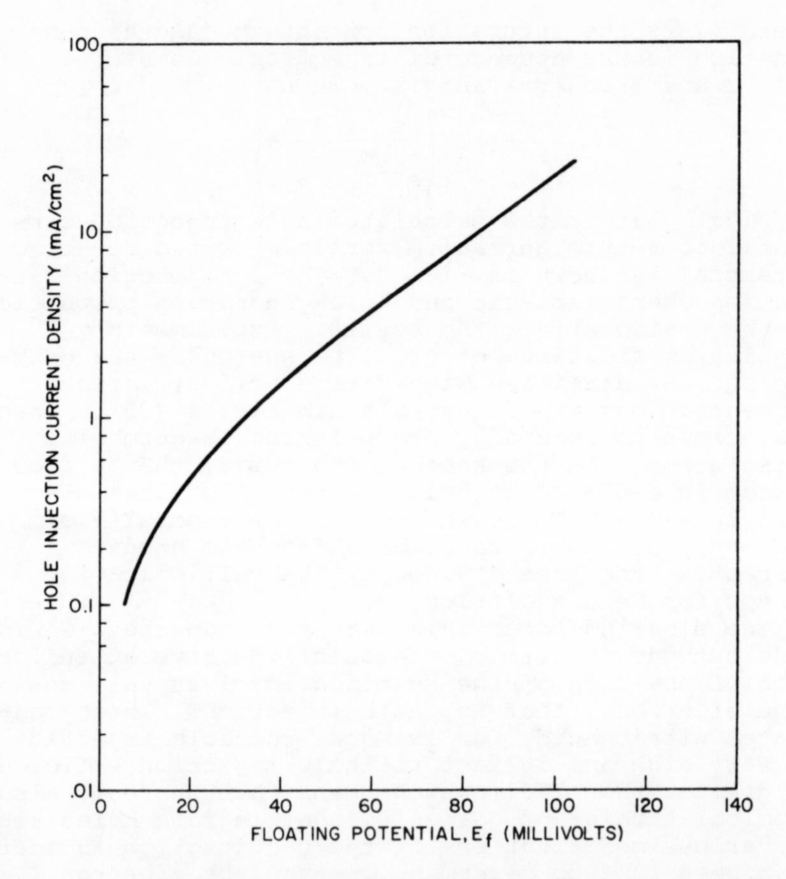

Fig. 3. Calculated hole injection current versus floating potential across a Ge p-n junction, where $i_s = 0.36$ mA/cm^2 and $E_f^{min} = -60$ mV.

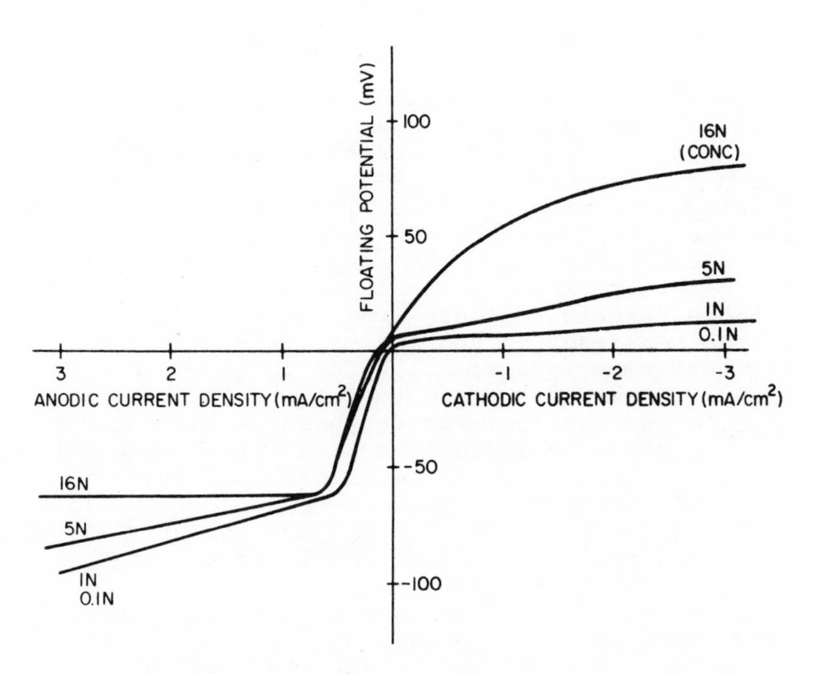

Fig. 4. Germanium p-n junction floating potential versus electrolysis current density at n-side of Ge in nitric acid at various concentrations.

A. Participation of Valence Electrons in Cathode Reactions

Transfer of valence electrons from a semiconductor to ions in solution means that holes form in the valence band of the semiconductor. When charge transfer process in reduction of ions at a semiconductor involves valence band electrons, holes are injected into semiconductor and these nonequilibrium holes diffuse to the p-n junction where they influence the floating potential across the junction.

Hole current through the Ge-electrolyte interface during the cathode reaction is given by

$$i_p = i_p^S \frac{e^{\beta E_f}}{e^{\beta E_f^{min}} - 1} \tag{5}$$

where i_p^S is the saturation current of the p-n junction in the absence of injection.

Another method of measuring the fraction of valence electrons in the reduction current was developed by Pleskov (27). The p-n junction is reverse biased so that the current through the junction is determined by the rate of hole generation in the n-type region. Holes injected into Ge by the cathode reaction diffuse from the electrode surface to the p-n junction and increase the saturation current. The loss of holes due to recombination in transit from the surface to the p-n junction is negligible, so the increase in junction current practically equals the injected hole current. The ratio of the hole current to the total cathodic current, ν, gives the fraction of valence-band electrons involve in the cathodic reaction. The concept of ν is quite analogous to the injection coefficient of the emitter of a transistor.

Hole injection efficiencies during the reduction of various oxidizing agents on n-type germanium with a p-n junction hole indicating electrode are given in Table 4.

IX. BIELECTRODE METHOD FOR STUDYING KINETICS OF ELECTRODE REACTIONS

The technique was first used by Pleskov (27). As shown schematically in Fig. 5, the cell consists of a thin disk of n-type Ge with a ring ohmic contact masked from the solution. Two auxiliary electrodes are used, each one in a separate electrolytic cell compartment. The germanium electrode is common to both cells. Thus one side can be biased anode or cathode and the nature of the electron transfer process occurring, whether it be hole or electron injection into or extraction from the semiconductor, can be detected as the result of the departure of the minority carrier concentration from equilibrium at the opposite side of the electrode. Pleskov (27),

Table 4. Hole injection efficiences during the reduction of oxidizing agents on n-type germanium with a p-n junction hole indicator electrode

Oxidizing Agent	Oxidizing Agent Conc. M/Liter	Indifferent Electrolyte	Apparent Hole Injection Coefficient	Cathode Current at Measurement mA/cm^2	p-n Junction Indicator Electrode	Ref.
O_2	Solution air sat.	0.1N KOH	0.1	8		2, 25
O_2	Solution O_2 sat.	0.1 N KOH	0.4-0.1	1-2		25
HNO_3	16	None	2.6	7.2	Floating	25
HNO_3	7	None	0.4	5.0	potential	25
HNO_3	1	None	0.1	8.0	measured	25
HNO_3	0.1	None	0.1	8.0		25
K MnO	0.12	1 N H_2SO_4	0.78-0.88	13.5		
$K_3Fe(CN)_6$	0.28-0.56	1-2 N KOH	0.66-0.80	8.6	3 V reverse	
KI_3	0.1-0.33	1 N KI	0.42	22	bias -	
Quinone	0.4-0.12	1 N H_2SO_4	0.38	3.3	measured	27
$K_2Cr_2O_7$	0.04-0.12	1 N H_2SO_4	0	20	Δi_S	

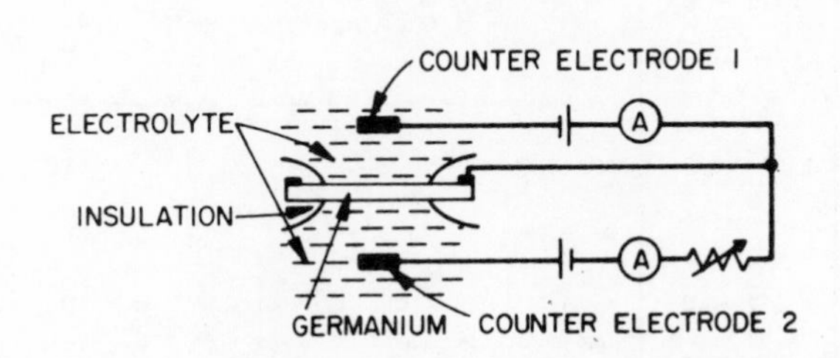

Fig. 5. Thin germanium bielectrode cell arrangement used by Pleskov (27).

for example, biased the opposite side anodically under conditions of limiting current so that it served as an indicator of the hole concentration in the Ge. He used this arrangement to determine what fraction of the current involving reduction of ions resulted in hole injection into the germanium. These holes diffuse to the opposite side of the electrode (assuming that few are recombined along the way), where they increase the limiting current in proportion to the number reaching the surface. Two factors influence the accuracy of this method: (1) recombination of injected holes in the germanium and (2) variability of the current multiplication factor a at the indicator side of the electrode.

X. CAPACITANCE

Impedance measurements, particularly the differential capacity between an electrode and an electrolyte, make it possible to determine the charge distribution at the phase boundary (28). Knowledge about the charge distribution is helpful in understanding the mechanism of electrochemical reactions at the electrode-electrolyte interface. The capacitance at an electrode-electrolyte interface is inversely proportional to the charge separation distance. On the solution side of the phase boundary, the charge is divided into two regions: (1) a surface charge that is formed by adsorbed ions (Helmholtz double layer) and (2) a space-charge region in which the charge may be distributed many ion diamters into the electrolyte (Gouy or diffuse double layer). The width of the diffuse or space-charge region in the electrolyte becomes negligible if the bulk ion concentration is about 1 N or higher. On the electrode side, with metal electrodes, all the charge is located within a few angstroms of the surface. Capacity measurements with metal electrodes in electrolytes therefore give information regarding the charge distribution on the solution side of the phase boundary.

The density of current carriers, holes and electrons, in semiconductors such as germanium and silicon is relatively low as compared to metals. A space-charge region may be formed in the electrode to a considerable depth. If moderately concentrated electrolytes are used, ∿1 N, Bohnenkamp and Engell (29) have shown that variations in the capacity of germanium electrodes in electrolytes are determined only by the space-charge distribution in the semiconductor. This technique has been used to verify the existence of depletion and enhanced carrier layers on both Ge (29-31) and Si (32-34) n-type electrodes biased anodically and p-type biased cathodically in appropriate solutions. Figure 6 illustrates the charge and potential distribution though the electrode-solution interface with a typical semiconductor electrode made anode in an electrolyte. Oxide films are assumed to be absent, although that condition can be illustrated with a slightly more complicated model. The measured capacitance through the semiconductor-solution interface may be represented as several capacitances:

$$\frac{1}{C_{meas.}} = \frac{1}{C_{S.C.} + C_{S.S.}} + \frac{1}{C_H} + \frac{1}{C_D} \qquad (6)$$

where $C_{S.C.}$ is the semiconductor space-charge capacitance, $C_{S.S.}$ is the capacitance due to charge trapped in semiconductor surface states, C_H is the Helmholtz double-layer capacitance, and C_D is diffuse (Gouy) double-layer capacitance. At moderate electrolyte concentrations (∿1 N) all the charge on the solution side is located at or very close to the Helmholtz

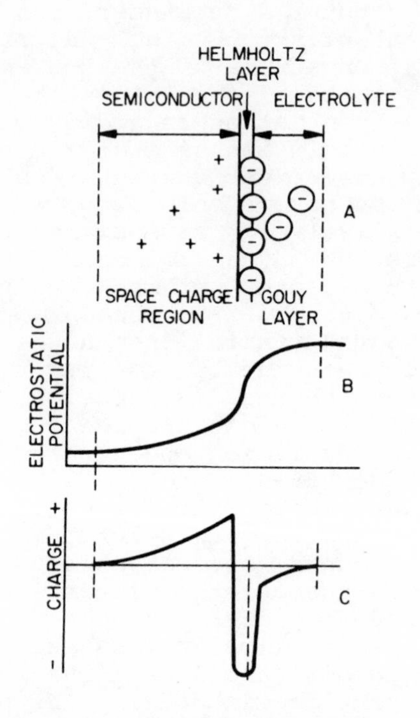

Fig. 6. Structure of the double layer (A), distribution of the potential (B), and distribution of charge (C) at a semiconductor-electrolyte interface.

plane. Since the Helmholtz region is very narrow, ~3 Å, C_H is large (~20 μF/cm^2) and $1/C_{meas.} = 1/C_{S.C.} + C_{S.S.}$. If surface states are absent or insufficient time is allowed to trap some charge on the states, then $C_{meas.} = C_{S.C.}$. Therefore, if impedance measurements are made in 1 N electrolytes at sufficiently high frequencies such that no charge is transferred to surface states and in the absence of faradaic processes, the observed capacitance should be that for the semiconductor space-charge region (35). At lower frequencies, where charge trapped in the surface states changes during measurement, the observed capacitance is the combined effect due to the space charge and surface states.

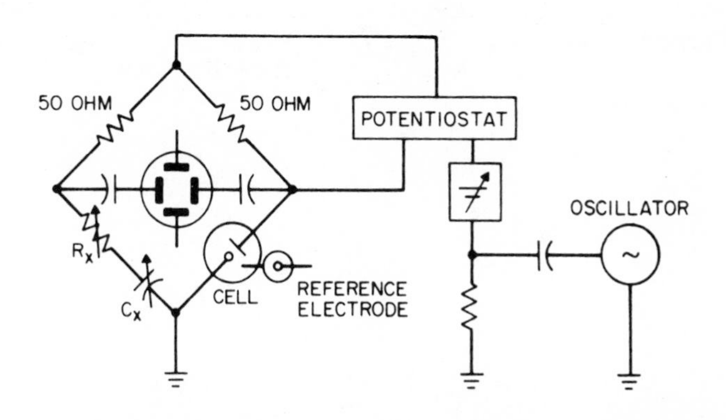

Fig. 7. Circuit diagram of a.c. bridge with potentiostat for d.c. bias for capacitance measurements (36).

Space-charge capacitance is measured by a.c. methods or a d.c. pulse method. The a.c. method employs a bridge such as shown in Fig. 7((36). Typical results obtained by using an a.c. bridge are given in Fig. 8. The d.c. pulse method has been used successfully on germanium (37, 38) and silicon (39). The electrode potential is observed as a function of time in response to a square current pulse as illustrated in Fig. 9 (35). If measurements can be made in a sufficiently short time such that only charging

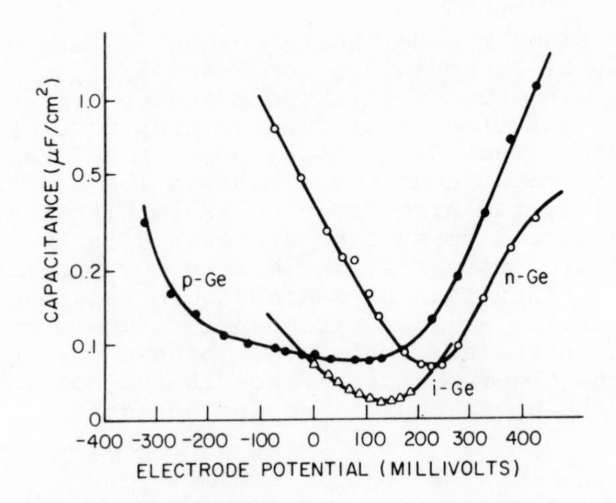

Fig. 8. Capacitance-potential curves for three semiconductor electrode types: p-type Ge, n-type Ge and intrinsic Ge in 1M $NaClO_4$ at pH = 2 (36).

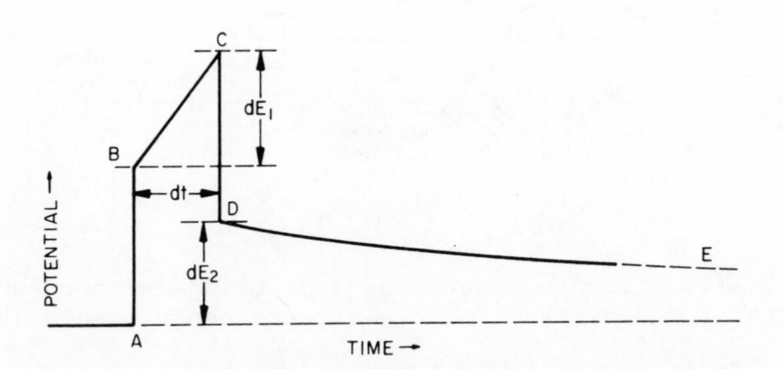

Fig. 9. Response of the semiconductor-electrolyte interface to a rectangular current pulse (35).

of the space-charge region occurs, then

$$C_{S.C.} = i \frac{dt}{dE_1} \tag{7}$$

and the capacitance is determined from the reciprocal of the slope of the charging curve BC in Fig. 9. At very short pulse times (∿1 μsec), oscillations may occur in the measuring circuit to make slope measurement difficult. This problem is avoided if the electrode potential is monitored for a time after the pulse is turned off. The potential decays quickly to a value equal to dE_2 which is the same as dE_1 provided that dt is small compared to the time constant RC for the interface. Since dE_2 decays relatively slowly, the curve can be extrapolated back to D for an accurate measurement.

Boddy (35) has shown that analysis of the data is simplified if the pulse width is short compared with the time constant for decay of the overvoltage. BC in Fig. 9 must be in the linear part of the charging curve. On germanium the time constant is several milliseconds at small currents and reduces to a few tens of microseconds at an anodic polarization of ∿1 mA/cm^{-2}.

The double-pulse technique was used to estimate the Helmholtz capacitance at a germanium electrode-electrolyte interface (40). Large-signal charging experiments have also been reported (41).

Typical results obtained by Brattain and Boddy (15) of differential capacitance measurements on germanium by the d.c. pulse method are given in Fig. 10. The theoretical minimum agrees with experiment assuming a normal surface roughness factor (real area/apparent area) of 1.3.

XI. SURFACE CONDUCTANCE

A surface conductance measurement of a semiconductor provides information about the surface concentration of charge carriers (holes and electrons). By using very thin samples, the surface conductance can be made an appreciably fraction of the whole. Changes in the total conductance of the sample may

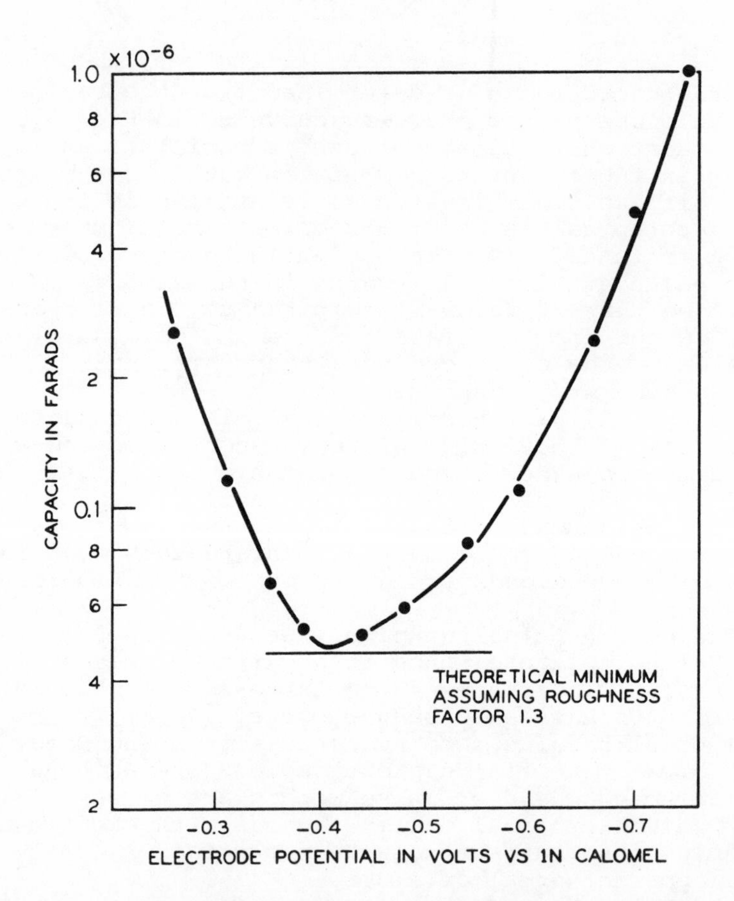

Fig. 10. Measured capacitance for Ge electrode in neutral K_2SO_4 solution (42.2 ohm-cm, n-type (100) orientation) (15).

be observed when the surface carrier concentration is varied relative to the bulk either by electrochemical bias or by changes in the solution or gaseous environment.

(a)

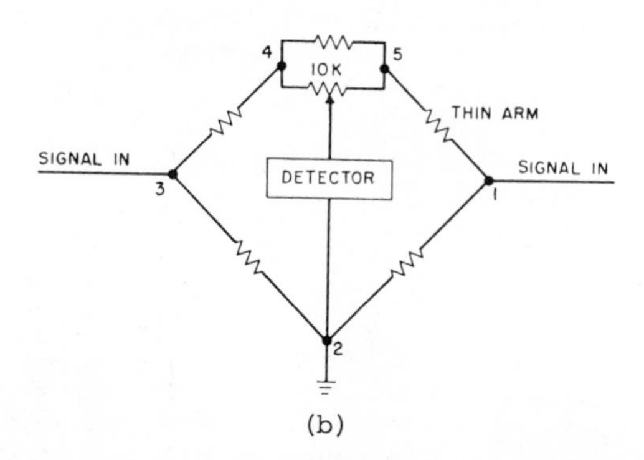

(b)

Fig. 11. (A) Germanium bridge. (B) Electrical schematic of germanium bridge for surface conductance measurements (15).

Brattain and Boddy (15) reported measurements on germanium using an electrode in which all arms of a bridge arrangement were cut in a continuous pattern from the same single crystal; see Fig. 11. The conductance bridge made this way is automatically temperature compensated, but it is complicated to make. Both d.c. and a.c. measurements have been made and they give the same results.

The particular advantage of the conductance measurement is that it is not influenced by the presence of surface states as is the capacitance measurement. However, it is severely limited by parallel conductance through the solution.

XII. ELECTROPLATING ON SEMICONDUCTORS

A. Electrodeposition of Germanium

Early attempts at electrodepositing germanium were done from aqueous alkaline germanate solutions (42-44). Only very thin layers of germanium were deposited before the hydrogren overpotential at the cathode was lowered to the point where 100% of the current went to the discharge of hydrogen ions. Furthermore, no germanium deposited on platinum cathodes whereas copper cathodes would receive a germanium deposit. Ostroumov (45) repeated much of the early work and found that copper cathodes corroded in the aqueous solutions and germanium deposited as an alloy with other metals, notably nickel and cobalt. This has been used to develop an analytical method for determining germanium electrolytically (46). The maximum content of germanium in the alloy is limited since too high a ratio will lower the hydrogen overpotential so that no alloy is deposited. For example, in the nickel alloy the limit is 53%.

Anhydrous organic solutions were tried when aqueous solutions failed to produce thick germanium deposits. The absence of water eliminates it as a source of material that can be reduced preferentially to germanium deposition. Germanium tetrachloride and germanium tetraiodide were used as electrolytes in ethyelene glycol or propylene glycol (45, 47, 48). Fink and Dokras (47) reported thick deposits of germanium from GeI_4-ethylene glycol solutions, but Ostroumov (45) could not repeat their results and he concluded that their solution must have been contaminated with metals.

Deposits of pure bright germanium up to 20 μ thick were obtained from $GeCl_4$-ethylene glycol or propylene glycol at 60° C (45, 48). The deposition rate is slow since the maximum cathode efficiency is about 2% in 5-7 vol. % $GeCl_4$ solutions at 60-100°C. The cathode efficiency also tends to decrease with plating time. At 200 mA/cm^2, about 6 μ of germanium deposit in 2 hr. The germanium plating solution ages with use as the result of chlorine liberation at the graphite anode and formation of ehtylene chlorohydrin (45). Ethylene chlorohydrin in the solution turns the germanium deposits dark.

B. Deposition of Electrodes

Ohmic and rectifying contacts may be deposited by electroplating. However, one must first remove an oxide layer that is usually present on the surface of the semiconductor and that may interfere with the formation of an adhering metal film (6). The best solutions for electroplating on germanium are those that cathodically reduce the oxide layer before the plating sction starts. Cyanide solutions would be suitable for plating on germanium, since their plating potential is larger than the potential for the cathodic reduction of the oxide. However, other electrolytes may be preferred because of their chemical stability. Table 5 shows the platable materials and the corresponding aqueous solutions. Chromium is plated from a chromic acid-sulfate bath. Bismuth may be plated from a perchlorate solution. All the metals of Table 5 that were deposited on germanium, except antimony, formed rectifying contacts on n-type material and ohmic contacts on p-type germanium. Antimony behaved in the opposite manner. Other formulations for plating solutions used by Borneman, Schwarz and Stickler (49) are listed in Table 6.

The simultaneous deposition of two different elements is usually difficult because their deposition potentials may be different. Thus, In and Ga deposit at different potentials from aqueous solutions; however, their halides in nonaqueous solutions have nearly equal deposition potentials (50). The slight difference in the deposition potentials permits control of the In to Ga ratio, a higher voltage favoring a higher proportion of Ga in the deposit.

Cathodic reduction of silicon surface oxide cannot take place before plating. However, the plating bath may contain HF, which attacks the oxide. The crystal is immersed for about 1 min in the solution before

Table 5

Metals plateable from aqueous solutions

	Sulfate and Chloride Solutions				Cyanide Solutions			Fluoborate and Fluoride Solutions		
VI B	VII B		VIII		1B	11B	111B	IVA	VA	VIA
24	25	26	27	28	29	30	31	32	33	34
Cr	Mn	Fe	Co	Nl	Cu	Zn	Ga	Ge	AS	Se
	43	44	45	46	47	48	49	50	51	52
	Tc	Ru	Rh	Pd	Ag	Cd	In	Sn	Sb	Te
	75	76	77	78	79	80	81	82	83	84
	Re	Os	Ir	Pt	Au	Hg	Tl	Pb	Bl	Po

the plating current is turned on. Plating baths that have been used for depositing contacts on silicon are listed in Tables 7, 8, and 9. The adhesion of Cu and Zn on silicon from the cyanide solutions is improved by "Strike" plating, which consists of starting at a high current density that is reduced after a few seconds to the normal plating current density.

When the contacts are alloyed into either germanium or silicon, a better wetting is obtained if a layer of gold is deposited first on the semiconductor. During heating, the gold forms a eutectic with the semiconductor, which results in an excellent mechanical bond. The electrical properties of the gold contact can be adjusted by the addition of suitable impurities.

The shape of the electrodeposited material may be defined by any of several techniques. Masking will, of course, block the deposition over the regions where it is not wanted. A suitably shaped anode in close proximity to the semiconductor surface will replicate its configuration in the deposit. However, to obtain a good resolution by this method, the

Table 6

Composition of solutions used to plate various metals on germanium (49)

Plated Metal	Solution Components	g/liter
Indium	$In_2(SO_4)_3$	8.6
	H_2SO_4	2.45
Zinc	$AnSO_4 \cdot 7H_2O$	50
	NH_4Cl	5
	$NaC_2H_3O_2$	5
Tin	$SnCl_2$	6.4
	HCl	6.1
Lead	$Pb(OH)_2 \cdot 2PbCO_3$	150
	HF	120
	H_3BO_3	105
Copper	$CuSO_4$	8
	H_2SO_4	4.9
Silver	$AgNO_3$	0.34
	$NaNO_3$	17
Gold	$AuCl_3$	10
	NaCl	2.9
Rhodium	$RhCl_3 \cdot 4H_2O$	8.5
	H_2SO_4	4.9
Nickel	$NiCl_2 \cdot 6H_2O$	11.8
	HCl	2.5
Platinum	$Pt(NH_3)_2(NO_2)_2$	5
	NH_4NO_3	50
	$NaNO_2$	5
Chromium	CrO_3	16.6
	Na_2CO_4	0.24
Tungsten	WO_3	32
	Na_2CO_3	48.5
	$NiCO_3$	0.09
Arsenic	As_2O_3	3.3
	Na_2CO_3	5.3
Antimony	SbF_3	6
	HF	4
Bismuth	BiOCl	20
	HCl	73.2

Table 7

Metals electroplated on Si with good adhesion and their electrical properties as contacts (6)

Metal	Solution Type	Current Density mA/cm^2	Relative Adhesion	Electrical properties of Contact	
				n-Si	p-Si
Cu	Cyanide	200-75	Very good	Ohmic	Rectifying
Au	Cyanide	15	Good	Rectifying	Ohmic
Ni	Fluoride	40	Good	Rectifying	Rectifying
In	Cyanide	30	Fair	Rectifying	Rectifying
Sn	Stannate	20	Fair	Ohmic	Rectifying
Zn	Cyanide	100-30	Poor	Ohmic	Rectifying

Table 8

Composition of solutions used to plate various metals on silicon (51)

Metal	Solution Components	g/liter	Solvent
Zinc	$ZnCl_2$	13.6	
	$SnCl_2$	0.016	
	$PbCl_2$	0.002	Ethylene glycol
Indium	$In_2(SO_4)_3$	17.26	
	H_2SO_4	4.52	H_2O
Tin	$SnCl_2$	22.32	
	NaF	14.93	
	$NaHF_2$	7.73	
	NaCl	4.93	
	$Na_4Fe(CN)_6$	0.36	H_2O
Lead	$Pb(BO_2)_2 \cdot H_2O$	180.3	
	H_3BO_3	180	
	HF	460	H_2O
Cadmium	$CdCl_2$	18.32	Ethylene glycol
	NH_4Cl	2.66	
Antimony	SbF_3	10.0	Ethylene glycol
	HF	57.5	
Copper	$CuSO_4$	8.0	
	H_2SO_4	16.36	H_2O
Bismuth	BiOCl	20.0	
	HCl	73.2	H_2O
Nickel	$NiSO_4 \cdot 7H_2O$	105	Ethylene glycol
Cobalt	$CoSO_4 \cdot 7H_2O$	151	
	NaF	4.7	H_2O
Iron	$Fe(NH_4)_2(SO_4)_2 \cdot 6H_2O$	29.4	H_2O
Silver	$AgBF_4$	10.66	
	NaF	0.07	H_2O
Gold	$AuCl_3$	10.0	
	NaCl	2.9	H_2O
Rhodium	$RhCl_3 \cdot 4H_2O$	4.0	
	H_2SO_4	34.80	H_2O
Platinum	$Pt(NH_3)_2(NO_2)_2$	6.0	
	NH_4NO_3	2.0	
	H_2SO_4	0.174	Ethylene glycol

Table 9

Comparison of rectification for metals plated on 2.5 ohm cm n- and p-type silicon (51)

Metal	n-Type Silicon	p-Type Silicon
Aluminum	Ohmic	Rectifying
Zinc	Ohmic	Rectifying
Indium	Ohmic	Rectifying
Lead	Ohmic	Rectifying
Tin	Ohmic	Rectifying
Cadmium	Rectifying	Rectifying[a]
Antimony	Rectifying[a]	Rectifying[a]
Cobalt	Rectifying[a]	Rectifying
Bismuth	Rectifying[a]	Rectifying
Silver	Rectifying[a]	Rectifying
Copper	Rectifying[a]	Rectifying
Nickel	Rectifying	Rectifying
Iron	Rectifying	Ohmic
Gold	Rectifying	Ohmic
Rhodium	Rectifying	Ohmic
Platinum	Rectifying	Ohmic

[a]Lower saturation current observed on this type silicon.

electrolyte must be more resistive than the semiconductor. An alternative approach is the use of a virtual anode (52). In this case, the actual anode may be far from the crystal but the plating current is confined almost all the way to the work surface by an insulating wall around that portion of the electrolyte that carries the current.

The deposition can be enhanced over the desired region by increasing locally the conductivity of the semiconductor, for example, by optical injection. Thus, an image on the semiconductor can be developed as a metallized pattern.

C. Decoration of pn Junctions

A reverse bias across a pn junction makes the p-type region negative with respect to the n-type region. Hence, the n-type region can act as the anode of the electroplating system that deposits metal onto the p-type region. A sharp demarkation will then appear at the boundary between the plated p-type region and the anodic n-type region. If the electrolyte is not very conducting, most of the plating current will crowd at the edge of the pn junction. One can also force the crowding of the plating current to the edge of the pn junction by coating the semiconductor surface with a thin layer of the plating solution instead of immersing the crystal in a plating bath. The technique increases the resistance of the solution away from the junction.

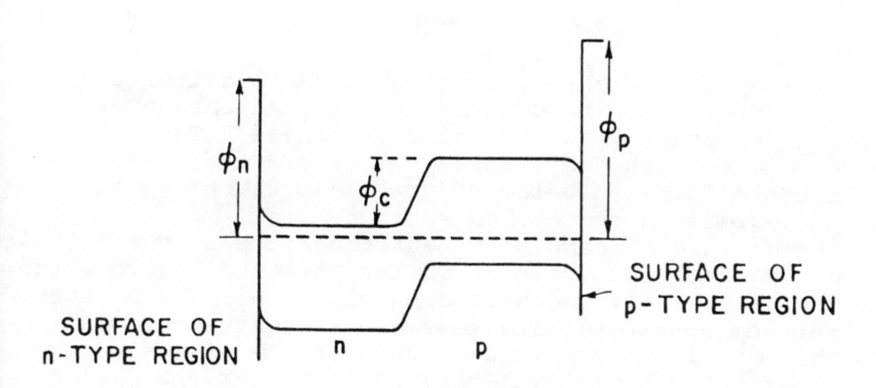

Fig. 12. Energy-level diagram illustrating how internal contact potential can render p-type surface more electronegative than n-type surface.

Preferential plating on the p-type region can sometimes be obtained without applying a bias across the pn junction (53). As shown in Fig. 12, the contact potential ϕ_c at the pn junction, resulting from the impurity gradient, causes the work functions ϕ_p and ϕ_n of the p and n regions to be different: the p-type region is more electronegative than the n-type region. In extreme cases, the electronegative difference may be sufficient to allow preferential plating by galvanic action where the crystal is simply immersed in the plating solution without attaching any electrode.

Illuminating a pn-junction with light having a photon energy greater than the band-gap energy, generates a photovoltage: the built-in field drifts the excess electrons to the n-type region and the excess holes to the p-type region. If the junction is immersed in an electrolyte, the photogenerated carriers can leak out through the electrolyte, which effectively short circuits the pn junction. Since the electrons are accumulated in the n-type region, which then becomes the cathode, it is the n-type region that is preferentially plated. Thus, the n-type side of a silicon pn junction was plated with silver (54), copper (55), or gold (56).

XIII. ELECTROLYTIC ETCHING OF SEMICONDUCTORS

Compared to chemical etching, the electrolytic method of removing material has the advantage of greater control and better cleanliness. The latter virtue results from the fact that electrolysis usually forms soluble products that therefore leave no residue on the etched surface.

The rate of removal of surface atoms is controlled by the etching current. On n-type semiconductors the rate may be limited by the supply of holes. However, this rate can be further accelerated by increasing the hole density at the surface of the crystal, such as by illumination. A variety of techniques can be used to concentrate or localize the etching current.

A. Electropolishing

Electropolishing is a treatment for removing work-damaged material from the surface of the crystal. Atoms in the neighborhood of crystalline imperfections are removed first. The remaining surface, although nearly uniform, is not necessarily smooth. The more rapid etching at the site of an imperfection results in an etch pit. The weakening of bonds by the imperfection is equivalent to the presence of a locally increased hole concentration, which accelerates the local etching. This lack of homogeneity can be compensated by uniformly increasing the hole concentration. This can be done by flooding the specimen with intense light or by electrical injection from a pn junction.

Actually, polishing consists of removing material from the surface nonpreferentially from high spots. Since the electric field is maximum at sharp irregularities of the surface, these are the regions that are removed most rapidly. In homogeneous materials, this refults in a very smooth surface. From the knowledge of the area of the work and the distribution of current flow (a plane parallel geometry is the simplest and most common case) the thickness of the material removed can be measured by coulometric techniques. Of course, the fact that the process can be measured implies that control of the process may be possible. It is found that in most cases germanium and silicon are removed as tetravalent atoms (i.e., one needs to transfer four electrons to liberate one atom).

Excellent results have been obtained by Kein et al. (57) on germanium and silicon. The semiconductor to be polished is placed very close to a rapidly rotating gold-plated cathode. The intense stirring action assures a flat mirror-like surface for n-type material, holes being injected by illumination through suitable slots or aperatures in the cathode. A surface roughness of $\pm$ 25 Å was obtained. A variety of etchants can be used for germanium; dilute KOH, HNO_3, HCl, and H_2SO_4 have all given a polish that is flawless on the microscopic scale.

p-Type silicon has been electropolished in an aqueous solution of about 5% HF using a current density in the range of 65 to 100 mA/cm^2 at 25° C. The crystal should be cooled by a sufficiently voluminous bath to prevent a thermal runaway. It is found that the critical current density at which electropolishing takes place increases upon stirring, or when increasing either the HF concentration, or the temperature or the viscosity of the electrolyte. A low concentration of HF (less than 2% causes etching instead of polishing while a high concentration of HF (over 10%) gives rise to excessive heating due to the high current density.

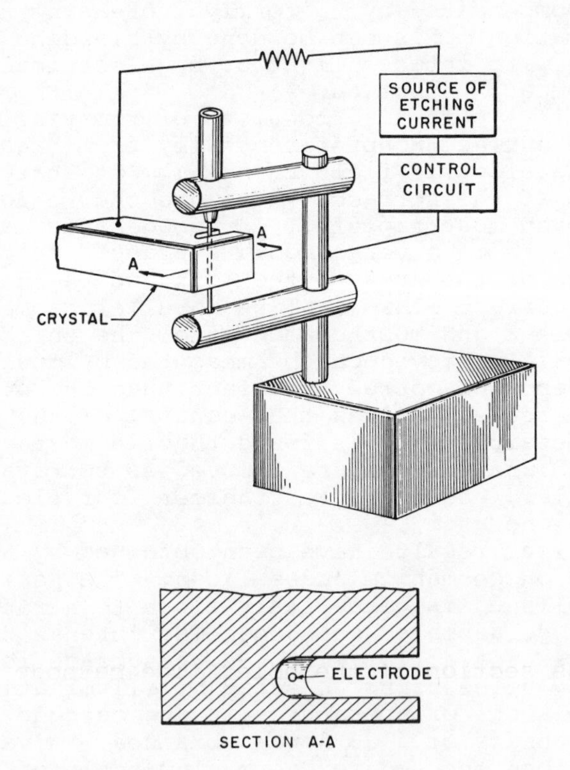

Fig. 13. Slicing by stream etching [after Barry and Seeley] (11).

B. Selective Etching

Selective etching allows cutting, drilling, or photoengraving of the semiconductor without the introduction of work damage.

Electrolytic slicing is certainly the most remarkable shpaing process. The crystal can be cut into smooth thin slices without straining the material. One-mil-thick wafers of germanium have been obtained by stream etching technique (58). In this method, shown in Fig. 13, the electrolyte flows along a stretched wire that forms the cathode. The stream of electrolyte also wets the crystal where etching is to take place. The material is removed from that region which is closest to the cathode. As etching proceeds, the cathode is advanced at a uniform rate inside the herf. The wire, not being subjected to wear, lasts indefinitely. It is desirable that the electrolyte have a higher resistivity than the semiconductor and that the spacing between anode and cathode be small. This condition results in a uniform cut because of a negative feedback action: as the gap between the crystal and the electrode tends to increase, the current density decreases, which results a reduction of the gap.

By way of example, Barry and Seeley (58) suggest a solution of 0.002% KOH by weight in water, with a flow rate of 10 ml/min a current of 28 mA, the 3.3-mil-diameter tungsten wire being fed at the rate of 6.5 mils/min. The width of the kerf is 7 mils. If the electrolyte does not wet the crystal readily, a wetting agent may be added.

To drill a hole or a well into the semiconductor, the etching current can be concentrated in the desired region by the close proximity of the cathode, as shown in Fig. 14. The shape of the cathode defines the cross section of the hole. The cathode is insulated on the periphery, while its bare end nearly touches the crystal. The current is confined near the end of the cathode. The area etched is only slightly larger than the cathode. The rapid evolution of bubbles stirs the electrolyte vigorously over the active region. An annular depression would be formed by a tubular cathode. Other shapes can be

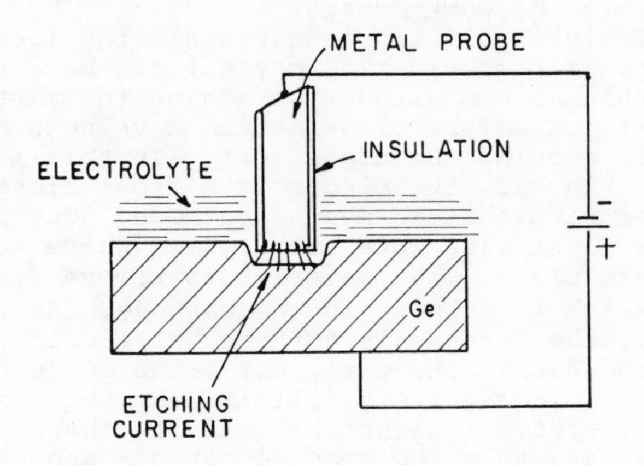

Fig. 14. Etching a well with a probe.

carved, such as delicate structures that should not be touched by any tool. In the case of microcarving, it is desirable that the electrolyte have a higher resistivity than the semiconductor. These close-proximity techniques lend themselves to automation. Since the etching current is a function of the probe-to-crystal spacing, a servo system can provide the automatic feed of the electrode into the cut at a rate equal to the etching rate until the desired depth is reached.

A virtual cathode was developed by Uhlir (52) for macromachining (see Fig. 15). That portion of the electrolyte which conducts the etching current is confined by an insulating wall all the way to the specimen. The insulating tube is opened very close to the crystal where etching is desired. The diameter of the bore may be controlled along the penetration of the virtual electrode by varying the current and the rate of tool advance. In all the close-proximity techniques, the maximum etching rate is limited by excessive heating of the electrolyte. This limitation can be alleviated somewhat by using electrolytes

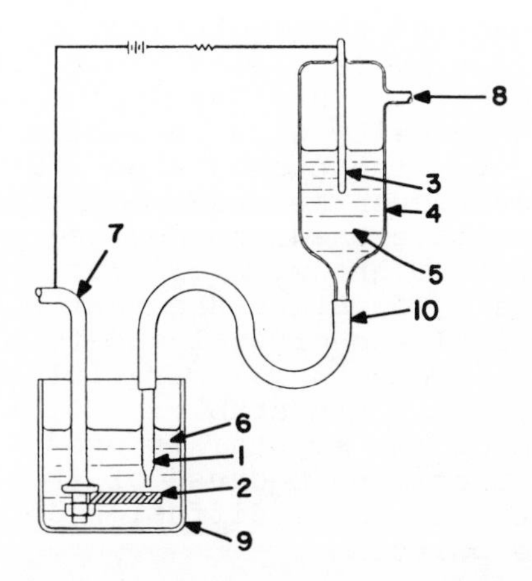

Fig. 15. Apparatus for virtual cathode micromachining. (1) Nonconducting partition (microelectrode). (2) Work. (3) Actual cathode. (4) Cathode compartment. (5) Electrolyte. (6) Electrolyte outside of microelectrode. (7) Support arm for work. (8) Connection for applying air pressure (optional). (9) Container with flat glass side for observing work. (10) Flexible tubing. From Uhlir (5).

under pressure to increase solution flow rate.

In the jet-etching (59) technique, the electrolyte has the shape of a fine jet which may be several mils in diamter. The electrolyte is projected through a glass or a metal nozzle (the latter forming the cathode). The jet carries the current. Hence, the current density is maximum over the area where the jet strikes the crystal. Beyond the impact area, the electrolyte spreads out in a thin sheet, the high resistance of which causes the current to drop off rapidly away from the edge of the jet. The resulting depression is a shallow well with a smooth bottom. By way of example, a depression 40 μ in depth and 250 μ in diameter can be etched in about 100 sec with a current density of 3 A/cm^2. Heating does not present a problem because of the cooling action of the jet. If two coaxial jets attack opposite surfaces of the crystal, it is possible to reduce the crystal thicknesses to the order of a few microns. By choosing the proper electrolyte it is possible to use the same solution for etching and then for plating by reversing the polarity of the current in the jet. Thus, an indium sulfate solution was used to anodically etch germanium and then to plate indium over the etched area (59).

Photoengraving can be obtained by projecting on the surface to be etched the desired pattern using for illumination light with greater than band-gap energy photons. To obtain the best contrast, it is desirable to cool the crystal and to eliminate ambient light. For deep etching, a parallel beam, such as from a laser, it is preferable. Illumination from the back side of the crystal is useful only for etching thin specimens because of the loss of resolution due to the latteral diffusion of photogenerated electron-hole pairs.

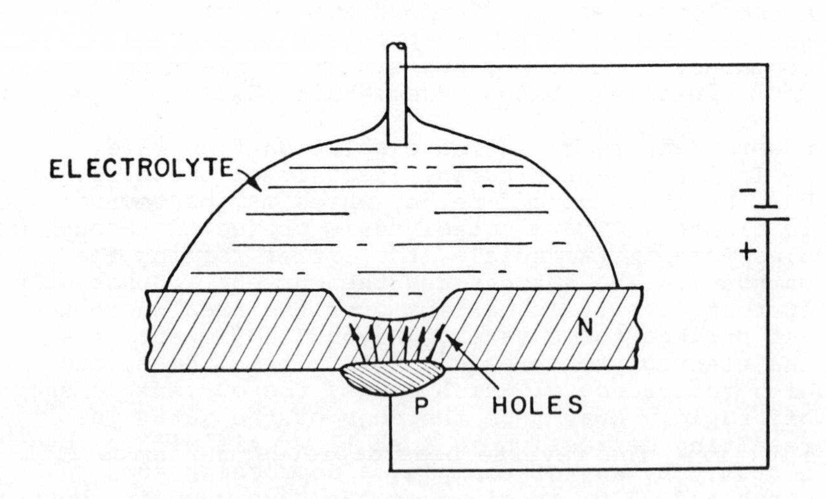

Fig. 16. Injection-accelerated etching.

Electrical injection may be used to locally accelerate the etching process. Thus, a pn junction or a Schottky contact having the desired configuration can inject holes into a thin n-type semicondctor (see Fig. 16). When the injected holes reach the opposite surface in contact with a negatively biased electrolyte, etching occurs in a configuration that replicates the distribution of the injected current.

On application of the above hole-injection technique is the formation of a very thin n-type layer adjacent to a pn junction. This thin layer can be the base of a high-frequency transistor, ior it can be the "window" of a sensitive photocell. Many other devices require a thin n-type layer or a close spacing between two junctions.

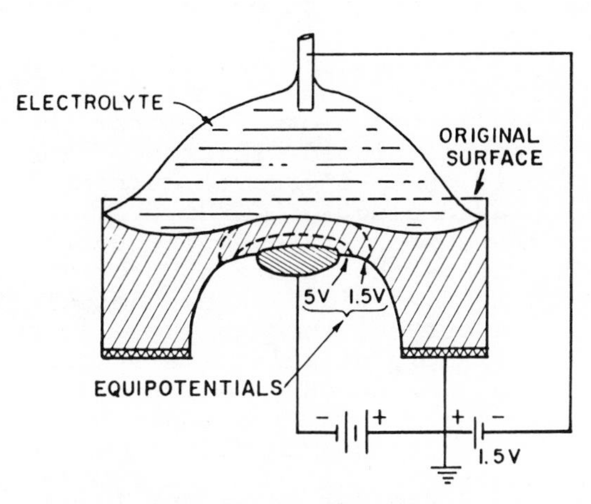

Fig. 17. Field controlled etching.

If hole injection due to forward bias can accelerate the motion of the etched surface toward the injection electrode, it is conversely possible to use hole extraction to stop the etching action near the same electrode. This is done by reverse biasing the pn junction (or a Schottky contact) as shown in Fig. 17. The reverse bias depletes the holes from the n-type layer and retards the etching process. Furthermore, a set of negative equipotential surfaces appears in the n-type region (two are shown in dashed lines in Fig. 17). One of these equipotentials is at the same potential as the electrolyte. Etching stops at this equipotential surface since no current can flow through it to the electrolyte (60).

The position of this terminal equipotential can be controlled by adjusting the potential of the reverse-biased electrode. Obviously, the extent of the depletion region depends on the resistivity of the material, increasing as the square root of the resistivity. Thus, 5-Ω cm n-type germanium can be etched to a thickness of 4 μ if a 10-V reverse bias is applied across the etch-limiting junction. In homogeneous materials, the equipotentials are parallel to the control electrode.

It is interesting to note that the current through the reverse biased pn junction may be used to monitor the thickness of the reamining n-type layer. When the electrolyte reaches the limiting equipotential, electrical "punch-through" is achieved: suddenly, the conventional current can flow from the electrolyte to the reverse-biased electrode. Punch-through manifests itself as an increased current flow through the reverse-biased electrode. However, now the polarity of the current is opposite to that needed for etching: hence the anodizing process stops. Punch-through is also signaled by the polarity reversal of the current through the electrolyte.

As a final example of selective etching, consider the case of GaN, a wide-gap semiconductor that is best grown by chemical vapor deposition on sapphire (61). The problem is to remove the GaN from its poorly matching substrate. GaN is almost as resistant to chemical etchants as sapphire; hence, preferential etching of sapphire is not practical. However, undoped GaN is n-type and very conducting, while the An-doped material can be insulating. If an insulating layer is grown on a conducting layer, itself grown on sapphire, it is possible to selectively etch away the conducting layer, thus separating the insulating layer of GaN from the sapphire substrate. An ohmic connection to the sectioned edge of the composite layer allows the etching current to flow along the intermediate conducting layer. An aqueous solution of NaOH was used as the electrolyte (62). A similar technique was used to preferentially remove the n^+ silicon substrate from an epitaxial high resistivity n-type layer of silicon (63). In this case, the electrolyte was an aqueous solution of HF.

C. Junction Delineation and Inhomogeneity Display

When a reverse bias is applied across a pn junction immersed in an electrolyte, the n-type region forms the anode and the p-type region forms the cathode. The anode etches off while the p-type region is not affected. A pronounced step down develops from the p-type to the n-type region. Furthermore, because of the differential etching, the two regions acquire different textures. In the case of germanium, if this technique is employed at 80°C with a high current density, the p-type region becomes covered with a dark film, probably amorphous germanium. Then, the pn junction appears in color contrast as shown in Fig. 18. This color contract technique works only after the electrolyte has been saturated with germanium ions by electrolytic etching; 1% aqueous solutions of NaOH, KOH, and HCl have produced the dark film.

Fig. 18. Color contrasted p-n junctions. The dark nonshiny surface indicates the p-type region; the shiny pitted surface marks the n-type region.

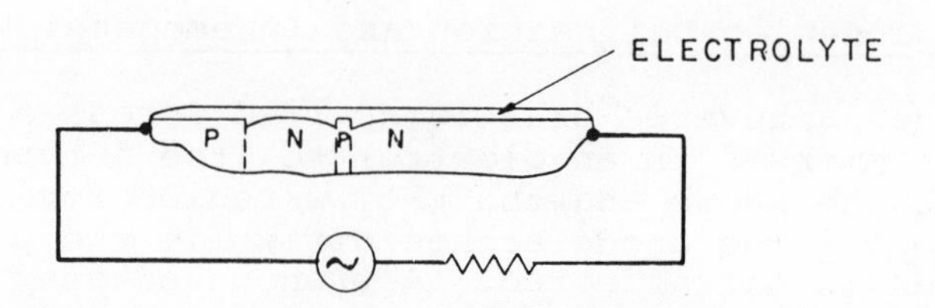

Fig. 19. Method of etching a plurality of junctions.

A multiplicity of pn junctions on the same crystal can be revealed without making an individual connection to each region. An a.c. bias is applied across the ends of the crystal as shown in Fig. 19. The crystal is coated with a layer of electrolyte. During each half cycle, all the reverse-biased junctions form an etching cell. Because of the high lateral resistance of the electrolyte, the etching action is strongest at the n-type edge of the junction. The p-type regions acquire a dull finish, whereas the n-type regions are bounded by lightly pitted zones that are dark or shiny depending on the illumination (Fig. 20). Even when the pn junction does not extend across the thickness of the crystal, as in the case of diffused junctions, it is possible to reveal the boundary of the "floating" junction. This is because a floating junction in an electric field assumes charge neutrality by distributing the current flow through the junction in such a way that the current which flows through reverse-biased portions of the junction emerges through another portion of the same junction where the bias is in the forward direction (64). In the presence of an electrolyte, some of the reverse-bias current fringes into the electrolyte causing the boundary of n-type region to be etched (see Fig. 21).

An extremely sensitive technique can be devised for revealing inhomogeneities in the crystal. This technique is so sensitive that one cannot distinquish between a pn junction and an abrupt change in impurity concentration. However, once the junctions have been located by one of the above techniques, it is possible to recognize them among the newly formed etch patterns. After the surface of the crystal has been carefully polished, a longitudinal

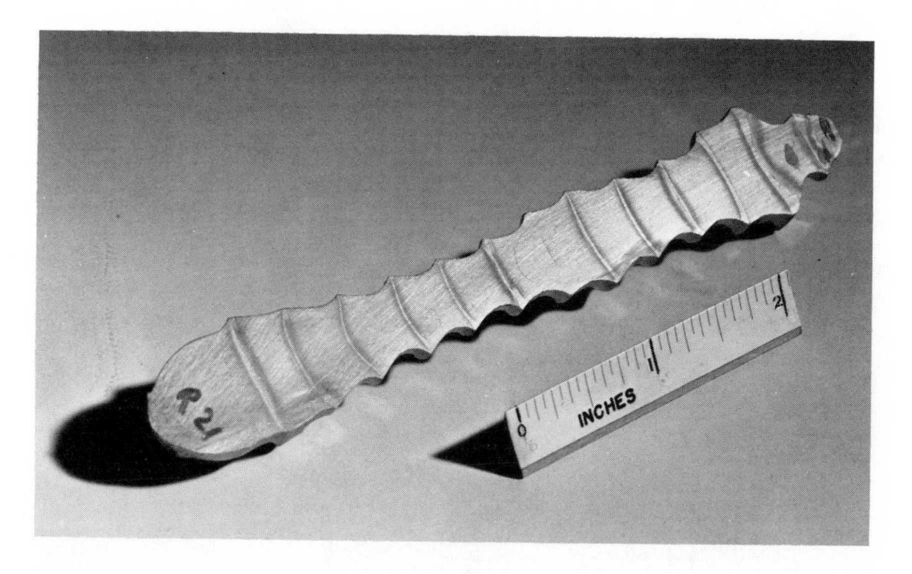

Fig. 20. Rate grown Ge crystal. The high-contrast regions represent pn junctions.

current is passed through the crystal as in the technique described above for revealing multiple junctions. A film of electrolyte wets the polished surface. The crystal is made positive with respect to a mobile cathode which spans the width of the crystal and which is located very close to the polished surface (Fig. 22). A mechanical arrangement causes the cathode to slowly scan the crystal while it is etched. When the cathode is over a region of low resistivity, that is, high impurity concentration, the etching current increases and leaves a more profoundly disturbed surface (for example, the etch pits become coarser; see Fig. 23.) The longitudinal field sets up potential gradients along the crystal depending on the nature of the inhomogeneity (pn junction, p^+p or nn^+ transitions) so that the more positive or the more conductive regions tend to etch faster. The longitudinal field must be reversed in a second scan in order to set up a potential gradient at those rectifying transitions that were biased in the forward direction during the first scan.

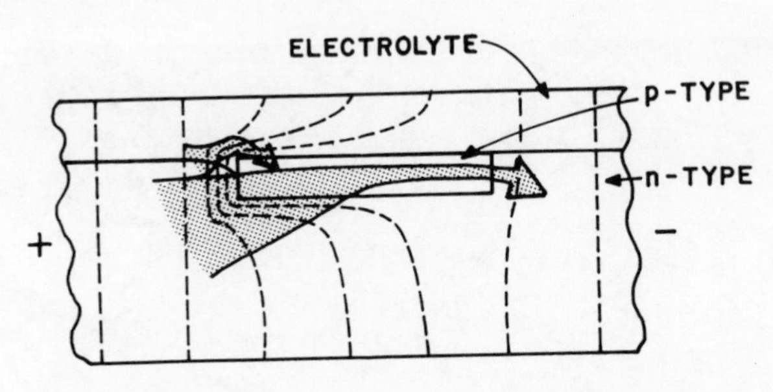

Fig. 21. Etching the edge of a floating junction.

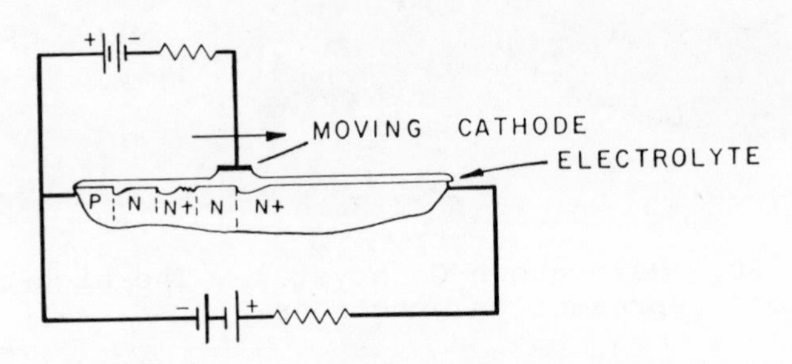

Fig. 22. Method for exposing inhomogeneities.

D. Control of Surface Recombination

Special attention must be given to the surface of semiconductor devices. A variety of centers, such as adsorbed chemical impurities or defects due to mechanical damage, greatly affect the performance and life expectancy of the device. Minority carriers can be lost by surface recombination, or unwanted carriers can be generated at the surface by these undesirable centers. In a transistor, surface recombination results in reduced amplification due to the loss of carriers injected by the emitter (65). Furthermore, the same centers that are responsible for high recombination and generation rates, in the vicinity of the collector are responsible for a high saturation current that limits the operational range

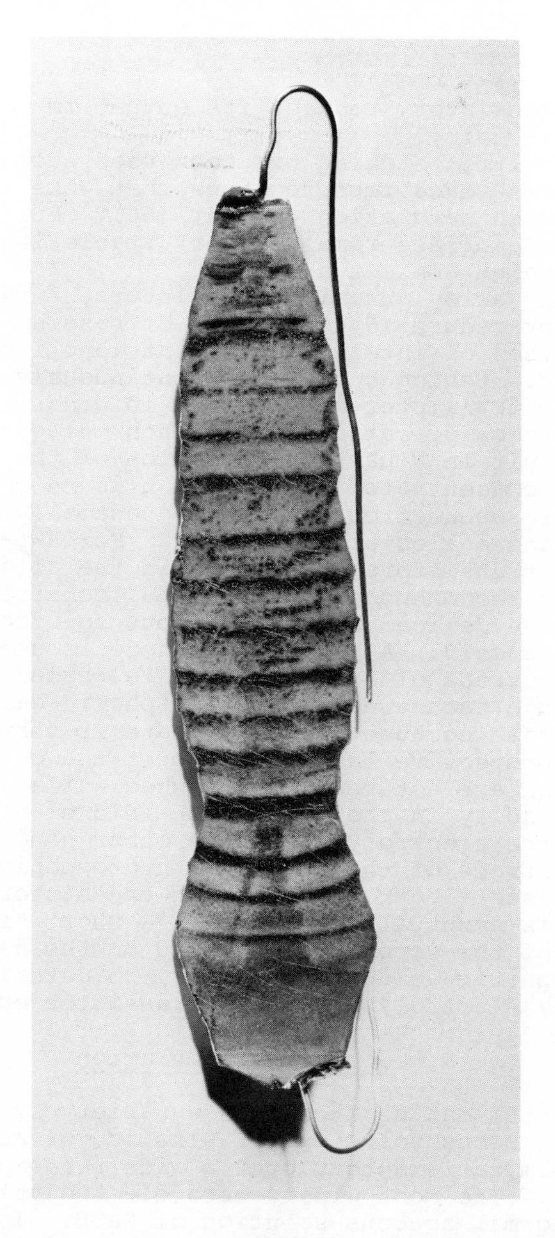

Fig. 23. Rate grown Ge crystal. Pitted bands represent regions of high impurity concentration.

of the transistor, reduces its output impedance, and renders it noisy.

Electrolytic etching has been used successfully to remove surface imperfections that were responsible for the high saturation current (66). Furthermore, the etched surface remained very stable in the presence of thigh ambient humidity.

For pnp alloy junction transistors, a recommended etching procedure (67) consists of passing a few short pulses of intense current at long intervals; a low d.c. etching current is continuously flowing while the transistor is immersed in order to avoid the diffusion of cations after each pulse since this would result in a partial depletion of the electrolyte. A concentrated KOH solution is recommended; the anodic connection is made sequentially to the emitter and collector electrodes. For n-p-n alloy junction transistors using lead in the alloy, a procedure recommended by Amaya (68) consists of etching the device in a 25% aqueous solution of phosphoric acid. A platinum cathode is used; an etching current of 500 to 800 mA is passed for 12 sec through each electrode. Phosphoric acid is used in this case because lead phosphate is very insoluble. Hence a protective layer forms over the electrodes, which then are not readily attached while germanium etches readily in the phosphoric acid electrolyte. An alkaline electrolyte, on the other hand, may form on the surface of the crystal a hydroscopic residue, which causes electrolysis during transistor operation; this eventually results in a short circuit because of the deposition of lead on the base. With the phosphoric acid electrolytes, no deterioration occurs by electrolysis during transistor operation.

E. Solution Formulation

Germanium can be etched in a variety of electrolytes. Aqueous solutions of alkalis, of acids, or of salts seem satisfactory over a wide range of concentrations. The most widely used electrolyte is perhaps a normal aqueous solution of NaOH. In particular cases, as discussed above, the resistivity of the electrolyte needs to be higher than that of the germanium. The choice of electrolyte, usually, is guided so as to avoid a reaction with other metals that may be immersed at the same time as the semiconductor.

Silicon appears somewhat difficult to etch electrolytically in aqueous solutions. This is believed to result from the formation of an oxide layer (69). Satisfactory results are obtained with an electrolyte consisting of a mixture of hydrofluoric acid and a hydrophylic organic compound such as alcohols, glycols, acetic acid, ether, and so on. Faust (70) recommends an electrolyte consisting of about 10% concentrated hydrofluoric acid in ethanol containing less than 17% water.

Uhlir (69) points out that electrolytes which give silicon an effective valence of 4 yield satisfactory results whereas at an effective valence of 2, p-n junctions may be shorted out and the surface may become unstable. A minimum current density for smooth etching is suggested as 0.01 to 0.1 A/cm^2 depending on the composition of the electrolyte.

Gallium arsenide has been etched and electropolished in 20% perchloric acid - 80% acetic acid solution (71). This solution must be cooled and kept away from carbonaceous matter. The specimen is made the anode, and a much larger area stainless steel cathode is used. Electroetching, to reveal etch pits and macrostructural details such as twins, grain boundaries, and other inhomogeneities, is best done at low current densities in the range of 0.3 to 0.4 A/cm^2. p-Type GaAs and $GaAs_{1-x}P_x$ alloys grown on an n-type GaAs substrate have been selectively removed from the n-type crystal by Nuese and Gannon using a three-molar aqueous solution of NaOH (72). Note that the selective removal of the p-type region cannot be achieved by chemical etching. The electrolyte is forced as a fast stream onto the specimen to flush away flakes of arsenolite, As_2O_3, which are a by-product of the anodizing reaction and which would tend to block the etching process.

InP, In $Ga_{1-x}P$ and $Al_xGa_{1-x}As$, all p-type, have also been successfully removed from their n-type GaAs substrates by the above Nuese-Gannon method.

Gap requires different electrolytes depending on the conductivity type (73); p-type GaP is anodized in a 3:1:1 solution of H_2SO_4, H_2O_2, H_2O, while the n-type material can be etched anodically in a 7:1 solution of NaOCl, HCl. Both solutions use a stainless steel cathode.

InSb has been electropolished by Dewald (74) in the following solution: $HClO_4$, 10 parts; acetic anhydride, 40 parts; H_2O, 2 parts. This electrolyte is used at 5°C with a current density of 50 mA/cm^2. This results in a solid black film of oxide that is tripped off in water. In this way, a mirror bright surface is obtained. Anodization of InSb is also possible with a 0.1 N KOH solution.

GaN has been anodized in a 0.1 normal solution of NaOH (62).

REFERENCES

1. C. Kittel, Introduction to Stolid State Physics, 4th ed., John Wiley and Sons, New York, 1971.
2. W. H. Brattain and C. G. B. Garrett, Bell Syst. Tech. J. 34, 129 (1955).
3. J. I. Pankove, in Electrochemistry of Semiconductors, P. J. Holmes, Ed., Chapter 7, Academic Press, New York, 1962.
4. D. R. Turner, in Electrochemistry of Semiconductors, P. J. Holmes, Ed., Chapter 4, Academic Press, New York, 1962.
5. V. A. Myamlin and Yu V. Pleskov, in Electrochemistry of Semiconductors, Chapter V, Plenum Press, New York, 1967.
6. D. R. Turner, J. Electrochem. Soc. 106, 786 (1959).
7. M. V. Sullivan and R. M. Warner, in Transistor Technology F. J. Biondi, Ed., Vol. 3, Van Nostrand, Princeton, 1958, p. 163.
8. J. O'M. Bockris, Chem. Rev. 43, 526 (1948).
9. R. Piontelli et al., Z. Elektrochem. 58, 86 (1954).
10. S. Barnartt, J. Electrochem. Soc. 106, 722 (1959).
11. D. R. Turner, J. Electrochem. Soc. 103, 252 (1956).
12. S. Barnartt, J. Electrochem. Soc. 99, 549 (1952).
13. R. Piontelli et al., Z. Elektrochem. 56, 86 (1952); 58, 86 (1954).

References

14. F. Beck and H. Gerischer, Z. Elektrochem. 63, 943 (1959).
15. W. H. Brattain and P. J. Boddy, J. Electrochem. Soc. 109, 574 (1962).
16. D. J. G. Ives and G. J. Janz, Reference Electrodes Academic Press, New York, 1961.
17. H. Gerischer and F. Beck, Z. Phys. Chem. [Frankfurt] 13, 389 (1957).
18. D. R. Turner, J. Electrochem. Soc. 107, 810 (1960).
19. V. G. Levich, Acta Phys.-Chim URSS 17, 257 (1942).
20. V. G. Levich, Physicochemical Hydrodynamics, English translation, Prentice-Hall, Englewood Cliffs, N.J., 1962.
21. Yu. V. Pleskov, Dokl. Akad. Nauk SSSR 130, 362 (1960).
22. Yu. V. Pleskov and B. N. Kabanov, Dokl. Akad. Nauk SSSR 123, 884 (1958).
23. A. C. Riddiford, in Advances in Electrochemistry and Electrochemical Engineering, P. Dekahay and C. W. Tobias, Eds., Vol. 4, Wiley-Interscience, New York, 1967.
24. B. Miller and R. E. Visco, J. Electrochem. Soc. 115, 251 (1968).
25. J. F. Dewald and D. R. Turner, unpublished results.
26. E. A. Efimov and I. G. Erusalimchik, Dokl. Akad. Nauk SSSR, 130, 353 (1960).
27. Yu. V. Pleskov, Dokl. Akad. Nauk SSSR 126, 111 (1959).
28. R. Parsons, in Modern Aspects of Electrochemistry, J. O'M. Bockris, Ed., Chapter 3, Butterworth, London, 1954.
29. K. Bohnenkamp and H. J. Engell, Z. Elektrochem. 61, 1184 (1957).
30. E. A. Efimov and I. G. Erusalimchik, Dokl. Akad. Nauk SSSR 122, 632 (1958).
31. E. A. Efimov and I. G. Erusalimchik, Zh. fiz. Khim. 33, 441 (1959).
32. E. A. Efimov and I. G. Erusalimchik, Dokl. Akad. Nauk SSSR 124, 609 (1959); 128, 124 (1959).
33. M. Seipt, Naturforsch A14, 926 (1959).
34. R. L. Meek, Surface Sci. 25, 526 (1971).

35. P. J. Boddy, J. Electroanalyt. Chem. 10, 199 (1965).
36. M. Hoffman-Perez and H. Gerischer, Z. Elektrochem. 65, 771 (1961).
37. W. H. Brattain and P. J. Boddy, J. Electrochem. Soc. 109, 574 (1962).
38. P. J. Boddy, J. Electrochemical Soc. 111, 1136 (1964).
39. R. M. Hurd and P. T. Wrotenbery, Ann. N. Y. Acad. Sci. 101, 876 (1964).
40. P. J. Boddy, Extended Abstracts of the Electrochemical Society Meeting, San Francisco, 1965.
41. Yu. V. Pleskov, Elektrokimiya 3, 112 (1967).
42. J. T. Hall and A. E. Koenig, Trans. Electrochem. Soc. 65, 215 (1934).
43. R. Schwarz, F. Heinrich, and E. Hollstein, Z. anorg. allg. Ch. 229, 146 (1936).
44. I. P. Alimarin and B. N. Ivanov-Emin, Zh. Prik. Khim. 17, 204 (1944).
45. V. V. Ostroumov, Zh. Prik. Khim. 37, 1483 (1964).
46. P. R. Subbaraman and J. Gupta, J. Sci. Ind. Res. 15B, 306 (1956).
47. C. G. Fink and V. W. Dokras, Trans. Am. Electrochem. Soc. 59, 80 (1949).
48. G. Szekely, J. Electrochem. Soc. 98, 318 (1951).
49. E. H. Borneman, R. F. Schwartz, and J. J. Stickler, JAP 26, 1021 (1955).
50. Philco Australian Patent No. 44, 774 (1959).
51. E. C. Wurst and E. Borneman, JAP 28, 235 (1957).
52. A. Uhlir, Jr., Rev. Sci. Instr. 26, 965 (1955).
53. G. W. Davis and M. C. Waltz, U. S. Pat. 2,694,040 (1954).
54. P. A. Iles and P. J. Coppen, JAP 29, 1514 (1958).
55. P. J. Whoriskey, JAP 29, 867 (1958).
56. S. J. Silverman and D. R. Benn, J. Electrochem. Soc. 105, 170 (1958).
57. D. L. Klein, G. A. Kolb, L. A. Pompliano, and M. V. Sullivan, J. Electrochem. Soc. 108, 60C (1961); Bell Lab, Rec. 39, 107 (1961).
58. J. F. Barry and N. C. Seeley, U. S. Pat. 2,827,427 (1958).
59. J. W. Tiley and R. A. Williams, Proc. IRE 41, 1706 (1953).
60. W. E. Bradley, U. S. Pat. 2,846,346 (1958).
61. J. I. Pankove, J. Luminescence 7, 114 (1973).

References

62. J. I. Pankove, J. Electrochem. Soc. 119, 1118 (1972).
63. R. L. Meek, J. Electrochem. Soc. 118, 1240 (1971).
64. A. R. Moore and W. M. Webster, Proc. IRE 43, 427 (1955).
65. A. R. Moore and J. I. Pankove. Proc. IRE 42, 907 (1954).
66. M. V. Sullivan and J. H. Eigler, J. Electrochem. Soc. 103, 132 (1956).
67. L. D. Armstrong and P. Kuznetzoff, U. S. Pat. 2,850,444 (1958).
68. A. Amaya, U. S. Pat. 2,890,159 (1959).
69. A. Uhlir, Jr., Bell System Tech. J. 35, 333 (1956).
70. J. W. Faust, U. S. Pat. 2,861,931 (1958).
71. J. G. Harper and M. S. Astor, Electrochemical Society, Electronics Division, Philadelphia Meeting, May 1959, Abstract No. 93.
72. C. J. Nuese and J. J. Gannon, J. Electrochem. Soc. 117, 1094 (1970).
73. W. H. Hackett, Jr., T. E. McGahan, R. W. Dixon, and G. W. Kammlott, J. Electrochem. Soc. 119, 973 (1972).
74. J. F. Dewald, J. Electrochem. Soc. 104, 244 (1957).

IV. PRIMARY BATTERIES

R. J. Brodd and A. Kozawa

Battery Products Division
Parma Technical Center
Union Carbide Corporation
Parma, Ohio

Primary Batteries

I. INTRODUCTION

The development of a viable primary battery requires the application of all of the tools possessed by electrochemists, from the basic understanding of electrode reaction kinetics to the engineering practices for producing a precision power plant. Many techniques utilized in investigating and developing primary batteries apply to secondary batteries and vice versa. The reader should refer to the chapter on secondary batteries to obtain the full picture of the experimental study of batteries.

II. PHYSICAL METHODS

A. Calculations

Calculations of the available energy, energy density, expected performance, and so on, serve to define useful areas of experimental studies. Energy-density calculations identify battery systems for development as well as areas for improvement of existing battery systems. Where possible, kinetic calculations indicate the expected performance from a given system.

The thermodynamic quantities of a system define the maximum ability of that system to produce electrical energy from chemical energy. Thermodynamic calculations serve to establish a guideline for the selection and optimization of battery systems. The free energy, ΔG, or net available work relates the voltage, E, and capacity, Q, of a system by the equation

$$-\Delta G = QE \tag{1}$$

where $Q = nF$, n is the number of electrons transferred in the overall reaction, and F is the Faraday. The ΔG value for any reaction can be calculated utilizing N.B.S. Circular-500(1) and Latimer (2) as reference sources of ΔG for most materials. If values for ΔG cannot be found, then the open-circuit voltage of a system can serve as a basis for estimation of the thermodynamic quantities used in the calculations.

Equation 2 defines the electrochemical efficiency, E, of a battery system:

$$E = \frac{\int Eidt}{\Delta G} \tag{2}$$

where $\int Eidt$ is the integral of the actual power delivered by the system. The energy loss or decreased efficiency occurs primarily as a voltage loss or polarization. The various sources of energy loss are discussed in Section 6.

It is worth noting in passing that Eq. 1 applies to most systems, since they are reduced or oxidized in a two-phase system. When materials are reduced (or oxidized) in a one-phase system, such as MnO_2 or RuO_2, Eq. 1 needs some modification since the equilibrium potential changes with the depth of discharge.

In comparing the various battery systems, thermodynamic calculations usually express the energy density as watt-hours per unit volume or per unit weight rather than in calories or joules. Equation 3 converts the free energy in calories/mole to watt-hours per pound (Wh/lb):

$$\text{Wh/lb} = \Delta G(\text{cal/mole}) \times \frac{454\ (\text{g/lb})}{\Sigma\ \text{atomic wts. of reactants (g/mole)}} \times \frac{\text{Wh}}{800\ \text{cal}} \tag{3}$$

Using the equation for average density of reactants, ρ_T, the energy density per volume, is given by

$$\rho_T = \frac{\rho_{cathode}\ \rho_{anode}\ (1 + \text{cathode/anode})}{\rho_{cathode} + \rho_{anode}\ (\text{cathode/anode})} \tag{4}$$

where ρ_{anode} and $\rho_{cathode}$ are the densities of the anode and cathode reactants, respectively, and (cathode/anode) is the cathode-to-anode weight ratio. Then

$$\text{Wh/in.}^3 = \text{Wh/lb} \times \text{lb/in.}^3 \tag{5}$$

Some investigators prefer to use the metric units of kilogram and cubic centimeter to the English units.

While thermodynamic calculations establish the maximum performance to be expected from a system, electrode polarization or inefficiencies resulting from kinetic or rate processes degrade the performance of a system. As a rough rule of thumb, the actual performance of a primary system approaches 50% of the theoretical Wh/lb and 20% of the theoretical Wh/in^3.

Where values of the kinetic parameters of electrodes exist, Greene and Greene (3) report a method for predicting overall system performance in terms of the exchange currents, I_0, and transfer coefficient, α. Unfortunately, kinetic parameters of most battery cathode systems remain unknown. The investigator frequently must rely on his own experimental program to determine the needed kinetic parameters to establish battery performance.

If diffusion processes are important, then concentration polarization and the limiting current I_0 must be considered in the performance of the battery systems. Equation 6 predicts the polarization, η, resulting from internal resistance, R_i, activation processes, and diffusion processes:

$$\eta = E_{OCV} - E_T = R_i I + \sum \frac{RT}{\alpha nF} \log \left(\frac{I}{I_0} \right) + \sum \frac{RT}{nF} \log \left(\frac{I - i_\ell}{i_\ell} \right) \quad (6)$$

B. Radiography

Radiography, the use of radiation other than light to produce a picture of an object of interest, reveals details of internal structure or defects of an otherwise opaque body. The technique of radiography is a nondestructive test method, in contrast to most other test methods. Radiography has had wide utilization in in metallurgy, biology, and medicine but has not been extensively applied to electrochemical problems. It has been used to study the shape and position of a meniscus inside a silver tube electrode and the electrolyte penetration into a porous gas electrode as well as catalyst distribution and homogeneity (4, 5). It has been suggested as a technique to determine the state of charge of mercury batteries (6). Figure 1 illustrates the difference between a fresh and a discharged mercury battery. The technique has been widely used to examine mercury batteries for heart pacer applications.

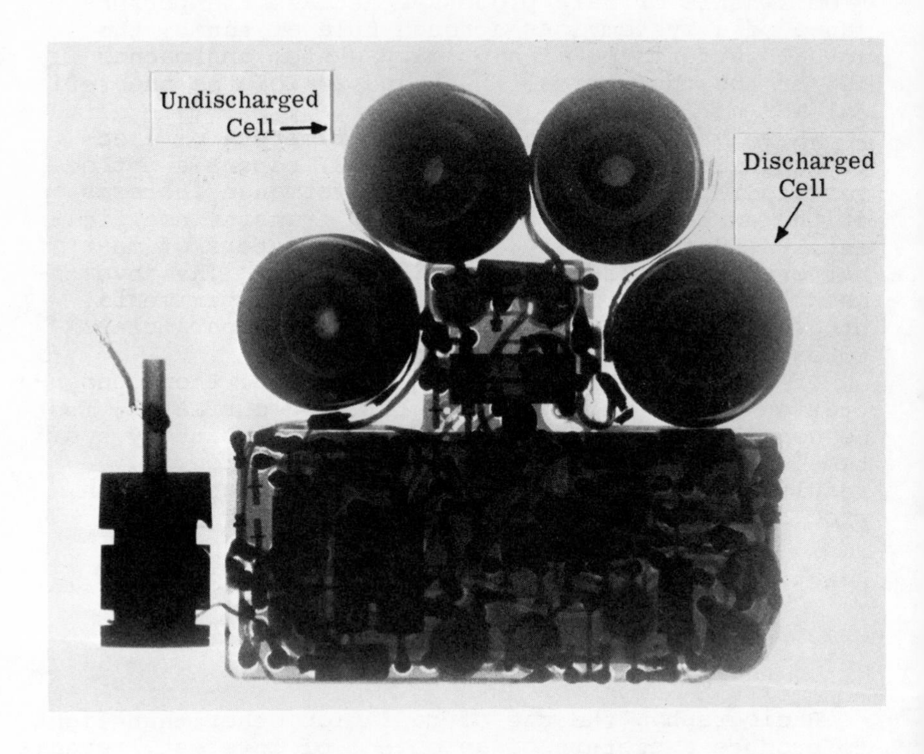

Fig. 1. X-ray Radiograph of mercury batteries illustrating different states of charge (Courtesy of Medcor, Inc.).

Figure 2 details a schematic of a typical radiographic experimental apparatus. The source depends on the type and intensity of radiation to be used, that is, X-ray, γ-ray, or neutron. The collimator, a series of diaphragms, or absorptive materials may or may not be used. X-ray sources are small and the sample-to-source distance is large compared to the sample detector distance. Photographic film is the most common radiation detector; a wide variety of

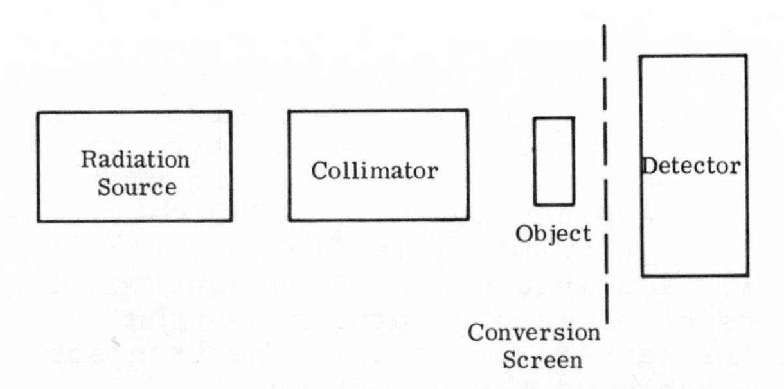

Fig. 2. Schematic diagram for the determination of the internal structure of an opaque object using radiation.

film has been used. It is also possible to use a radiation-sensitive fluorescent screen for the detector. In this case, a still camera or orthicon tube records or transmits the picture for viewing. These last techniques produce pictures instantaneously. Lack of collimation in the beam (i.e., non-parallel beams) produces a hazy or fuzzy picture. The amount of backscattering from the interaction of the radiation with the materials behind the detector controls the quality of the radiographs.

X-rays (γ-rays) are most commonly used in radiography; the type depends on the source of the radiation. A cathode tube produces X-rays, while γ-rays originate from radioactive sources. The X-ray tube is convenient to operate and control, although γ-ray sources may be preferred for their steady high-energy characteristics. The picture produced by X-rays depends on the absorption characteristics of the materials in the object under study. The absorption of X-rays by a material increases with the atomic number of the element as the result of increased density of the orbiting cloud of extra nuclear electrons. Materials composed predominately of the light elements, that is, H, Li, O, and so on do not absorb X-rays and may excape detection; on the other hand, the more strongly absorbing heavy elements, that is, Fe, Pb, Au, Hg, and so on show up nicely by X-ray radiography. A good source of information concerning X-ray radiography has been written by Berger (7).

Neutrons may also be used as the exciting radiation (8). This experimental technique is very similar to that for X-ray radiography. Three neutron sources are a nuclear reactor, a charged-particle accelerator, or a radioactive source such as ^{252}Cr. Reactors produce the largest flux of neutrons but are not portable. Radioactive sources provide a very low particle flux and require long exposures. Accelerators produce neutrons from a nuclear reaction such as proton-neutron reactions using a ^{7}Li target. Generally, neutron sources are not as portable as X-ray sources.

Since neutrons are not charged, they may pass through photographic film used as the detector without interaction. Therefore, a conversion screen in front of the detector converts the neutron either into charged particles or into light which easily exposes the film or activates the screen and camera.

Unlike X-rays, neutrons are absorbed in the atomic nucleu. This interaction bears no relation to the atomic number but depends on the characteristics of the nucleus itself. The neutron absorption will depend on the energy of the neutron as well as the particular isotopic composition of the object under study. Also unlike X-rays, neutrons can pass through heavy elements such as Pb without appreciable absorption. On the other hand, neutrons are absorbed by thin layers of H, B, Li, and so on, which do not absorb X-rays. Therefore, neutron radiography permits study of many materials that cannot be differentiated by X-ray studies. Neutron radiography should not be considered as a competitor for X-rays and γ-rays for simple radiography but rather as a complement in the study of internal structure.

Autoradiography used radioisotopes to study the location and changes in location of materials under the influence of experimental stresses. It has been widely used to study corrosion reactions (9), adsorption (10), and metallurgical and biochemical studies (11). It has also been used to determine exchange currents of electrode reactions, as well as to study sulfation and the effects of phosphate additives on lead-acid battery plates (12).

The half-life, its radiation characteristics, and its relationship to the study, e.g., ^{35}S for sulfate reactions, ^{65}Zn for zinc reactions, and so on determine the isotope selection. The isotope-containing solution should be purified before use. The radioactive isotope usually constitutes only a portion of the total element involved in the study. For instance, H_2SO_4 is enriched with about 5% of radioactive $SO_4^{=35}$ to give about 1-2 mC of activity per unit of electrolyte. The sensitivity of the detection equipment determines the amount of activity required in the experimental solution, although counting equipment can be used to determine total activity. The use of film permits the identification of both the location and the amount of the activity.

C. Surface Area

The surface properties of electrodes determine the ability of the electrode to deliver useful power. Whenever the exchange currents for the electrode reaction are low, the surface area of the electrode must be enlarged for the electrode to function satisfactorily in battery environments. The understanding and developing of battery electrodes requires a knowledge of the surface area of the electrode components and electrode structures.

The total surface area and pore volume characteristics can be obtained from BET gas adsorption and mercury porosimeter techniques. Sometimes the mercury of the mercury porosimeter interacts with the adsorbent, for example, MnO_2, making comparison of surface area and pore volumes invalid. X-ray line broadening and optical and electron microscopic techniques give information on the crystallite size of reacting substances or fuel cell catalysts. The electron probe microscope depicts the distribution of elements on the surface when coupled with the scanning electron microscope, and gives a detailed surface structure even for rough and porous electrode surfaces. This tool is very valuable in developing a "picture" of electrode surfaces.

While the BET nitrogen adsorption method is accepted as a method for measuring total surface area of porous solid, adsorbates other than nitrogen assist in characterizing porous battery matrix materials. Adsorption of a dye, for example, methylene blue, from a dilute solution often provides a convenient method for determining surface areas of materials with highly developed surfaces.

Another simple method utilizes water, carbon tetrachloride, and similar liquids (13, 14). Since adsorbents may interact differently with water or carbon tetrachloride than with nitrogen, these adsorbents will provide information on the nature of the surface. Adsorption curves for water and carbon tetrachloride adsorption are obtained gravimetrically by a desiccator method at room temperature. The samples are equilibrated in a desiccator at various vapor pressures supplied from saturated salt solutions and weighed on an analytical balance. Often, only the adsorption at $P/P_O = 1$ (pure adsorbent) is obtained. This value gives the relative volume of water adsorbed and is proportional to a combination of surface area and pore volume. The comparison of water and carbon tetrachloride adsorption gives an indication of the interaction of the surface with polar and nonpolar adsorbents. A complete adsorption isotherm can be obtained with appropriate computer analysis for surface area and pore volume, thereby providing a check on the BET technique. (C.F. Vol. 1, chapt. 4 of this series (79).

The zinc ion adsorption method measures the surface of a material by adsorption of zinc ions from a standard solution. Kozawa (15, 16) established an empirical relation that 14 m^2 of BET surface area compound to 10^{-4} mole of adsorbed zinc ion. The suggested procedure follows:

1. Add 5.00 ml of 0.11M zinc solution (2M NH_4Cl+ 0.11M ZnO) to a given sample weight (w), for example, 0.4 g for 20 m^2/g material, in a stoppered glass tube.
2. Agitate or rotate the tube to ensure good mixing, for example, 20 hr at room temperature or 3-4 hr at 70°C.
3. Remove 2.0 ml of solution and titrate the zinc with 0.02M EDTA (b). If desired, the hydrogen ion released can be determined by HCl titration.
4. Titrate original zinc solution with 0.02M EDTA (a).

5. Calculate surface area using

$$\text{Surface area} = \frac{7\ (b-a)}{w}\ m^2/g$$

Frequently, the Gardner Oil Adsorption or its modification provides a rapid and relatively reproducible method for comparing the surface areas of powdered materials. A small sample of powder (e.g., 1 g of carbon black, MnO_2, or other high-surface-area material) is placed on a glass plate. Standard linseed oil, kerosene, dibutyl phthalate, or other oil is added dropwise to the sample, with thorough mixing or working of the oil into the material after each addition of oil. A steel spatula may be used or the mixture mulled to ensure uniform mixing. The endpoint is taken when the mixture just coheres to form a putty or paste of a predetermined consistency. This endpoint is somewhat subjective in nature. Nevertheless, the oil adsorption test is accepted by industry and is reasonably reproducible. Some recognition of the operator bias may be necessary to compare results from various calculations. Sljaka and Pie (17) suggested a technique to improve the accuracy wherein the change in torque to operate the muller indicates the endpoint. A sharp increase in torque to operate the muller occurs at the endpoint. This eliminates operator judgment in determining the endpoint.

The oil adsorption test yields a number (e.g., grams of oil per grams of material) that relates to the total surface area and pore volume of the sample. In the case of carbon black, the oil adsorption test provides an indication of the "structure" or chain-like arrangement of the primary particles of the black.

X-ray line broadening indirectly yields the surface area of small crystallites such as catalysts, carbons, and so on. From the line broadening, the dimensions of the crystallites can be calculated. Then the surface area of a given material can be calculated from the particle size. This method can be used to study changes in particle size and heating or reaction time. A relationship of particle size (surface area) and conductivity of carbon to the ability of the carbon to function as an oxygen electrode has been reported (22).

The capacitance of the electrical double layer at the electrode-solution interface can be used to evaluate the surface area of an electrode structure. The electrochemical method (18 - 21) for determining surface areas relates the electrochemically active area rather than the total surface area measured by the gas adsorption or BET. The electrochemical method applies best where electrodes are porous or involve mixing and reforming nonconductive binders and wetproofing materials along with the electroactive powders. Kronenberg (23) developed a method for use with powders. The powders are suspended in an oxygen-saturated electrolyte by vigorous stirring. The particles collide with a large-area collector electrode held at a steady-state potential. The current is interrupted with a mercury-wetted relay. The potential decay is followed with an oscilloscope. The capacitance is calculated from

$$C = I \left(\frac{\Delta T}{dE}\right) \tag{7}$$

where I is the current flowing at t= 0. The prime use of the capacitance values compares the available surface areas of various materials, for example, carbons and powders.

Morley and Wetmore (24) showed that the decay of EMF near the reversible potential follows the equation

$$\frac{dE}{dt} = \frac{b}{2.303}\left(\frac{1}{t + \theta}\right) \tag{8}$$

where b is the slope of the Tafel equation, and θ is the time for the EMF to decay from a hypothetical potential to the actual potential at the time of interruption of the polarizing current, assuming a constant double-layer capacity. Integrating Eq. (8) yields

$$E = E_i - b \log (t + \theta) \tag{9}$$

Then θ can be evaluated semiempirically from EMF-log time plots as the value of time which when added to the time t gives an initially linear plot. The capacitance of the double layer is then obtained from

$$\theta = \frac{b}{2.303} \frac{C}{i} \tag{10}$$

where i is the current at the moment of interruption. This technique also yields the Tafel slope for the rate-determining process of the electrode reaction.

The electrode capacitance can be measured directly from charging curves, impedance measurements, and square-wave techniques.

The total surface area is obtained from

$$\text{Area} = \frac{\text{Measured electrode capacitance}}{\text{Capacitance/unit area}} \tag{11}$$

Unfortunately, the capacitance per unit area for many electrodes varies from 20 to 40 $\mu F/cm^2$ for clean electrodes in the absence of an electrode reaction. For many electrodes near the hydrogen potential, $C_{DL} = 20\ \mu F/cm^2$, while for many electrodes near the oxygen potential, $C_{DL} = 40\ \mu F/cm^2$. The C_{DL} varies with potential with a minimum near the zero point of charge so that these numbers should be used with care. For an oxide-covered electrode, the C_{DL} drops to 3 to 6 $\mu F/cm^2$.

D. Phase Diagrams

Phase diagrams predict the thermodynamically stable species to be found in a given situation. These diagrams are especially useful for predicting the long-term stability of a system. A special form of the phase diagram, the pH-potential or Pourbaix diagram, finds extensive use for correlating the activity of MnO_2 samples.

The phase diagram for Leclanché electrolyte (shown in Fig. 3) illustrates the usefulness of this type of diagram (25, 26). The system contains two stable solid species, $ZnCl_2 \cdot 3NH_4Cl$ and $ZnCl_2 \cdot 2NH_4Cl$. Suppose point P represents the equilibrium composition of a cell; then discharging the cell (also corrosion of the can) introduces $ZnCl_2$ into the electrolyte. As the $ZnCl_2$ accumulates, the composition follows the line P-$AnCl_2$ to the point where solid $ZnCl_2 \cdot 3NH_4Cl$ appears. This salt precipitates

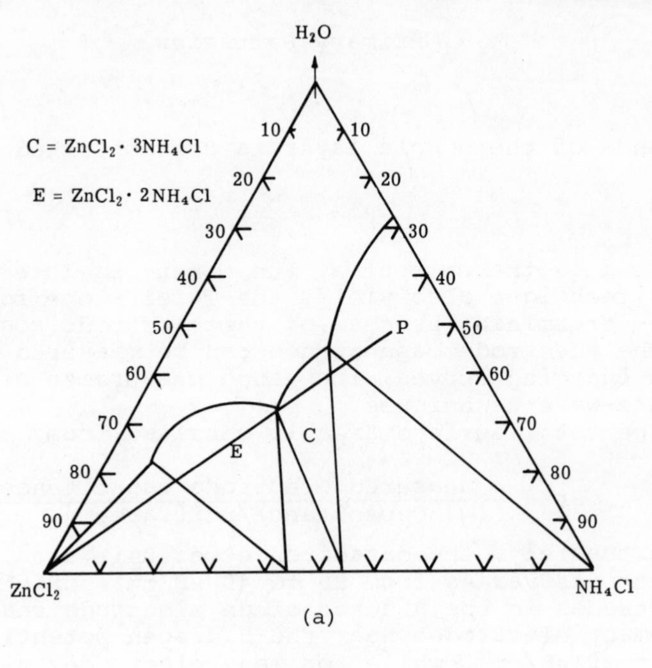

(a)

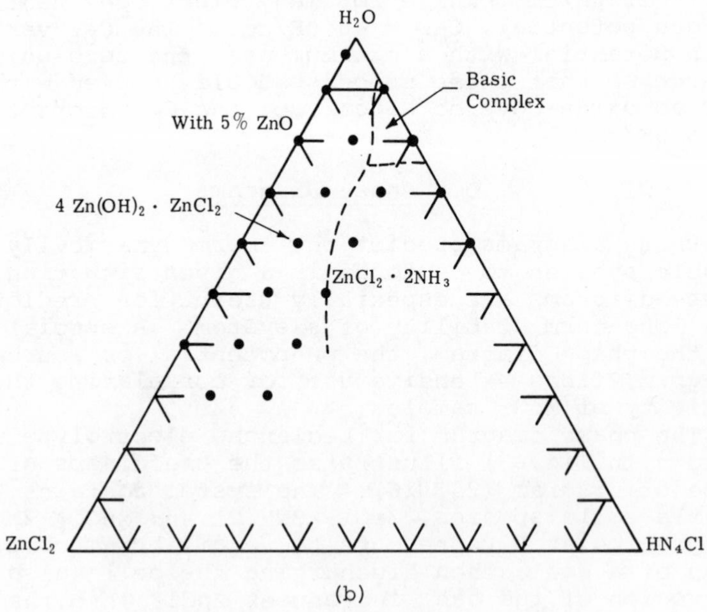

(b)

Fig. 3. (a) Phase diagram of $ZnCl_2$-H_2O-NH_4Cl (25); (b) as modified by 5% ZnO Addition (26).

in the form of long needles or whiskers, which can seriously disrupt the internal structure of a cell and inhibit proper electrode operation. Commercial Leclanche cells frequently contain excess solid NH_4Cl in part to retard the formation of the double salt.

In poorly sealed Leclanche cells, which dry out during discharge, or in the more alkaline regions of the cathode, zinc oxychloride salts may form, as predicted from the equilibrium with $ZnO-NH_4Cl-H_2O$. In alkaline-zinc cells, $KOH-H_2O$, $NaOH-H_2$, and $ZnO-KOH-H_2O$ diagrams find use; for example, low-temperature electrolytes, compound formation, and so on.

Two examples of the pH-potential diagrams shown in Fig. 4, 5a and 5b illustrate the usefulness of this type of phase diagram. Figure 4 depicts the various phases present in the silver-zinc system as a function of pH. The pH determines the soluble species present as well as the potential of the system.

Figure 5a depicts the variation of MnO_2 potential with pH in the absence of Mn^{++} ions. The equilibrium behavior corresponds to the reaction with a theoretical slope of 0.59 V/pH,

$$MnO_2 + H^+ + e \rightarrow MnOOH \qquad (12)$$

The value of the potential at a fixed pH, say 6, is used to compare the activity of various MnO_2 materials. The higher the potential, the greater the activity because the free energy of the material is less negative. Since MnO_2 possesses considerable ion-exchange properties, various ions influence the slope of the pH-potential plot.

In the presence of Mn^{++}, a theoretical slope of 0.12 V/pH (see Fig. 5b) corresponds to the reaction

$$MnO_2 + 4H^+ + 2e \rightarrow Mn^{++} + 2H_2O \qquad (13)$$

Two methods (27, 28) are described below for measuring the pH-potential behavior of MnO_2; one is for research studies, and the other is for quality control measurements. Both methods are capable of yielding useful results with careful control of the experimental conditions.

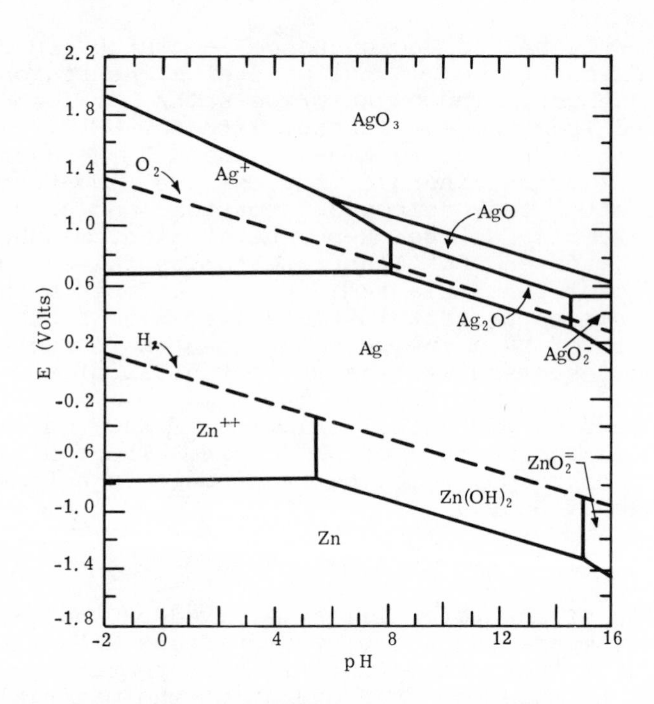

Fig. 4. The pH-potential diagram of zinc and silver as related to the Zn-Ag battery.

The MnO_2 should be of the particle size to be used for electrode construction. Various size fractions differ slightly in pH-potential behavior, primarily in the time to reach equilibrium. Nylon sieves are preferred since they avoid contamination of the ore. A typical sample would pass 150/in. sieve but would be retained on the 200/in. sieve.

Solutions common to battery use are normally employed, but analytical grade reagents are preferred for careful work. Typically, NH_4Cl or KCl solutions constitute the test electrolyte, adjusted in the acid range with HCl or $ZnCl_2$, and in the alkaline range

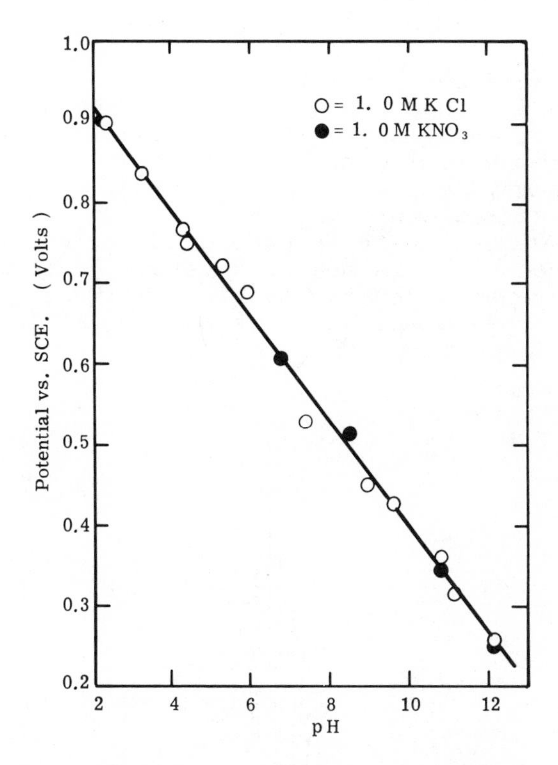

Fig. 5. The pH-potential behavior of MnO_2 (27).

with KOH or NH_4OH. Buffer solutions may be employed, but because of the ion-exchange capacity of MnO_2 small differences of pH-potential behavior occur between various buffer systems, for example, phosphate, acetate, and so on. Constant temperature should be maintained. The pH of the solution should be determined at each potential reading. The potential measurements should be made with an electrometer or potentiometer to ± 0.001 V. Potential measurements should be taken until constant values are obtained. The time to reach equilibrium varies (some say 3 wk), but a 24-hr equilibration is common. Care should be exercised to prevent disturbance of the collector-ore contact during equilibration. Platinum contacts find frequent use, although MnO_2 attacks platinum over a period of time. For this reason, carbon contacts may be preferred. It does not appear necessary to exclude oxygen from the experiment except for strongly acid solutions.

A simple method employs small flasks or glass jars. A sample of MnO_2, say, 1-5 g, is placed in the flask along with 50 to 100 ml of test solution. Buffer solutions may be used. The MnO_2 sample and the solution are stored overnight or for 24 hr. A platinum spade or wire contact is placed in the MnO_2 slurry and the potential is measured versus a calomel electrode. By using several samples and test solutions, a complete pH-potential curve can be obtained with a minimum of effort.

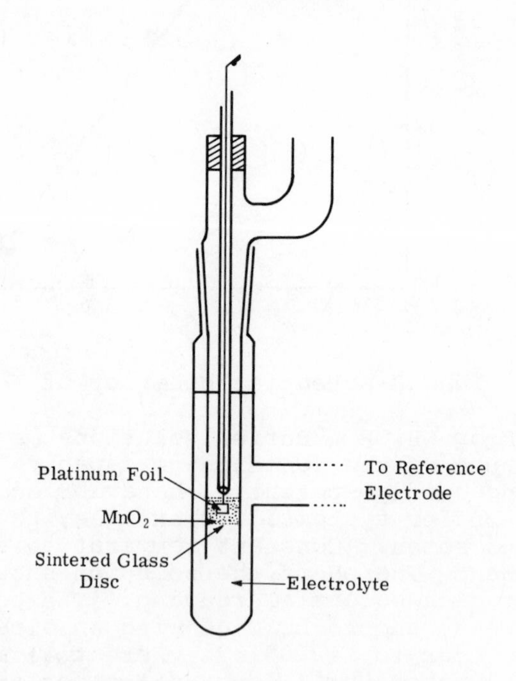

Fig. 6. Apparatus for determining pH-potential behavior for battery active materials (28).

The electrode compartment used for precision measurements is shown in Figure 6. The fritted disk permits washing to the MnO_2 with the test solution under suction to compact the MnO_2 about the platinum contact. The platinum contact may be cleaned with aqua regia. The electrode vessel may contain several MnO_2 electrode assemblies along with a reference. The reference electrode should be reversible to anion of the solution. Calomel electrodes are almost always used as chloride is the common electrolyte for the Leclanché system.

E. Reference Electrodes

Reference electrodes are electrodes with a well-defined potential that is invariant with time under experimental stresses. Properly constructed and used, they are insensitive to reasonable current flow and behavior predicted by thermodynamics. Reference electrodes provide a point of reference by which to refer the behavior of other electrodes whose potential may change as a result of the experimental situation (29, 30).

The hydrogen electrode is the primary standard to which all electrode potentials are referred. However, the hydrogen electrode is not convenient to use and is easily poisoned in practical application of battery electrode studies. The hydrogen electrode is not used extensively in battery studies. More commonly used reference electrodes are the saturated calomel and mercuric oxide reference electrodes. The reference electrode is selected to have a common ion with the test environment and to be in a stable pH range for good operation. A salt bridge, whether an asbestos wick or agar bridge, generally isolates the reference electrode from the test environment. A list of reference electrodes used in battery studies is given in Table 1.

Figure 7 shows several common constructions of reference electrodes. The electrodes can be assembled in the laboratory or purchased commercially. With reasonable care, electrodes can be constructed to give reproducible values of potential of 0.002 V from batch to batch. For greater precision, special care must be given to each detail of manufacture, purity of materials, and so on.

TABLE 1

Common Reference Electrodes

Ions in Solution	Reference Electrode	Solution Acidity
Cl^-; K^+	Saturated calomel	Acid, neutral, and basic (with asbestos wick)
OH^-; Na^+, K^+	Mercuric oxide	Basic
Cl^-; K^+, Na^+	Silver chloride	Acid and neutral
$SO_4^=$	Lead SO_4-PbO_2	Acid
$SO_4^=$	Mercuric sulfate	Acid and neutral
OH^-	Silver oxide	Basic
H^+, OH^-	β-Pd	Acid, neutral, and basic
	Cd	Acid and basic
	Zn	Neutral and basic
SO_4, $Cu^{++=}$	Cu-$CuSO_4$	Acid

Two electrodes, Zn and Cd, have been included in the list although they do not meet the requirements as true reference electrodes in a thermodynamic sense. However, they are used to infer voltage changes in situations where limited accuracy ±0.02 V is required. The zinc electrode is useful in alkaline and neutral solutions while the cadmium electrode has been used extensively in both acid and alkaline solutions.

For additional discussion of reference electrodes the reader is referred to R. Bates (80) chapter 1, Volume 1 of this series.

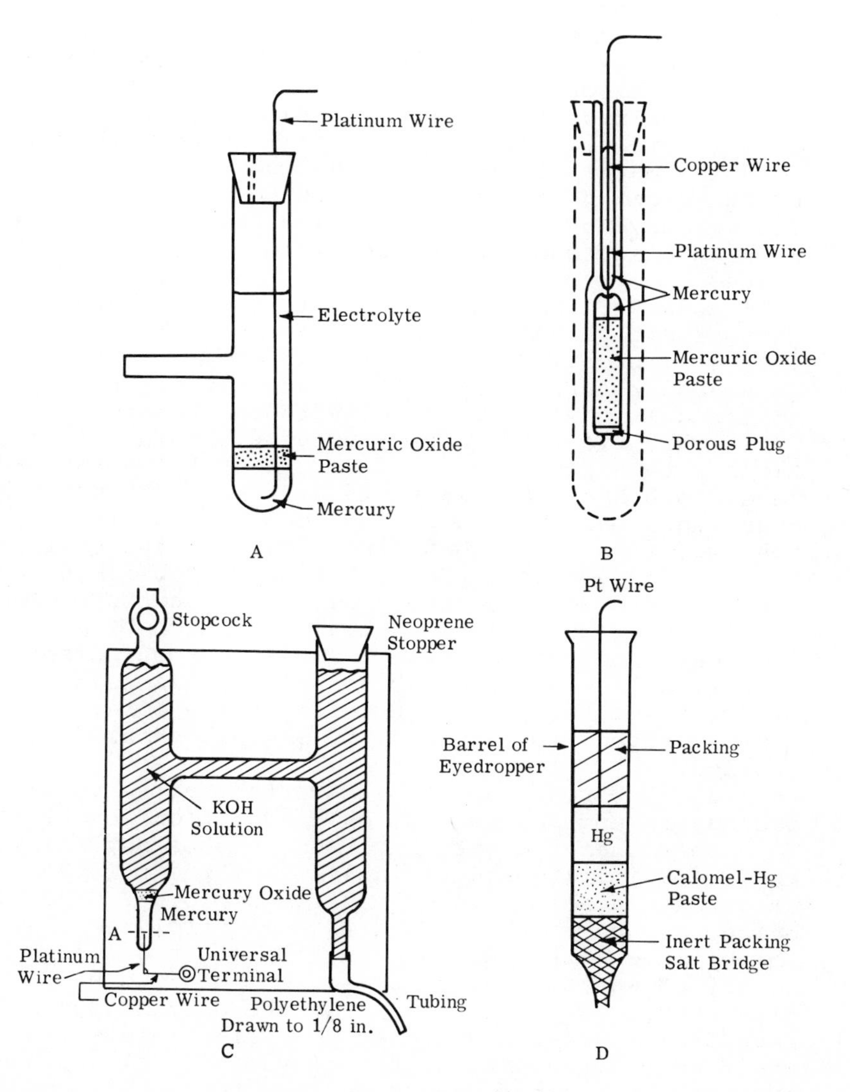

Fig. 7. Constructions of typical reference electrodes.

The calomel and mercuric oxide electrodes are made by similar techniques. The components, Hg-Hg_2Cl_2-Sat'd. KCl or Hg-HgO-KOH solution are mulled together in a mortar and pestle to give a pasty intimate mixture. The paste is placed into the tube on top of a mercury pool or drop. After packing in the paste, a wad of cellulosic or other separator material, for example, felted nylon, is forced into the tube to hold the materials in place and to act as a salt bridge from the reference to the test environment. The mercury-mercurous sulfate and lead sulfate-lead dioxide electrodes may be assembled by similar techniques.

The Ag-AgCl and Hg-Hg_2Cl_2 reference electrodes are photosensitive and must be protected from light for long-term stability.

A good reference electrode that has not been widely utilized is the β-palladium-hydrogen electrode (31, 32). This electrode behaves reversibly in all solutions with a potential 0.05 V more positive than the reversible hydrogen electrode. It is constructed by charging a palladium wire and bead to visible hydrogen evolution. Hydrogen dissolves into the palladium to form the stable β phase, thereby establishing a reversible electrode potential. The electrode can be easily restored to its original condition by a boosting charge before each use. The electrode possesses good stability and remains relatively insensitive to impurities although solutions of metals more noble than hydrogen, for example, Cu and Hg, should be avoided.

Many electrodes are sensitive to their environment and may undergo hydrolysis or dissolution reactions. To ensure that the test environment does not influence the reversibility and reliability of the reference electrode, a salt bridge is often employed. The salt bridge may be a tube filled with electrolyte sometimes immobilized with 1-2% agar-agar or other gelling agent, or an asbestos fiber sealed into the end of a glass tube. Whatever the form, the salt bridge insulates the reference electrode electrolyte from the test environment. Often a hydrostatic head between the reference system and the test environment provides a flow into the test environment and slows diffusion of the test electrolyte into the reference system. It may be necessary to discard the salt bridge from time to time to prevent contamination of the reference electrode.

The salt bridge introduces an error in potential measurements due to differences in concentration and composition of the two solutions. This junction potential remains constant as long as the environment does not change; it does not influence the measurement of changes in potential in the test electrode. The junction potential must be considered if the potentials are to be used to identify thermodynamic or reversible reaction characteristics. The junction potential can be reduced to 1 to 2 mV by the use of highly conductive solutions with ions in common, for example, saturated KCl for Cl^- or K^+ contained test environments.

Whenever potential measurements are made, the instrumentation draws current from the electrodes. The current flow introduces: (1) an IR drop in the salt bridge or solution between the reference and test electrode and (2) a slight shift in the reference electrode potential. Well-constructed reference electrodes reduce the second effect to less than 0.001-volt. The amount of current flow depends on the internal impedance of the measuring instrument and the voltage between the reference and test electrode. For instance, if the internal impedance of the voltmeter is 10^5 ohms and the voltage difference is 1 V, the current flow equals 10^{-5} A. If the resistance in the solution is 1000 ohms, then a 0.01-V error occurs fairly commonly in the asbestos-fiber-type of salt bridge. The internal impedance of the measuring instrument should be as high as possible to avoid errors due to the iR and reference electrode polarization. The use of low-impedance voltmeters is not recommended with reference electrodes.

Error-free reliable measurements require the proper placement of the reference electrode. If the tip of the capillary of the electrode itself is positioned in the current lines between the counter and test electrode, the iR drop associated with the current flow to the electrode will be included in the reference voltage measurement. Movement of the reference electrode in this situation changes the voltage reading. This provides a convenient experimental check on the proper position of the reference electrode. If the reference probe interferes with the current lines, a nonuniform current distribution occurs on the test electrode. This shielding of the test electrode can introduce serious errors in the polarization measurements, especially at high current densities. A rule of thumb predicts that placing the capillary end at the distance of four times the radius of the capillary eliminates the shielding effect. This does not eliminate the ohmic potential drop from the tip to the test electrode while current is flowing. Piontelli et al. (33) and later Barnartt (34) estimated the errors associated with various electrode designs and locations. The use of a "back-side" Luggin capillary has been recommended especially with flowing electrolyte (35). With proper design of the reference probes, there is good agreement between potential measurements made by the various techniques. The use of techniques to eliminate the ohmic potential drop is discussed in Section III d.

F. Resistivity

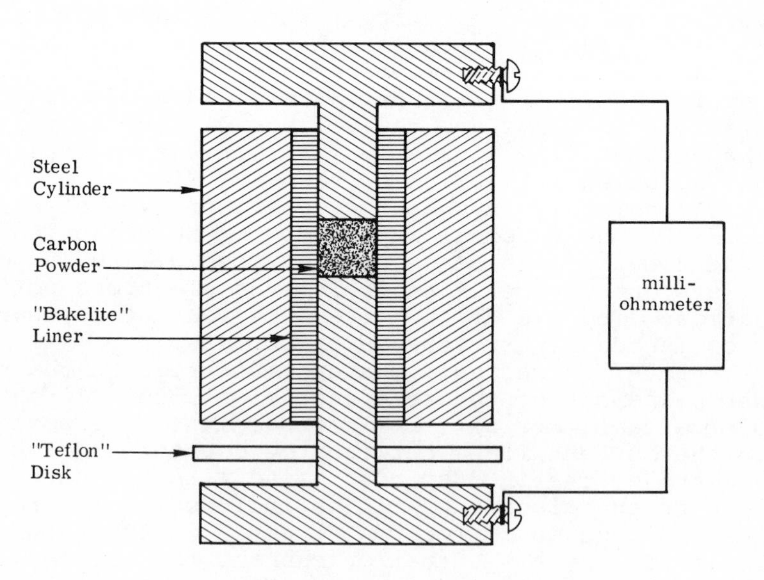

Fig. 8. Apparatus for determining the resistivity of a material as a function of applied pressure.

The apparatus for determining the resistivity of various materials is shown in Fig. 8. A known amount was weighed into the pressure chamber. The resistance of the sample is measured under pressure using a conductivity bridge or other resistance-measuring device. The measurement of solids should be independent of frequency and measuring current if electrode effects are absent. The thickness of the sample is determined using a traveling microscope or cathetometer. The volume of the sample may then be calculated from the measured thickness and the cross-sectional area of the pressure chamber.

The resistivity depends strongly on the measurement pressure, as shown in Fig. 9. The comparison of resistivity should not be made at a constant or fixed pressure as is commonly done. Comparisons of resistivities are valid only if comparisons are made at the same relative density.

A comparison of conductivities using a sort of "Swiss cheese" model is preferred. The particles are compacted in a somewhat continuous medium with holes in it. The effective cross-sectional area of the sample is reduced by the volume of holes multiplied by a constant because some material on each end of the hole is not carrying current. Expressed mathematically, this translates to

$$\sigma = \sigma_0 \frac{A_{eff}}{A} \tag{14}$$

$$= \sigma_0 (1 - a\theta) \tag{15}$$

Since $d/d_0 = 1-\theta$, Eq. 15 becomes

$$\sigma = \sigma_0 (a\frac{d}{d_0} - a + 1) \tag{16}$$

where σ = measured conductivity
σ_0 = bulk material conductivity
A = true area
A_{eff} = effective area
θ = relative volume of holes
d = measured density
d_0 = bulk material conductivity
a = constant

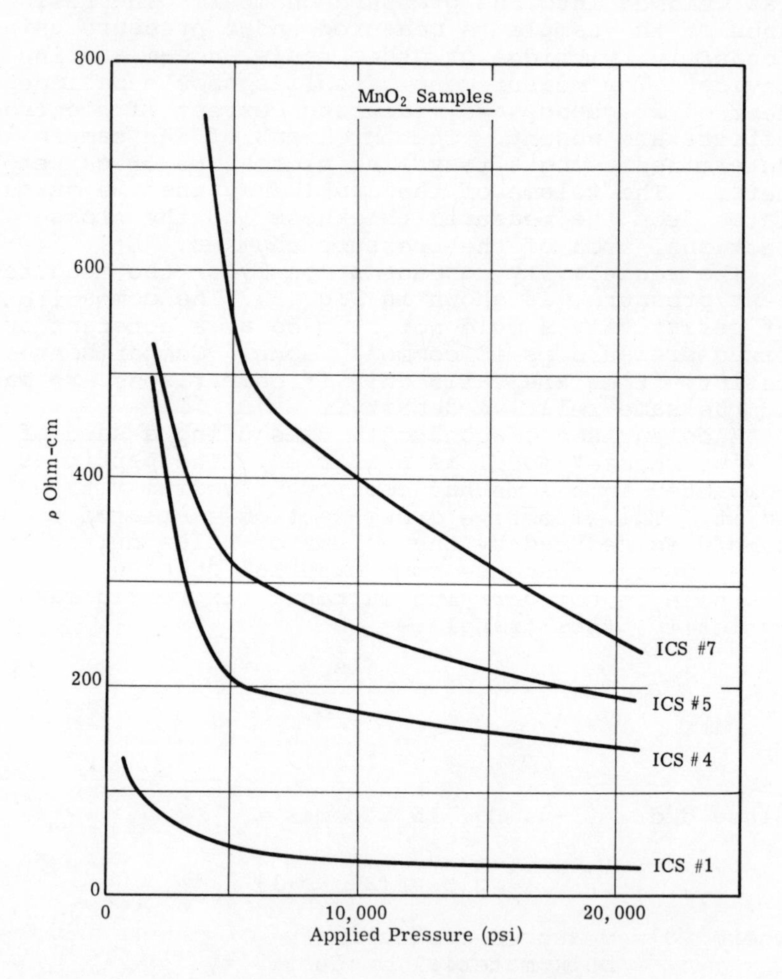

Fig. 9. Resistivity of MnO_2 as a function of pressure.

Using Eq. 16, the conductivity at the theoretical density of other selected density is easily determined. Typical resuts for several MnO_2 and active carbon samples are shown in Figs. 9, 10, and 11.

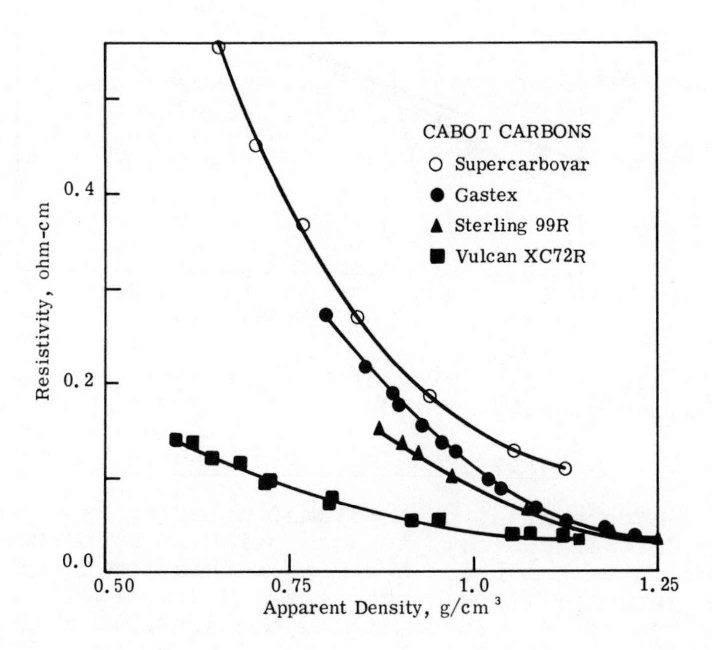

Fig. 10. Resistivity of active carbon as a function of relative density.

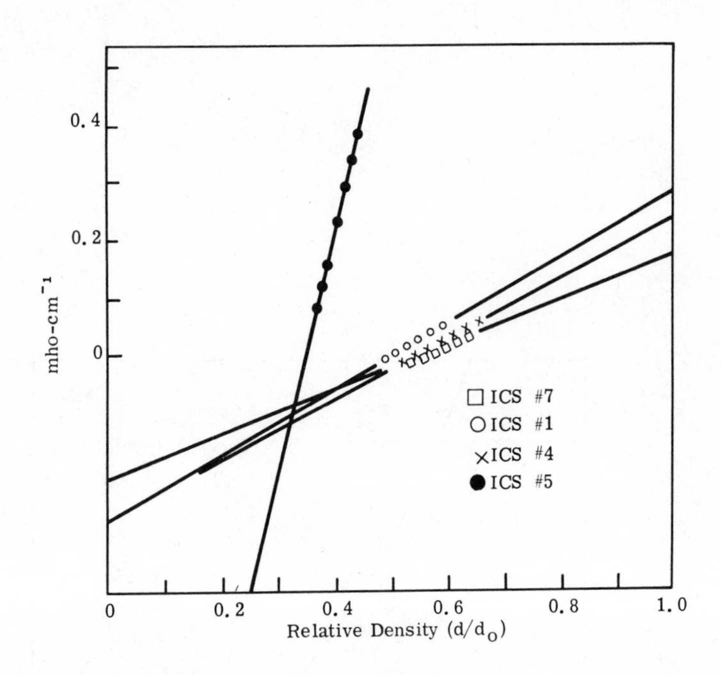

Fig. 11. Conductivity of MnO_2 as a function of relative density.

G. Porosity, Pore Size Distribution, and so on

The pore size, pore volume, porosity, and so on are determined from gas adsorption or by mercury intrusion techniques. A simple method for determining gas permeability and pore size is discussed in Section III D. In addition, considerable material applicable to all surfaces is given by Salkind (79) in Volume 1, chapter IV of this series.

III. ELECTROCHEMICAL METHODS

A. Corrosion

The study of corrosion in batteries follows closely the methods described in the chapter for the study of corrosion metals. Weight loss, gas collection, scanning electron microscope with micropore analysis, zero resistance ammeter, potentiodynamic techniques, X-ray, and so on find wide application to identify the extent of corrosion. In addition, one frequently encounters problems of stress corrosion in a conductive medium, intergranular attack and grain size effects, concentration cells (especially with oxygen or mercury-zinc alloys, for example), and trace impurities of heavy metals.

Gas evolution constitutes a particularly difficult problem in battery technology, especially in sealed cells. Gas evolution usually arises from corrosion processes in the cell, for example, hydrogen fron anode corrosion and CO_2 from cathode corrosion.

Figure 12 depicts a simple gas collection device for short-term studies. The test cell rests on a polystyrene rod. The long test tube covers the cell and rod arrangement. Oil is added to the beaker and nitrogen (or other gas) is introduced into the inverted tube to produce a controlled atmosphere for the corrosion test. A partial vacuum will draw the oil up into the inverted test tube. A high-quality vacuum pump oil, mineral oil, or other inert oil may be used. This test should not be used for long-term studies. The solubility of gases in the oil permits atmospheric gases to diffuse into the test chamber as well as diffusion of the corrosion product gas, for example, hydrogen, out of the gas space. Electrical leads can then be brought out and immersed in oil for measuring gas evolution on discharge, and so on.

Figure 13 depicts a gas collection device suitable for long-term studies (39). In this device, the gases evolved from the corrosion process displace mercury, which can be collected and weighed. If desired, the cell can be fitted with electrical leads for discharge of the cell in a controlled environment and measurement of gas evolution on discharge. The top of the cell is a 45/50 ground glass joint.

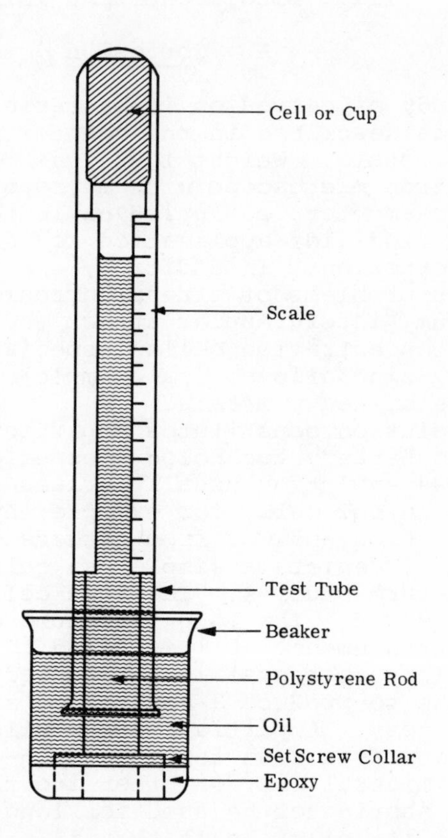

Fig. 12. Simple gas collection apparatus for corrosion tests.

In both methods the gas volumes should be corrected to STP condition, with due allowance for the difference in height of the containing fluid inside and outside the container and the vapor pressure of the solvent in the corrosion test.

The gas phase can be sampled using a syringe connected to a long, thin, flexible plastic tube. The plastic tube can be inserted into the gas phase through the containing fluid and the gas sample withdrawn into the syringe. The syringe can then be emptied directly into a gas chromatograph for analysis.

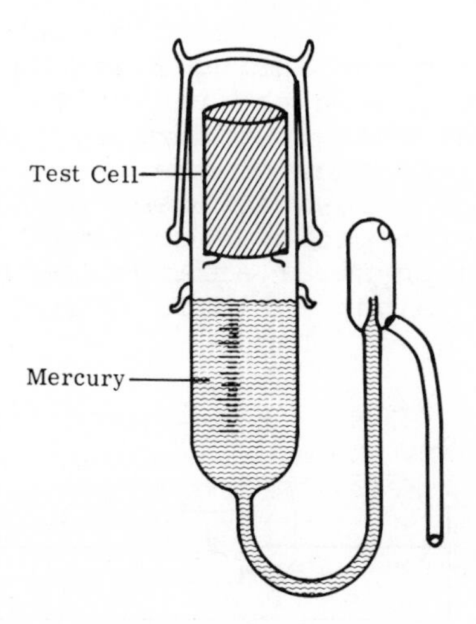

Fig. 13. Mercury displacement gas test apparatus for long-term corrosion tests (39).

B. Impedance Measurement Techniques

The internal impedance of a battery determines its ability to function as a power source. The impedance of a battery changes with the state of discharge, storage, and temperature (43-52). The measurement of cell impedance is one of the few available nondestructive test methods.

The total impedance, Z, of the battery is found by the vector addition of the resistive, R, and reactive, X, portions:

$$Z = R + jX \tag{17}$$

where $x = \omega L - 1/\omega C$ and $\omega = 2\ \pi f$ with f as the frequency of the applied alternating field in hertz (cycles per second). L is the inductance in henrys and C is the capacitance in farads.

The impedance of a cell may be divided into five componenets, as shown in Fig. 14. The resistances R_a and R_c are associated with the nature of the supporting anode and cathode electrode structure. Changes in R_a and R_c that occur during discharge (or charge) arise from changes in the resistivity of the electroactive material when it undergoes change in oxidation state. The resistance of the supporting structure does not change unless it loses mechanical integrity. An inductive or capacitive component may be present because of the physical arrangement of the electrode structure but is not included here.

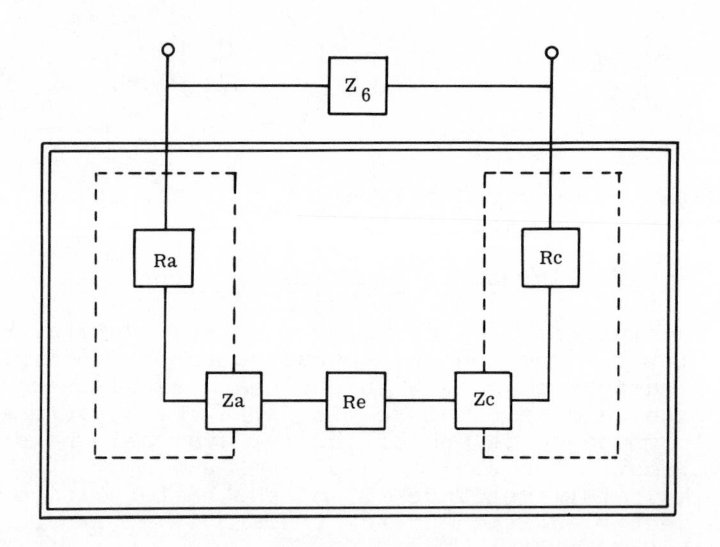

Fig. 14. Schematic of the components of battery impedance. The term, Z_6, corresponds to terminal geometry and constructional contributions to the impedance.

The electrolyte resistance, R_e, is determined by the ionic concentration of the battery electrolyte and the physical arrangement of the electrode structures. At frequencies below the threshold of the Debye-Falkenhagen effect (about 10^8 Hz for most battery electrolytes), R_e has the characteristics of a pure resistance. The electrolyte resistance is not necessarily invarient during discharge (or charge). Battery electrode reactions frequently involve the electrolyte as a reactant. The value of R_e increases or decreases, depending on the specific nature of the electrode reactions.

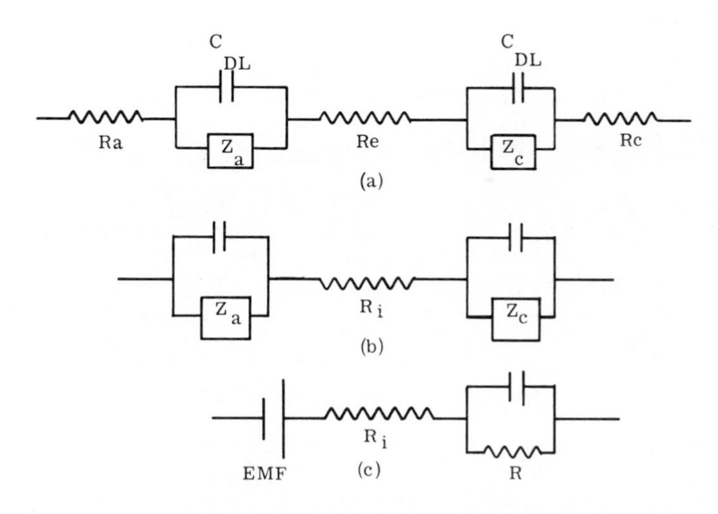

Fig. 15. Progressively simpler resistance-capacitance networks that simulate battery impedance. In (b) and (c), $R_i = R_a + R_e + R_e$. The source of EMF is not always shown but is implied.

The impedances, Z_a and Z_c, depend on the characteristics of the electrode reactions at the anode and cathode, respectively. The electrode impedance varies with frequency of the applied field and will change as the cell is discharged (or charged). Figure 15 shows the double-layer capacitance, $C_{d\ell}$, in parallel with the reaction impedance. The diffusional or Warburg impedance, W, is shown in series with the reaction resistance. The exact nature of these impedances depends on the specific nature and condition of the reactions at the electrode surface.

The simplified circuit combines many elements of the more complex circuit. The value R combines the resistance of the electrode structure and electrolyte, R_a, R_c, and R_e, into one unit. This combination really consitutes the "true" internal resistance of a battery. This portion is frequency insensitive but may change in value as composition and resistance of the electrolyte and electrodes change on discharge (or charge). The value R gives useful information concerning:

1. Maximum short circuit current;
2. iR loss during cell discharge;
3. Some information on composition of electrodes and electrolytes;
4. State of charge indication.

The values of R_2 and C_1 result from a combination of the electrode impedance Z_a and Z_c. The values are frequency sensitive to some extent. Often one electrode controls the values of the electrode impedance, for example, Zn in miniature mercuric oxide cells. It is often difficult to assess the effects of discharge, and the like. The values of R_2 and C_1 may be useful in obtaining information concerning:

1. Exchange current of cathode and anode reactions;
2. Electrochemically active surface area;
3. Kinetic limitations on the short circuit current;
4. Diffusional processes;
5. Short-term polarization predictions;
6. State of charge, especially at low frequencies.

The three main methods of measuring internal impedance may be classified as:

1. Volmeter-ammeter;
2. Pulse or square-wave methods;
3. Alternating current methods.

The voltmeter-ammeter method includes electrode polarization in the measurement and must be discarded as an unreliable method for measuring the internal impedance of a battery.

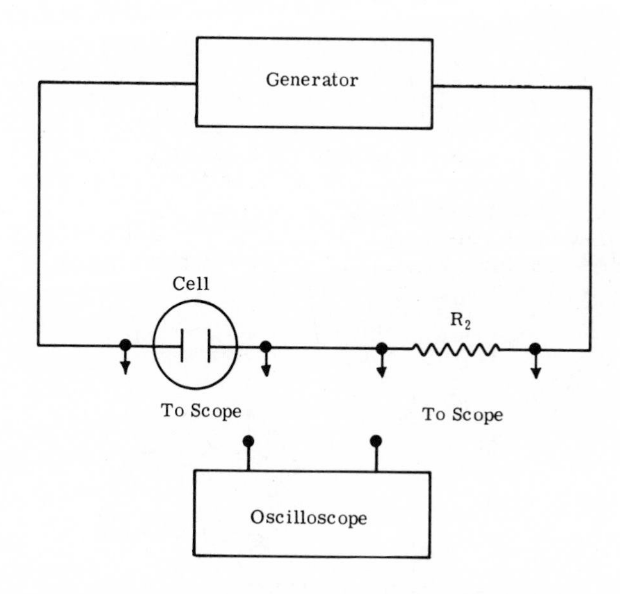

Fig. 16. Schematic diagram for the pulse or square-wave method.

The pulse method (40), shown schematically in Fig. 16, relies on the instantaneous appearance or disappearance of the ohmic polarization (iR drop) when current is turned on or turned off to measure the internal resistance (R_1) of the cell. It is essentially a very-high-frequency method for measuring the impedance. The voltage of the cell is observed with an oscilloscope as the current pulse flows through the cell. The instantaneous change in the cell voltage when the pulse travels through the cell and the voltage change measured across a standard resistor are used along with Ohm's law to determine the internal resistance. Either the leading or trailing edge of the pulse may be used.

$$R = \frac{E}{I} \tag{18}$$

The square-wave method (41, 42) is essentially a pulse method with a long-duration pulse. The internal resistance is determined from the instantaneous change in the leading or trailing edge of the square wave. In addition, some knowledge of the rate of polarization buildup is obtained by observing on an oscilloscope the voltage as a function of time. Since $t \approx 1/f$, the impedance of an electrode at a particular frequency can be found from the square wave. That is, the resistance computed via Ohm's law from the polarization and current flow at t = 0.01 sec compares to an impedance measurement made at 100 Hz.

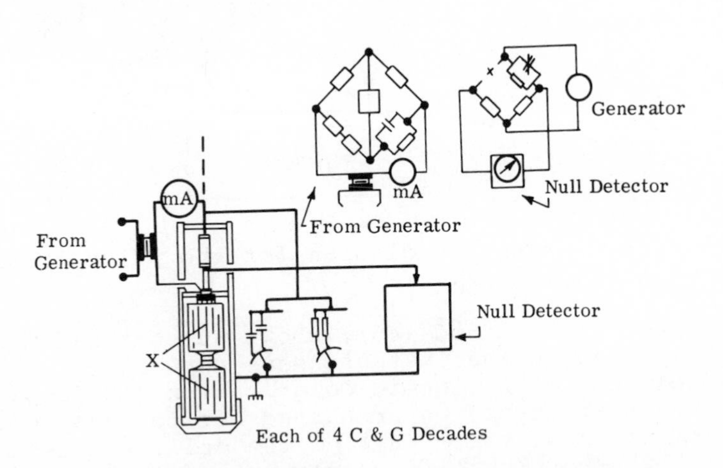

Fig. 17. Impedance bridge for measuring battery impedance (note that the two cells are in opposition to prevent discharge during measurement) (43).

An impedance measurement should be a relatively simple process. In the case of batteries, however, the low values of resistance and high values of capacitance (low values of reactance) make bridge

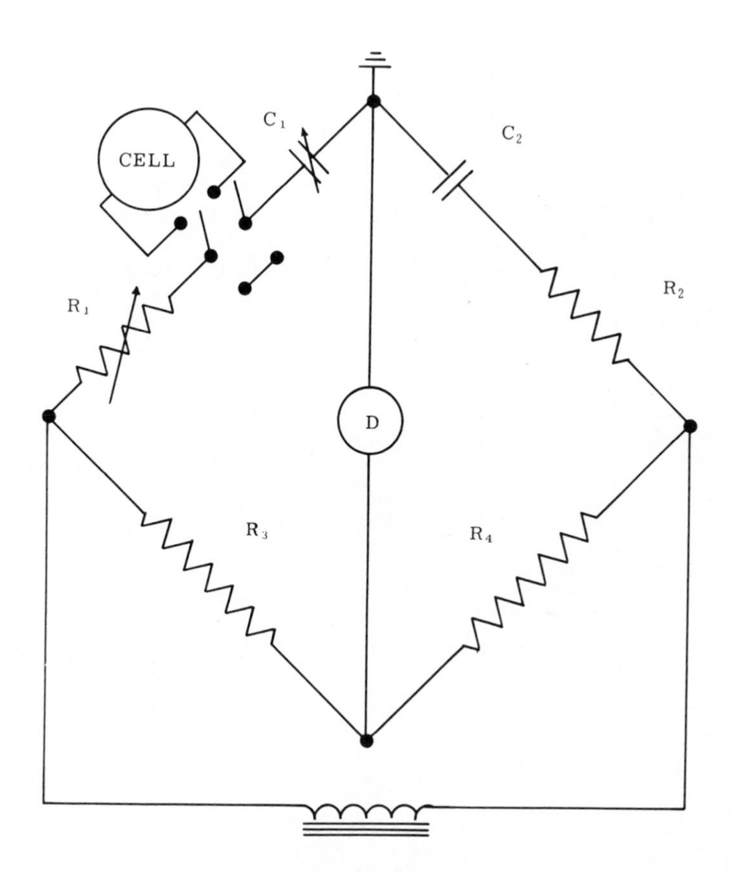

Fig. 18. Circuit diagram of substitution method impedance bridge.

measurements difficult. The EMF of the battery can also cause difficulty in the balance point. Two types of bridges find application in battery impedance measurement: the ratio arm resistance-capacitance bridge (43) shown in Fig. 17, and the substitution-type bridge (44, 45) shown in Fig. 18. Both bridges need a good resistive balance in order to secure a good null response in the detector. The substitution method relies on the differences in balance point for the standard components so that residuals which affect the balance point cancel out. The ratio-arm bridge may be difficult to balance when the ratio arms depart from equality as happens for large values of capacitance and low values of resistance. Also, provision must be made to prevent discharge of the battery through the resistance parts of the bridge. Some accomplish this by simultaneously measuring two cells in opposition, while others use a capacitive ratio arm. With good quality components, the resistance and capacitance elements remain constant over a wide range of frequency. It is always advisable to check the bridge with standard components.

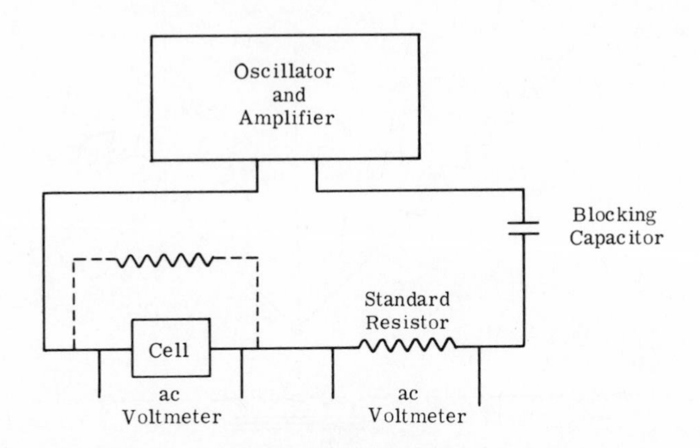

Fig. 19. Schematic of simple a.c. voltmeter impedance method. The impedance may be measured while discharging the cell as shown by the dotted lines.

A simple technique for measuring total impedance of a battery utilizes a simple a.c. voltmeter, as shown in Fig. 19. This method does not yield the detailed information that the a.c.-bridge technique supplies, but it serves adequately for many studies. Although the same voltmeter can be used to measure the a.c.-voltage drop across the standard resistor and the test cell, it is very convenient to have two voltmeters. The current is adjusted to a set value, say, 10 mA, 100 mA, and so on, by adjusting the output of the amplifier; then the voltmeter across the test cell becomes direct reading in total impedance. The total alternating current should be kept as small as is practical to avoid heating or exaggerating the rectifying effects of the battery and yet obtain reasonable operation of the voltmeter.

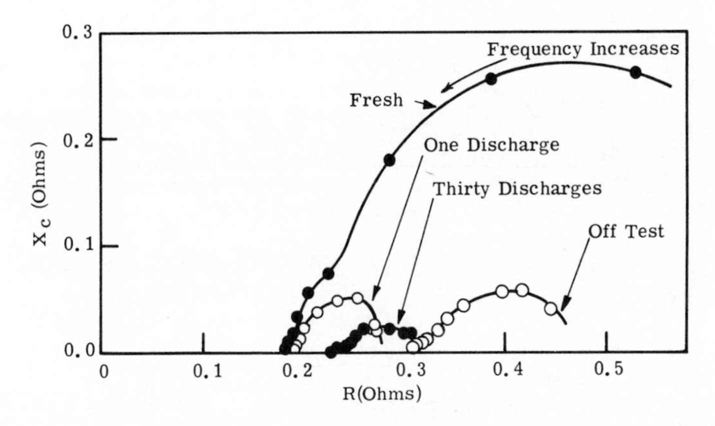

Fig. 20. Typical results of impedance of D cells as a function of discharge (4-ohm LIF) and frequency (50 H_2 to 50 kH_2) using an impedance bridge (45).

Typical results are shown in Figs. 20 and 21 for both the bridge and a.c.-voltmeter methods. The advantage of the Argard diagram representation lies in extrapolating to the high-frequency value for the internal resistance, R_1, and in the interpretation of the electrode reaction contribution from the frequency effects on resistance and capacitance.

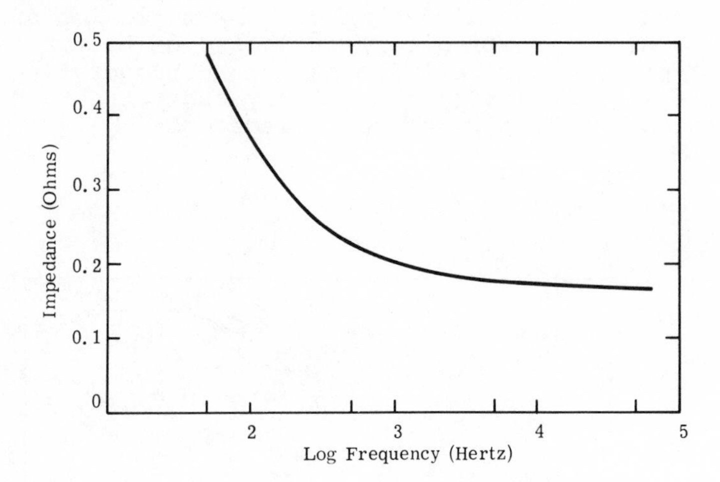

Fig. 21. Typical results of impedance of a D cell measured using the a.c.-voltmeter method.

C. Polarization Methods

The current-voltage curves of a battery system determine the intrinsic ability of a battery or electrode to produce useful power. The current-voltage characteristics (often termed "polarization characteristics") can be measured by a number of different techniques. The galvanostatic, potentiostatic, and linear-sweep voltammetry (LSV) experimental methods find extensive use in studying the kinetic behavior of individual battery electrodes. The application of these techniques to research studies of battery electrode performance follows a straightforward manner that is described elsewhere. The LSV technique is especially useful, as the regions of electrochemical activity at low sweep speeds correspond to the plateaus of the discharge curves for practical battery systems.

The simultaneous measurement of current and cell voltage is the classical method for studying the properties of electrochemical battery reactions.

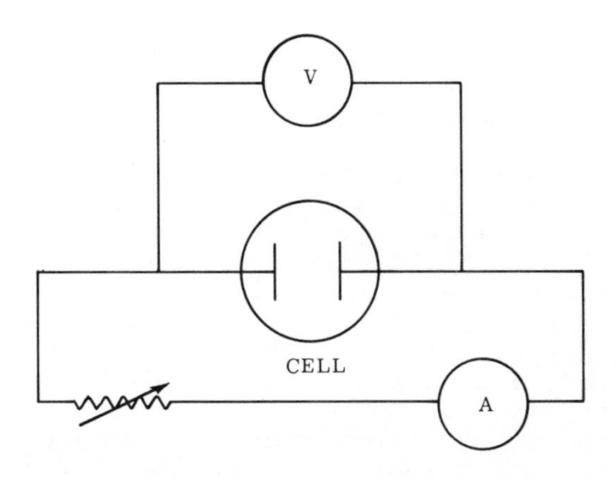

Fig. 22. Circuit diagram for determining the current-voltage characteristics of a cell or battery.

The usual experimental arrangement is shown in Fig. 22. A reference electrode may or may not be used, depending on the experimental situation. Although it is possible to use the cell terminal voltage readings to determine the service on a given test, it is difficult and almost suicidal to attempt to interpret individual electrode performance from terminal voltage measurements. In order to delineate individual electrode performance, it is essential to incorporate a reference electrode into the cell.

The voltmeter chosen to measure the EMF depends on the end use for the data and the characteristics of the electrode system. If one of the electrodes is easily polarized (i.e., changes voltage with small current flow) an electrometer or other high-input impedance (megohm or greater) voltmeter must be used. Precision measurements require a potentiometer for voltage measurements. The potentiometer balances with 10^{-6} to 10^{-8} A and is capable of 0.01% accuracy. Only where conditions tolerate appreciable current flow can a regular voltmeter (e.g., 1000 ohms/V) be used. Generally speaking, for research and development of individual electrode performance the electrometer or potentiometer-type voltage measurement is preferred, while for evaluation of practical cells the regular voltmeter is preferred.

The interrupter technique finds favor in electrode evaluation studies of a research and development nature. The use of an interrupter technique circumvents many of the uncertainties associated with proper reference electrode Luggin capillary placement and eliminates ohmic polarization from the voltage measurements. Careful experiments using either the Luggins capillary technique agree closely with the results of interrupter studies on the same system. The current interrupter utilizes the fact that ohmic polarization disappears instantaneously when current flow stops. The three types of interrupters are the electronic interrupter, the mercury relay, and the Kordesch-Marko bridge.

The electronic interrupter is a complex current supply with a fact-action electronic switch to start or stop current flow in less than a microsecond. When the current flow is interrupted, a potentiometer or other voltage measuring device is automatically connected to the terminals of the cell and the voltage measured instantaneously. The voltage measurement can also be made at definite time intervals after circuit interruption, so that the potential decay or buildup of activation and concentration polarization can be obtained. Normally, the duty cycle (ratio of time on-time current flow) is kept large, for example, 99%. The period of time of no-current flow is 10 to 100 μsec.

The elimination of both the disturbance of the current stream lines by reference electrode placement and the uncertainty of the iR-drop from solution resistance makes the interrupter technique a preferred experimental method. The complex nature of decay of activation and concentration polarization could interfere with the determination of the iR-free voltage.

Kordesch and Marko (51) developed an interrupter-type instrument with wide application for study of battery and electrode characteristics. Their interrupter-type circuit, shown in Fig. 23, is simple and reliable and does not require other complex electronic equipment or highly skilled personnel for proper measurement and interpretation. The Kordesch-Marko interrupter derives its power from the line voltage

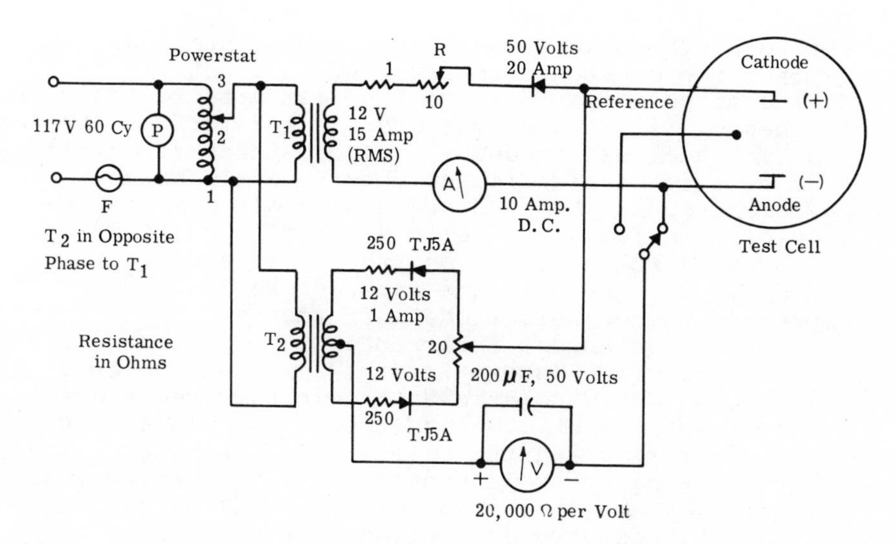

Fig. 23. Circuit diagram of the Kordesch-Marko interrupter.

for both charge and discharge of the battery or electrode. Reference electrodes can be used with this instrument. Since the instrument eliminates electrolyte resistance effects, the positioning of the reference is not critical as it is with other dc techniques.

The direct reading for iR-free and iR-included voltage measurements include activation, diffusion, and polarization.

The Kordesch-Marko interrupter measures the cell voltage only during the time when no current flows. By use of a suitable transformer and diode, current flows through the test cell during only one half of the cycle. The measurement of the average current and cell voltage or test electrode-reference voltage is made with meters have D'Arsonval movements. The meter reading for this condition gives the cell voltage or test electrode-reference voltage without the ohmic polarization or iR drop. If desired, the cell resistance may be obtained from the two voltage values and the average current,

$$R = \frac{E_{iR\ incl}E_{iR\text{-}free}}{I} \tag{19}$$

The Kordesch-Marko measurements are accurate as long as the time constants for the decay of activation and diffusion polarizations are long compared to the power frequency (52). This condition is met for most battery and fuel cell electrodes. The resistance-free voltage of a cell is of significant importance in battery research, since the iR drop in a laboratory cell bears little resemblance to the iR drop in a practical cell. In the research application, the effects of activation and diffusion are of prime interest in developing and improving battery electrode performance.

Another convenient interrupter construction, current interruption (shown in Fig. 24), is also used to measure the ohmic polarization. In this technique, a mercury-wetted or other suitable relay interrupts the the current flow and the cell voltage transients are recorded using an oscilloscope. The oscilloscope is triggered by the energizing of the relay as the relay opens or closes at a controlled time interval. This technique may be used with either constant current charge or constant resistance discharge studies.

In addition to the interrupter, constant-resistance discharge is the other most common method for studying batteries. It is convenient in practice and, more importantly, corresponds closely to most commercial battery applications. Many times the use of standard test equipment is cumbersome. Figure 25 shows a very simple test setup with multiple test positions that can be energized on the test program. It is essential to include the relay contact resistance and lead resistance in the total discharge resistance. The relays are electromechanical, and several can be activated simultaneously. A minimum of three and preferably six or more batteries should be tested simultaneously; otherwise the statistical analysis may suffer wide fluctuations.

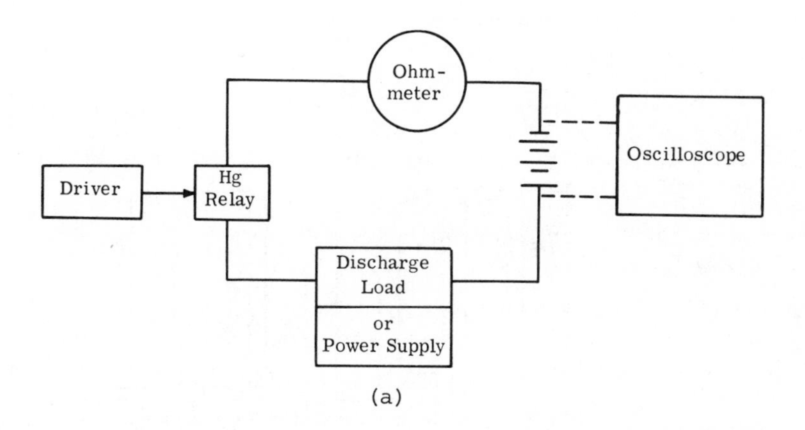

(a)

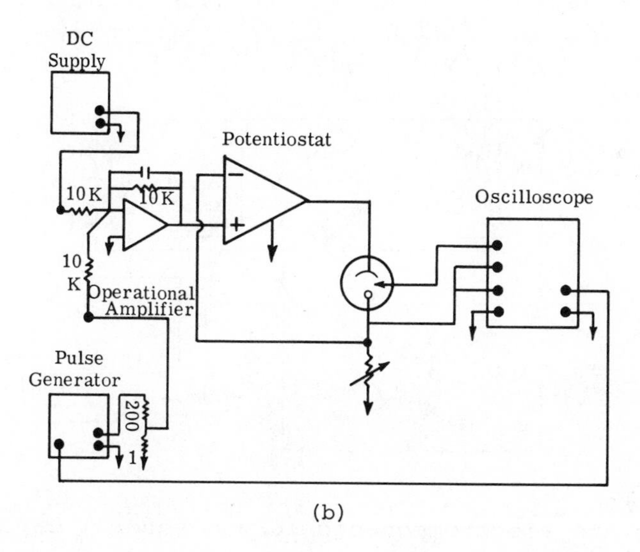

(b)

Fig. 24. (a) Schematic for mercury relay interrupter. (The mercury relay operates from a variety of inputs. The duration and regulator rate of the interruption depend on the relay. A power supply for constant current or voltage or a resistive load may be used. The oscilloscope records the interruption). (b) Electronic current interrupter circuit.

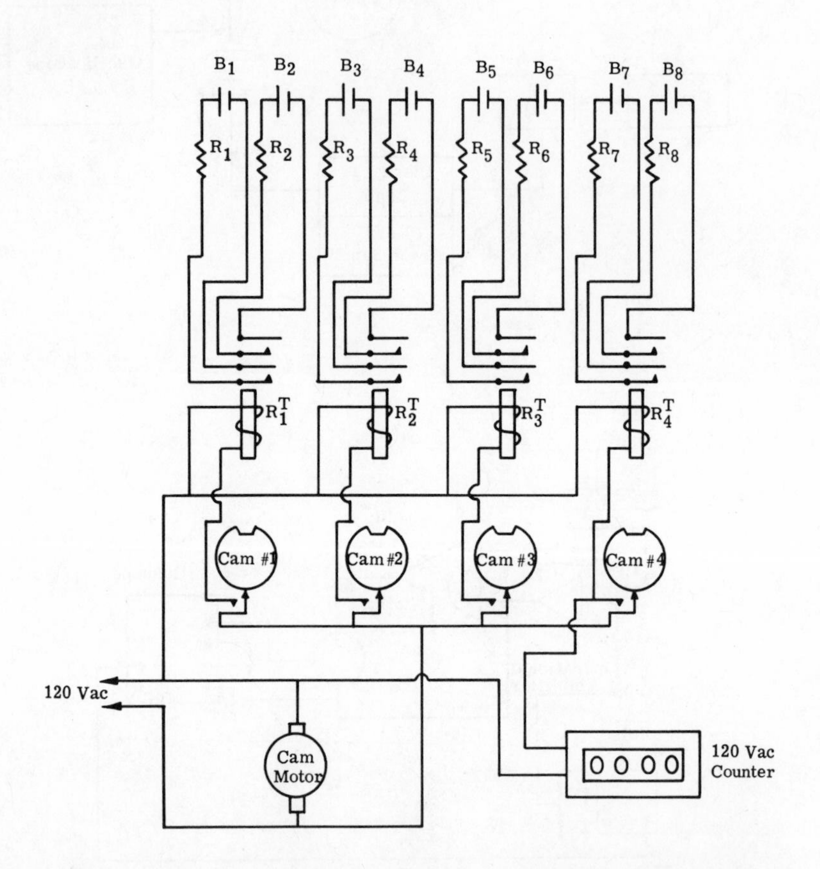

Fig. 25. Circuit diagram for a constant resistance discharge of a cell or battery. (The timing motor and cam permit discontinuous testing if desired.)

Constant resistance discharge tests that approximate various uses have been established. The comparison of the test results under various experimental conditions is essential to the continued development of a battery system. There are many different test configurations available, but all contain the same basic elements: a timer to control the discharge time on intermittent test loads; a discharge resistor of known value; a relay network activated by the timer to connect the discharge resistor at the proper time intervals; and a means of measuring the voltages of the test cells at fixed intervals. The apparatus, whether electromechanical or computer controlled, must be reliable and capable of operating unattended. The various standard tests and test methods are described in Section III F.

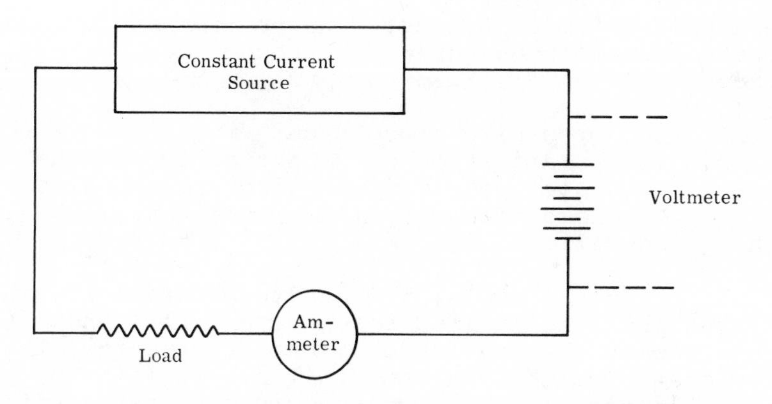

Fig. 26. Circuit diagram for a constant current discharge of a cell or battery.

Another testing method is by constant current discharge (Fig. 26). Here, the power supply may be electronic or a high-voltage battery with large dropping resistors in series to keep resistance fluctuation introduced by cell discharge to a minimum. The polariztion curves may be taken automatically and the results displayed directly on an x-y recorder. This requires sophisticated equipment such as described by Schreiber and Schwarz (53). The internal resistance in a cell can also be measured with an a.c. superimposed on the discharge. The resistance can be measured on open circuit or on load, in which case it is often called an a.c.-ripple, or a superimposed-a.c. method. This technique constitutes an impedance and has been discussed in the impedance section.

D. Electrochemical Activity of Battery Components

It is often inconvenient to construct full batteries (anode, cathode, etc.) to evaluate a material for electrochemical activity. Two types of test procedures are described that are designed to evaluate only one component for electrochemical activity. The search for new electrode materials, whether anodes or cathodes, or the selection of a better material from a number of samples of the same chemical nature, for example, MnO_2, continues. The preliminary evaluation rests primarily on three electrochemical characteristics:

1. The open-circuit voltage;
2. The polarization or departure from open-circuit voltage under various discharge conditions;
3. Discharge capacity (Ah/g) under various discharge conditions.

In practical cells, the active material is usually packed into the limited space of a cell. Under these conditions, the available discharge capacity often deviates from the maximum available capacity of the active material. Various factors such as insufficient electrolyte, poor electrical contact with the active material, nonuniform mixing of conductor and active material, and nonuniform current distribution contribute to the poor overall utilization of the active material. Several tests (54-62) described below attempt to measure the intrinsic activity of a material under standard reproducible conditions without complications of compaction, porosity, conductivity, and the like.

Figure 27 shows a typical test cell construction. A sample material (100 mg) was thoroughly mixed with 1.0 g of battery grade graphite, for example, Dixon "Air Spun" graphite, or the like, and 0.2 g of coke and placed in the cell. Some mixes may include acetylene black for its absorptive properties. Then, 0.2 to 0.3 ml of appropriate electrolyte solution was added to the mixture and patiently stirred with a

r

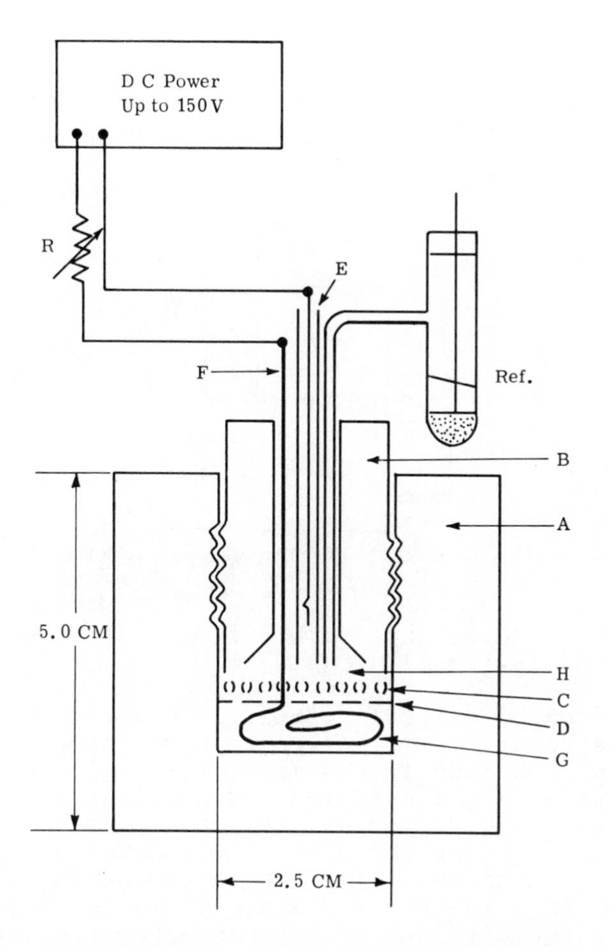

Fig. 27. (a) Discharge Cell. A, body of the cell (Lucite); B, threaded Lucite plug, with hole in center which is used to keep pressure on G; C, perforated Teflon disk; D, separator paper (made of synthetic fiber); E, anode compartment (glass tube with a fine glass frit at the end) containing a Pt wire electrode; F, gold wire (1 mm diameter), with a spiral at the end which was buried in the mixture-the top straight portion of the gold wire was covered with a polyethylene tube; G, a mixture of 0.100 g of MnO_2 + 2.0 g of coke + 1.0 g of graphite; H, electrolyte; R, resistance (25 to 400 kohm) (54).

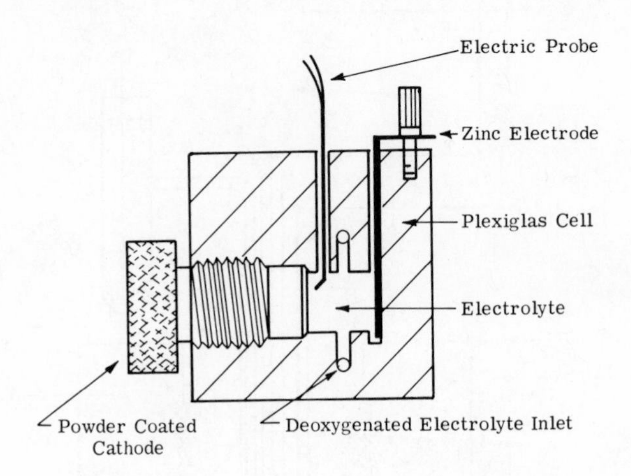

Fig. 27 (b) Discharge cell using thin powder-coated electrode (62).

Teflon rod or a metal spatula in order to mix the dry components and the electrolyte homogeneously. A gold wire was embedded in the mix. After placing a separator paper and a perforated plastic disk over the mix, the mix was prepressed with a plastic rod (having a hole in the center to accommodate the gold wire) from the top. Then, a threaded plug was screwed down tightly onto the mix. The electrolyte was usually squeezed out of the mix above the perforated disk. More electrolyte was added from the top to ensure a flooded electrolyte condition. A counter-electrode and the reference electrode completed the setup.

After assemblying the cell, it is advisable to leave the cell on open circuit overnight before measuring the first OCV (open-circuit voltage). The cell is then discharged at a constant current of 1.0 mA for 6 hr and the CCV (closed-circuit voltage) is measured at the end of the 6 hr discharge. Then the cell is left on open circuit overnight (18 hr) and the second OCV is measured immediately before starting the second 6 hr discharge. This process (6 hr discharge and 18 hr rest) is repeated daily as long as the cell is capable of discharge. The OCV values and the CCV values are plotted against the depth of the discharge (mAh). Typical results of such plots of MnO_2 samples in 9M KOH are shown in Fig. 28. Other test regimes may be used, for example, constant current, higher or lower currents, shorter discharge times, and so on.

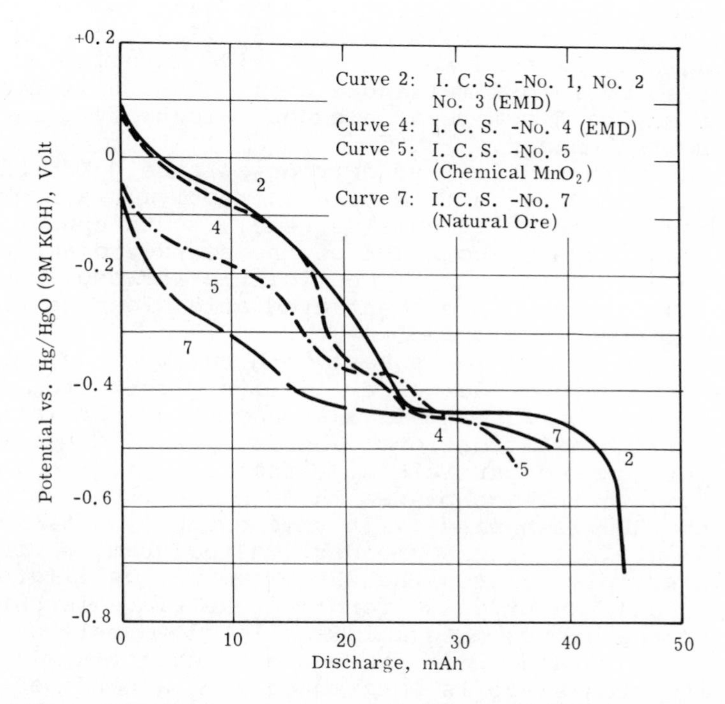

Fig. 28. Discharge curves of international common samples of MnO_2 at 1.0 mA/100 mg sample in 9M KOH using method shown in Fig. 27 (a).

When the electrolyte has a reasonably good conductivity, that is, KOH electrolytes (1 to 9M) and $NH_4Cl + ZnCl_2$ (2 to 4M), the iR drop associated with the CCV measurement is usually only 5 to 10 mV. The correction for the iR drop is not necessary for the CCV values in these electrolytes. However, if the electrolyte is a poor ionic conductor or if the discharge product accumulated on the cathode is a poor conductor material, the iR-free CCV voltage must be measured. To measure the iR-free CCV, the circuit is opened for 1 to 2 sec while the cathode voltage against the reference electrode is recorded with a chart speed of 20 inc/min or faster using a fast-response recorder. Carefully conducted tests show good correlation to cell discharge data (59).

Another quantitative method (60, 61) consists of embedding grains of the active oxide material or other solid into a malleable foil. This foil electrode may be discharged in excess electrolyte in a controlled manner.

The preferred foil or carrier is pure gold, although a conductive film of a soft polymer and graphite may be used. If gold is used, its thickness would not exceed about 40% of the average diameter of the powder under study. The foil is thoroughly degreased, cleaned in concentrated acid cleaning solution, rinsed, and dried.

This technique works best with uniform particle-size reactants. The powder is first freed from fines by sieving, by washing in a steady stream of water, or by an air classification. The dry powder is sieved and the desired particle size fraction retained, for example, that which passes through the 150-mesh but not the 200-mesh sieve. In this example, a 0.002 to 0.003-in. foil is recommended. A few grams of the oxide are placed into the 200 sieve and oscillated to distribute the oxide uniformly. The sieve is inverted and the excess material is allowed to fall out. The sieve retains a considerable amount of dry powder. The inverted sieve is positioned over a hardened-steel plate. Brushing lightly over the inverted sieve with a camel-hair brush deposits a uniform layer of particles on the hardened-steel plate. The weighed gold foil is placed over the powder and a polished 1-in.^{2} die is placed on top of the foil. The assembly is pressed at 20,000 psi or more in a hydraulic press to embed the powder into the gold foil. The process is repeated with a clean hardened-steel plate and die to lock the powder into the gold foil. The foil is rinsed in distilled water, dried, and weighed. The amount of powder is found by the difference. This technique produces an electrode of essentially single grains of active material with 3 to 7 mg powder/cm^2.

Softer powders necessitate the use of the polymer-graphite foil. The embedding process is carried out at lower pressures and the foil heated to near the softening point to embed the powders into the foil.

Care must be exercised in drying the electrode, because many powders react with oxygen from the air. Vacuum drying at slightly elevated temperatures, say, 60 to 80°C, for several hours is recommended.

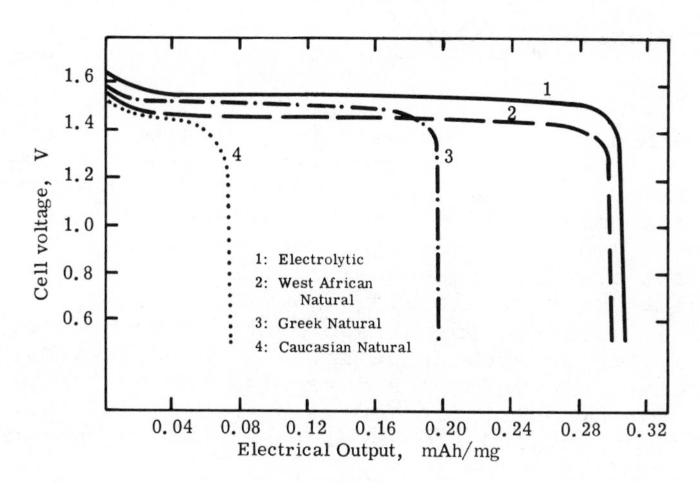

Fig. 29. (a) Discharge curves of (1) electrolytic; (2) West African natural; (3) Greek natural, and (4) Caucasian manganese dioxides. Discharge current 0.01 mZ/mg electrolyte 20% NH_4Cl, 10% $ZnCl_2$, 70% Water (62).

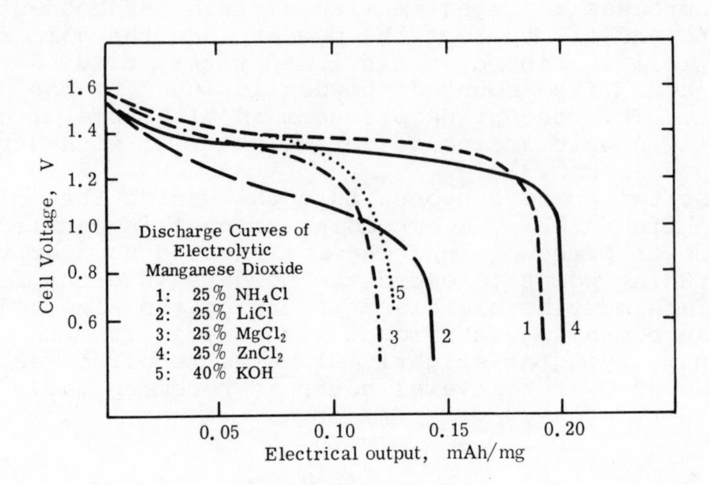

Fig. 29. (b) Discharge curves of electrolytic manganese dioxide in (1) 25% NH_4 Cl; (2) 25% LiCl; (3) 25 $MgCl_2$; (4) 25% $ZnCl_2$, and (5) 40% KOH. Discharge Current 0.1 mA/mg (62).

The foil electrode held in a plastic frame is immersed in the deaerated test solution. A Luggin capillary reference positioned near the foil monitors the discharge. If a large-area unpolarizable counter is utilized, cell voltage readings may be recorded instead of the reference-foil voltage. Typical results are shown in Fig. 29.

The discharge is carried out in a controlled atmosphere with or without stirring. The solution, test material, and reference determine the system of interest; for example, for MnO_2 use $ZnCl_2$-NH_4Cl solution, a zinc anode, and a calomel reference, as noted in Section II E.

E. Cell Construction Concepts

After single-electrode tests define the performance of an electrode or material, its performance in an actual cell environment provides the final test.

There are so many subtle influences of mechanical design or of construction on cell performance that extensive practical cell testing must be carried out under a variety of test conditions. A full description of practical cell construction is beyond the scope of this chapter.

Battery construction varies with the chemical system. The preparation of Leclanché cells has been described elsewhere in detail; other cell types have also been discussed elsewhere (63). The reader is referred to these sources for detailed methods.

The preparation of the anode or cathode mix for battery construction determines to a large extent the ultimate performance of the battery. While procedures vary depending on the particular cell construction, a general outline of mix preparation follows

1. Dry blend ingredients (e.g., carbon black, graphite, MnO_2, binder for the cathode, and binder or gelling agent, zinc powder for the anode, etc.) (care must be exercised to avoid overmixing that results in segregation of the mix components).
2. Electrolyte (or water) added, usually as a spray to assure uniform distribution of the electrolyte. (Care must be exercised to avoid overmixing.)
3. Adjust electrolyte to yield proper wetness and plasticity for molding.
4. Molding of the electrodes (cathodes) varies from about 200 psi for Leclanché style cathodes to over 20,000 psi for the mercuric oxide or button-style cathodes.

Two types of wetness tests find application in the battery industry. One utilizes a commercial "wetness" meter that measures the weight loss of the sample when heated in a controlled fashion.

The second type uses a setup similar to that used for resistivity measurements. A blotting paper or similar absorbent paper covers one electrode. As the pressure in increased, the sample compacts. At the point where liquid is expressed from the mix, the liquid wets up the blotting paper and the circuit is completed. Usually a simple buzzer is used so the buzzer sounds as the circuit is completed. The pressure at which the buzzer sound measures the wetness.

Generally, this pressure corresponds to compaction pressure in cell construction.

To accomplish the needed study of the effects of various parameters on battery performance, experiments should be conducted utilizing a statistical approach to experiment design (64).

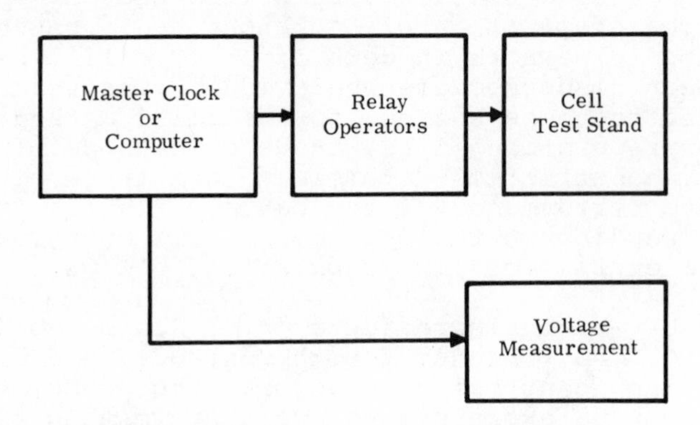

Fig. 30. Schematic diagram for standard test methods.

A schematic of the standard test methods is shown in Figure 30. The master clock controls the relay operators for cell discharge as well as the timing of the voltage measurements. The exact makeup of each test laboratory varies, but the data are collected as prescribed for each standard test. The testing essentially expands the simple setup of Figure 25 to provide precise control and voltage measurement.

When using this approach, it is essential to identify and control all of the experimental variables. The presence of an unidentified variable can be recognized by a large experimental error or by the lack of good reproducibility between experiments. However, this lack of preception in controlling and in planning the experiment can be costly.

Terms of statistics include:

Mean	Arithmetic average of a set of data
Median	Middle value of a set of data
Range	Difference between largest and smallest value
Variance	A measure of dispersion
Standard deviation	A measure of dispersion
3-Sigma limits	95% confidence that data falls between the 3-sigma limit
Coefficient of variation	A dimensionless number that permits comparison of dispersion among sets of data
Skewness	Measure of a lack of symmetry of a set of data

F. Standard Battery Tests

The performance of cells and batteries on standard tests that simulate actual use conditions constitutes an important criterion in battery research and development. The American National Standards Institute subcommittee C-18 periodically revises and issues specifications for the expected performance of dry cells and batteries (65). The International Electrochemical Commission through the technical committee on primary cell and batteries (TC-35) coordinates cell testing procedures and performance levels on an international basis (66). Most primary battery consist of constant resistance discharges with varying frequency and period of discharges. Table 2 lists several typical standard tests that simulate actual use.

TABLE 2

Summary of Typical Standard Tests

Test	Sample Size	Resistance (ohms/cell)	Duration (min)	Interval (hr)	Cutoff Voltage V/cell
Heavy intermittent, No. 6	8	2 2/3	60	6-16 [(a)]	0.85
Lighting battery	6 [(b)]	9 [(c)]	30	0.5-16 [(d)]	2.6
Safety flasher	6	15	0.25	(b)	0.9
General purpose C, AA	6	4	5	24	0.75
Light industrial flashlight D	6	2 1/4	4	1-16 [(e)]	0.65
Transistor battery	3	166 2/3	240	20	0.9
High rate (Alkaline)	6	1	Continuous		0.75

a. Two periods of 1-hr discharge, with 6 hr intervening with a 16-hr rest period at the end of the second discharge.
b. Two strings of 3 batteries in series.
c. Per 6 V nominal battery voltage.
d. Ten periods of 4-min discharges at hourly intervals followed by a 14-hr rest period; on the seventh day every other discharge period omitted.
e. Four-minute discharges at hourly intervals for 8 consecutive hr with a 16-hr rest period intervening.
f. Discharge 0.25 sec each second, 24 hr/day.

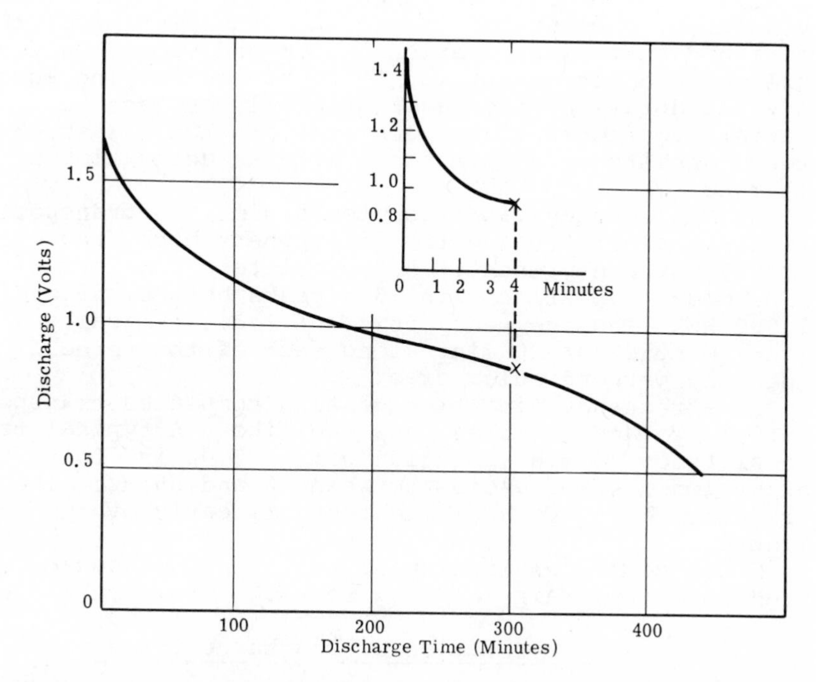

Fig. 31. Performance of a typical D Leclanché cell on the 2.25 ohm-LIF test. The inset shows one of the 4-min discharges, with the voltage read at 4 min as required by the standard test procedure.

A typical discharge curve is shown in Fig. 31. The test procedures specify the voltmeter to be used, for example, a 10,000 ohms voltmeter for small cells and 1,000 ohms voltmeter for larger cells. The scale size and meter accuracy are also specified.

Delayed or shelf testing constitutes an important aspect of the testing program. The ability to retain a sizable fraction of its original performance often determines the usefulness of a battery system.

Certain applications, for example, military and aerospace, require batteries to perform satisfactorily after being subjected to specified vibration and shock testing procedures. These environmental tests simulate the conditions (high temperature, shock, and vibration) encountered in storage and shipping of the finished batteries. Various shock and vibration tests are employed in an attempt to reproduce varying severity and duration; the tests generally subject the battery to shocks along each axis. A typical shock test consists of a 60-g shock with an activation of 11 msec.

High-frequency vibration tests simulate transportation via aircraft, and the like, where high-frequency vibrations are present. A typical test consists of sinusoidal vibration with 10-g peaks between 55 and 2,000 Hz. The frequency is swept logarithmically every minute for 30 min. along each of the three mutually perpendicular axes.

Low-frequency vibration testing simulates transportation by land vehicles, and the like. A typical test consists of 90 min vibration with a 0.06-in. total excursion at frequencies between 10 and 55 Hz. The frequency is swept back and forth linearly every minute.

Other common environmental tests include bounce, high-humidity, rain, and fog tests.

G. Interpretation of Discharge Curves

Figures 32, 33, and 34 show typical results for the discharge of several different oxide samples. The OCV curve for the discharge approximate the cell discharge when the cell discharges at a very low current. From the shape of the OCV curve determined by eliminating the iR contribution to the electrode polarization, the nature of the discharge process may be deduced based on thermodynamic principles. When the OCV decreases gradually with continued discharge, the electrode system consists of a homogeneous phase. Only one solid phase is present, and the reaction product and reactants form solid solutions (are mutually soluble). The first half of the discharge of MnO_2 or RuO_2 follows this type of discharge (54a).

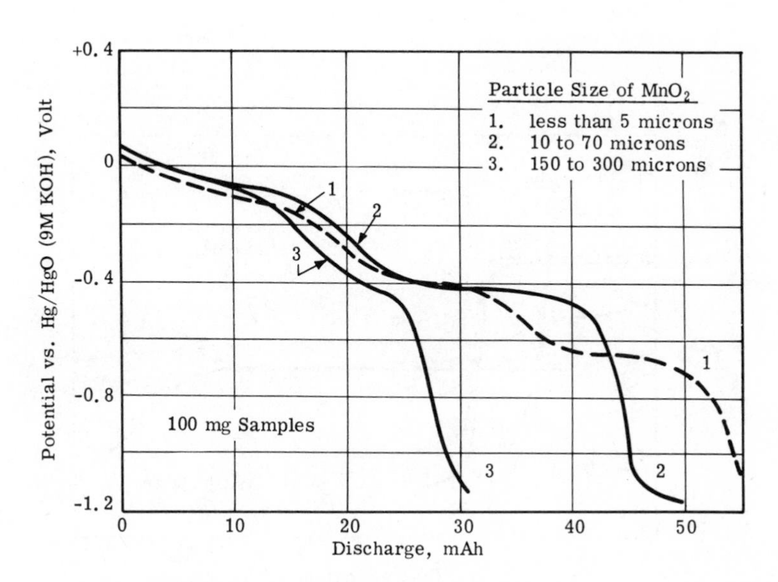

Fig. 32. Effect of particle size of MnO_2 samples discharged in 9M KOH Solution at 1.0 mA for 6 hr each day, as described by Ref. 54. The curves are closed-circuit voltage curves.

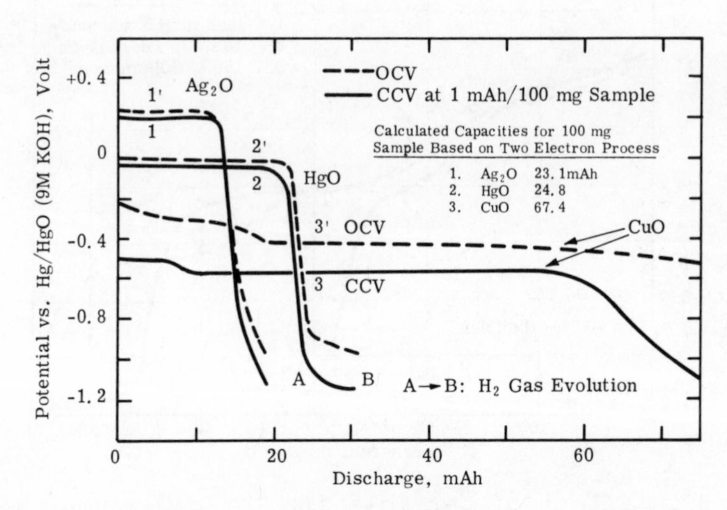

Fig. 33. Discharge characteristics of Ag_2O, HgO, and CuO in 9M KOH solution. Dotted lines are the open-circuit voltage values as described in Ref. 54.

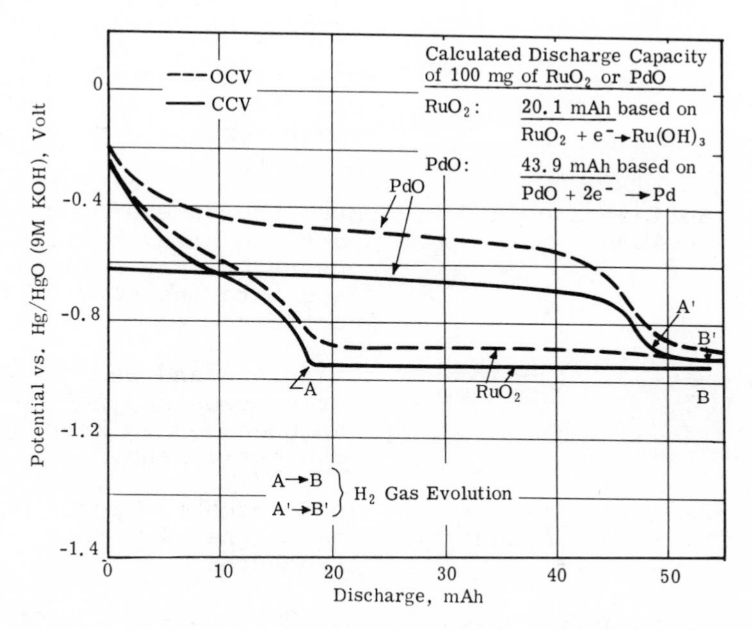

Fig. 34. Discharge curves of PdO and RuH_2 in 9M KOH solution, discharged at 1.0 mA for 6 hr/day as described in Ref. 54. Dotted lines are the open-circuit voltage values.

When the OCV remains essentially constant as the discharge continues, the system consists of two distinct or heterogeneous solid phases. The reactants and products retain their identity and no solid solutions form. The discharges of HgO, Ag_2O, and the second step of MnO_2 and RuO_2 discharge exhibit this behavior.

In each figure, the calculated discharge capacities to a given voltage cutoff are shown based on the most likely electrochemical reaction, that is, a one- or two-electron process. Here these calculated values agree with the experimental values. If poor agreement exists between the calculated and experimental values, several factors must be considered.

Among these are:

1. Purity	Water content or extraneous foreign material.
2. Particle size	Too large a particle with high resistivity and the inner portion may not be discharged
3. Surface area and pore volume	Much the same effect as particle size, but the porous nature may permit reaction into the core of the particle
4. Rate of discharge	Diffusion and charge transfer process may not be rapid enough to keep up with the demand.
5. Reaction of active material with conductor	Self-descharge reaction inside the electrode structures

The difference between the OCV and the CCV of the cell defines the polarization. The iR polarization should be separated out and identified, leaving activation and concentration polarization. The magnitude of the polaraization depends on the material under test. For instance, in Fig. 35, the gradual loss of the ability to produce power shows up if periodic checks (e.g., 1-min tests) of the polarization characteristics of the cell are made during the life of the battery. One electrode may be the limiting electrode in determining the useful life of the battery, but the other electrode may be the limiting factor in determining the ability to produce instantaneous power. The polarization of HgO is very small compared with that of other oxides such as MnO_2. The factors that often determine the magnitude of the polarization include:

1. Surface area of the powder;
2. The pH change in the fine pores of the powder (especially true for unbuffered electrolytes);

3. Dissolution rate of the reacting species when a soluble intermediate is involved in the electrode reaction.

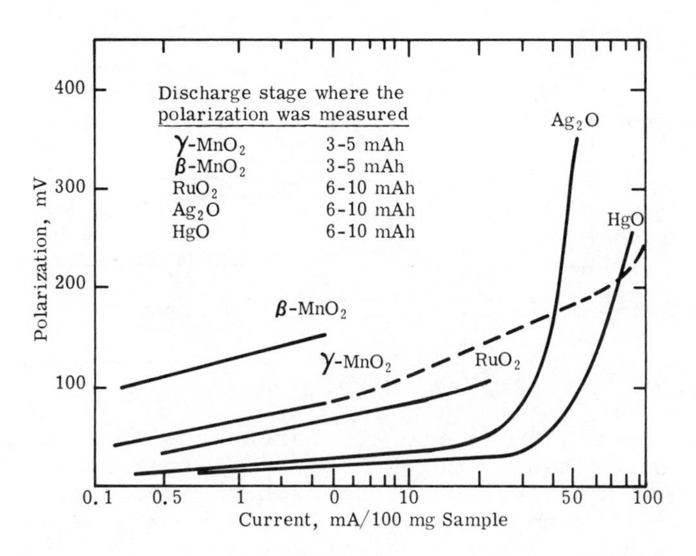

Fig. 35. Typical polarization curves.

Figure 36 shows the discharge curves of PbO_2 in 9M KOH at various current drains. In general, the discharges at higher currents show the greatest polarizations. High polarization when discharged at lower currents may indicate reaction of the active material with the conductor. Storing the cell at 45°C for 4 days lowers the capacity, perhaps because of reaction between the active material and the carbon powder during storage.

Whenever electrochemical reduction involves a powdered material, the particle size and true surface area of the material must be considered. Figure 30 shows the discharge of three MnO_2 samples with varying particle size. The first discharge step in which MnO_2 discharges in a homogeneous phase is not

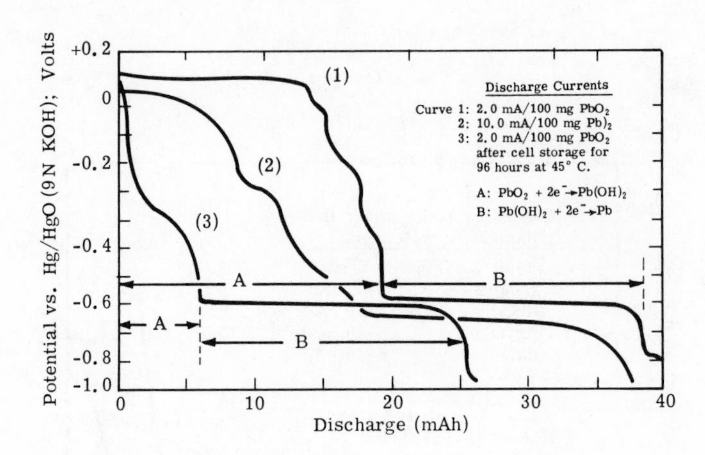

Fig. 36. Discharge curve of PbO_2 in 9N KOH using test method described in Ref. 54.

significantly influenced by particle size, whereas the second step in which MnO_2 discharges via a dissolved species varies considerably with particle size. The sample with the largest particle size and smallest surface area cannot be discharge (i.e., it exhibits large polarization) because of the slow dissolution or diffusion rate of the reacting species (Mn (III) ions).

In comparing test methods, a source of battery-active materials of known activity serves a useful purpose. The Cleveland Section of the Electrochemical Society maintains the International Common Samples of MnO_2 for research use.* These various samples of well-characterized MnO_2 provide a means to compare or standardize various test methods. Factors such as electrode geometry, cathode and anode mix formulation, electrolyte, separator material, cell construction,

*For further information write I. C. Sample Office, Attention: Dr. A. Kozawa, P. O. Box 6116, Cleveland, Ohio 44101.

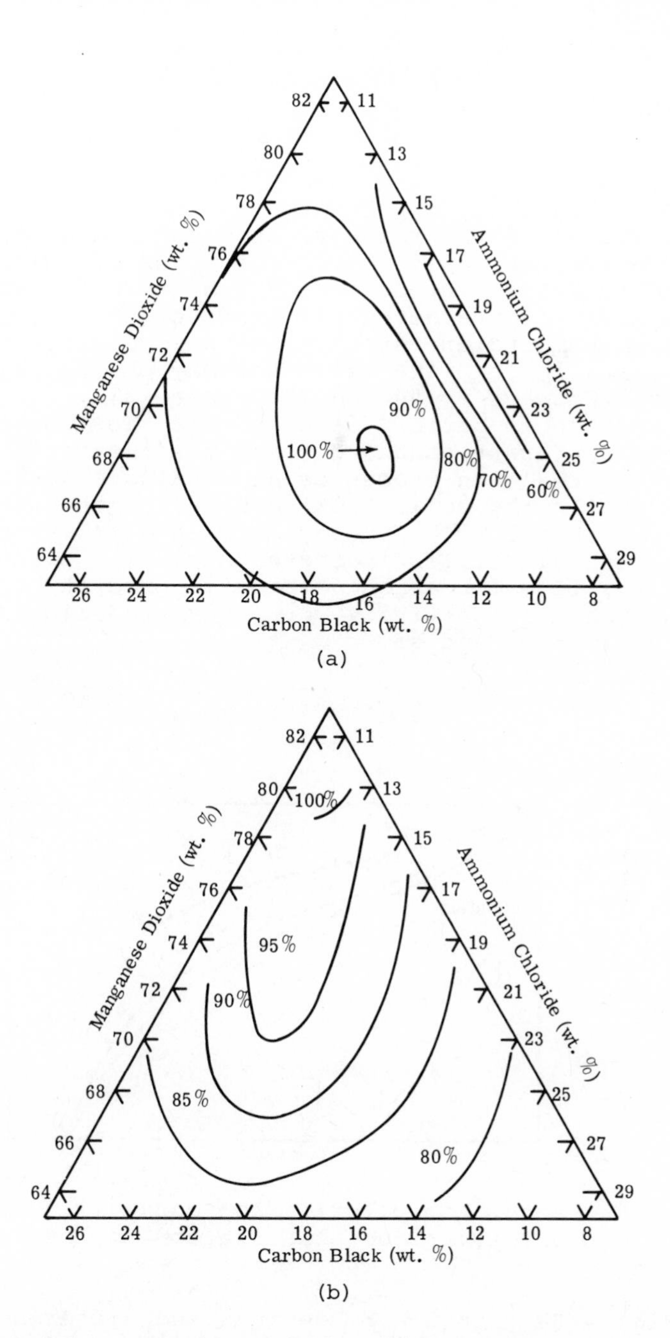

Fig. 37 (a) Relative service-life contours from pasted cells discharged through 5 ohms continuously to 0.75 V as a function of mix composition (57). (b) Relative service-life contours for pasted cells discharged at 100 mA constant current for 6 hr day at 0.75 V as a function of mix composition (57).

and so on have interrelated effects on cell performance. Figure 37 depicts the influence of the cathode composition on the relative service life on two discharge regimes (63). Representations of this type serve to assist the design of a battery for optimum performance for a given application.

High-temperature storage tests accelerate the aging processes in complete cells but may change failure modes. Storage at 54 and 71°C especially find wide application in battery development. Testing the cells on the different standard tests after storage helps to identify reaction processes and evaluate seal effectiveness in the cell. Low rate performance indicates changes in the total energy available from a system. High rate discharges indicate changes in electrode integrity, film formation, corrosion product buildup, electrochemical activity, or power-producing ability of the system, and so.

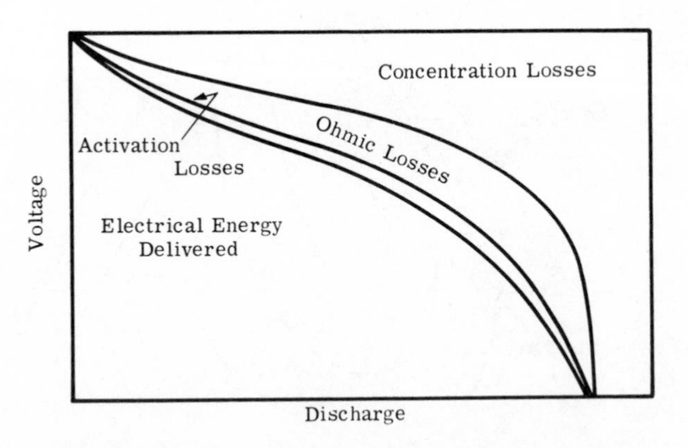

Fig. 38. Schematic polarization curve illustrating contribution of various polarization to total polarization.

Figure 38 graphically depicts the difference between a perfect electrochemical battery and a practical battery. In actual practice there are three sources for energy losses when chemical energy is converted into electrical energy. These are (a) concentration losses, (b) activation losses, and (c) ohmic losses. As a result of these losses, the actual power or useful energy delivered is only a fraction of the total available energy in a battery.

Concentration losses arise from (a) changes in the effective concentration (activity?) of the reactants during battery operation, including pH changes induced by electrode reactions, and (b) transport problems of the reaction sites. Concentration losses usually account for most of the losses in practical cells.

Ohmic losses occur when power is dissipated as i^2R heating due to internal resistance of the cell. The terminal voltage deviates from the open circuit due to iR losses such as occur from film formation on the surface of the electrode. Ohmic losses are probably the second largest source for energy losses in practical batteries.

Activation losses occur when electrode reactions proceed slower than the rate at which current is withdrawn. Activation losses are associated with kinetic factors of the energy-producing ractions in the battery and are usually small except for extremely high-rate batteries.

The dissipation of energy decreases the efficiency of the energy conversion process in a battery. The current efficiency is always close to 100% because of Faraday's law. Therefore, the energy loss results primarily from the deviation of the cell voltage from its theoretical value. This deviation, termed polarization, is defined by

$$\eta = E_i - E \qquad (20)$$

where E_i is the cell voltage with current flowing, and E is the theoretical cell voltage.

Figure 38 shows the various voltage losses for a typical battery. When the ordinate of the figure is expressed in units of current, the figure defines the polarization curve. When the ordinate is expressed in units of time, the figure defines the discharge curve.

In some batteries it is the positive electrode that limits high-rate performance, but the negative electrode limits the total capacity of the battery. The use of a suitable reference electrode is essential to determine the operation of each electrode. Many cells, especially sealed cells, require careful balance of the anode and cathode capacities.

Analysis of the rise time of voltage transients at several currents can be used to distinquish between the three types of polarization losses. Ohmic polarization appears or disappears instantaneously on current flow. It is the easiest of the three to identify and measure.

The rise time, T, for activation polarization is given by

$$\mathrm{T} = \frac{C}{i_o} \frac{RT}{\eta F} \qquad (21)$$

where C is electrode capacitance and i_o is the exchange current. The time T varies from about 10^{-5} to 10^{-2} sec, depending on the reaction. It is independent of the current, so rise time is not a function of current but is a constant quantity.

The rise time for concentration polarization is given by

$$\mathrm{T} = 4\pi D \ \eta F \left(\frac{C}{i}\right)^2 \qquad (22)$$

where D is the diffusion coefficient, C is concentration of reacting species, and i is the current. The rise time is a direct function of current; the higher the current the faster the rise time. This dependence on current distinguishes concentration from activation polarizations. The rise time for concentration varies from about 10^{-3} to several seconds.

The transient analysis in the time of 10^{-6} sec or less identifies the ohmic contribution to polarization. In the time 10^{-5} to 10^{-2} sec both activation and concentration polarization may esixt, but the activation polarization does not depend on value of current flow. Buildup of concentration can be rapid, but the rate of buildup is a function of the current. In times greater than 10^{-2} sec, concentration is almost solely responsible for observed polarizations.

The temperature coefficient of the observed polarizations can also serve to identify the source of irreversibility. Processes with concentration control have heats of activation of about 2 to 5 kcal/mole, while processes with activation control generally give heats of activation greater than 5 kcal/mole.

H. Failure Analyses

The study of the integrated structure and character of discharged and aged cells provides insight into the causes for battery failure and also identifies the proper direction for improving battery performance. Thus the "post mortem" analysis constitutes one of the most important part of any battery research and development program.

Each battery system has its principal chemical system but also contains a myriad of minor chemical constituents and components. A wide background of the basic chemistry involving all the components and their interaction is essential to fully realize the benefits of the post mortem. The influence of small amounts of impurities cannot be overlooked. The combination of deductive reasoning with an observant investigation provides the basic requirements for successful post mortem.

Suggestion for Post Mortem Procedures

1. Examine the outside of the cell to locate bulges, perforation, possible sources for leakage, and so on.
2. Dissect or open up the cell to reveal the electrode structure, etc., with a minimum of disturbance to the cell components.
3. Examine the anode for:

   Surface appearance and presence of compounds;
   Uniformity of utilization, location of reaction products;
   Integrity of surface;
   Corrosion, pitting, and so on.

4. Examine the cathode for:

   General mechanical appearance, that is, hard, mushy, and the like;
   Presence of insoluble reaction products;
   Change in structure; clogging by precipitation;
   Decreased or increased fluid content;
   Conductivity of matrix particle-to-particle contact;

5. Examine the separator for:

Structural integrity;
Chemical or mechanical attack;
Clogging up or precipitation of compounds;
Fluid content - drying out or increased fluid;
Presence of conductive particles such as carbon from the cathode mix;

General

Type of packaging;
Structure of seal - its integrity;
Cause of failure, anode, cathode, and so on;
Adherence of electroplating;
Can corrosion, adherence of electroplate.

I. Calculations of Performance

1. Current Distribution

Problems of current distribution arise in all practical battery systems. Maximum performance requires a uniform current distribution. Many excellent discussions of this problem exist in the literature for both solid and porous electrodes. The work of Shephard (67) on planar electrodes deserves special comment because of its simplicity. The terms S_h defines the uneven current distribution

$$S_h = \left(\frac{\rho_{anode} + \rho_{cathode}}{Re + Rp}\right) \quad (23)$$

where ρ_{anode} and $\rho_{cathode}$ are the resistivities of the anode and cathode, respectively, Rp is the resistance of the electrode reaction, and Re is the resistance of the electrolyte between the electrodes. Values of S_h larger than 1.2 indicate a significant departure from uniformity of current distribution on the surface of an electrode and reduced overall electrode performance.

Shephard (67) also points out the effect of construction and electrode configuration on current uniformity. The large exchange current desired of most battery electrodes for overally good performance works against the uniformity of current distribution for both solid and porous electrodes. In actual practice the properties of the electrode must be optimized for a given performance standard. In some instances, electrodes with lower exchange currents may be preferred over electrodes with superior kinetic characteristics.

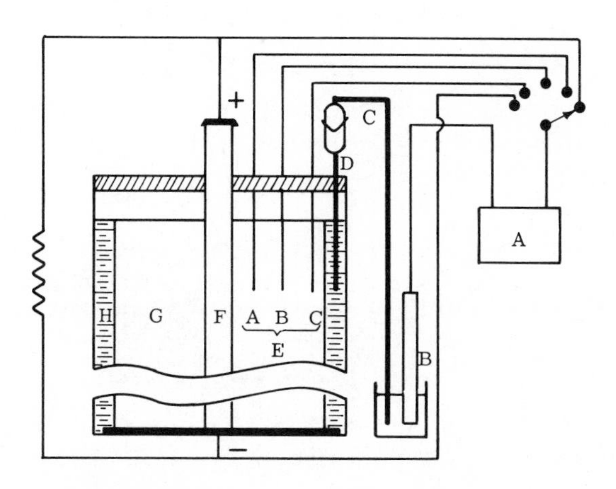

A Potentiometer
B Saturated Calomel Electrode
C Saturated KCl Bridge
D Glass Capillary Filled with Paste
E Small Carbon Electrodes or S.C.E.
F Carbon Rod
G Bobbin
H Paste Layer

Fig. 39. Schematic diagram test method for studying Leclanché-type Bobbin cells (55).

Reference electrodes and sectioning of partially discharged electrodes confirm the applicability of theoretical calculations to practical electrode situations.

For the Leclanché round cell, Hirai and Fukuda (55) developed a method for separating ionic and electronic components as well as concentration, activation, and ohmic polarizations. Figures 39 and 40 depict their experimental approach. The experimental technique also gives information on the current distribution in the MnO_2-carbon matrix. As the cell discharge, the reference and carbon electrodes reveal the dynamics of the electrode behavior. Using this technique, current

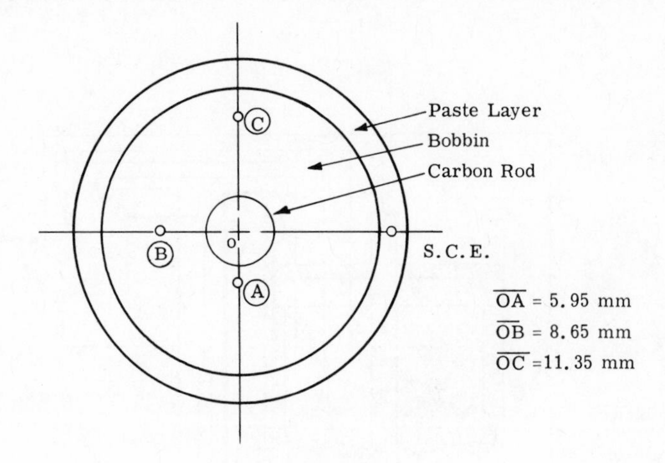

Fig. 40. Position of the auxiliary electrodes for studying D cells as described in Ref. 55.

interruption studies allow the separation of the ohmic from activation and concentration polarization, as has been described in Section III G.

Several investigators (68, 69) have calculated the current distribution in battery electrodes. These calculations serve to point out the effect of physical and electrochemical parameters on overall cell performance.

2. Discharge Battery Performance Predictions

One often needs the ability to calculate or to predict battery performance on a given test rather than to measure it directly. The discharge curves are frequently given for a fixed set of discharge conditions. The ability to calculate the performance for any operating condition saves considerable experimental effort. Several equations predict performance over a wide range, based on relatively meager amounts of data from several discharges at different rates. The equation developed by Shephard (67) gives excellent results, but requires the determination of several adjustable parameters,

$$E = E_s - k\left(\frac{Q}{Q - it}\right)i - Ni - Ae^{B\ it/-\ Q} \qquad (24)$$

where k, Q, A, and B are constants, E_s is the open circuit voltage, i is the current, and t is the time on discharge.

The empirical equation by Selim and Bro (70) also gives good results, especially for miniature cells,

$$q = q_o \frac{\tanh\ (i/R)^{1/A}}{(i/R)^{1/A}} \qquad (25)$$

where q and q_o are the capacity and maximum capacity of the battery, respectively, and A and R. are constants.

These equations permit reporting experimental data in a very condensed form.

IV. FUEL CELL TECHNIQUES

A. Introduction

Berger (71) and Liebhafsky and Cairns (72) have summarized the results and experimental techniques of the extensive fuel cell studies of the 1950 to 1960 era. The reader should consult these works for greater detail of the experimental methods for studying gas electrodes. Many methods for measuring and interpreting electrode polarization as well as for determining the physical properties of electrodes described in Sections II and III apply directly to the study of gas electrodes.

B. Gas Electrode Preparation

The patent literature contains a multitude of fuel cell electrode preparations. Most modern electrodes utilize Teflon as a binder for the catalyst and its supporting material and as a hydrophilic agent to control the gas-liquid interface. Several typical thin electrode constructions follow below as examples commonly encountered in modern fuel cell technology.

Niedrach and Alford (73) describe a convenient method for preparing gas electrodes in order to evaluate various catalysts, electrode substrates, etc. A Teflon suspension (DuPont's Teflon-30) serves to form the binding agent and to wetproof the electrode. The spraying of a diluted Teflon-30 suspension (7 volumes of water) with an air brush forms a Teflon film onto a heated aluminum foil (180 to 150° C). The film was sintered at 350° C for about 1 min between the open jaws (1/8 in.) of a press with heated platens.

Platinum black or other catalyst with a diluted Teflon-30 suspension was mixed to form a creamy slurry and spread over the Teflon-coated foil as uniformly as possible with a spatula, then dried on a hot plate at 250-350° C. A screen (platinum, nickel, silver, etc.) with suitable tabs was then centered on the spread 40 to 50-mesh screen with 5 to 8-mil wire.

A second spread of the Teflon-30 catalyst slurry was then placed on an aluminum foil, dried and heated to 250 to 350° C to dispel the nonionic wetting agent. The sheet of aluminum with the catalyst and screen is then combined with the other Teflon-coated catalyst to form a sandwich with the screen in the center.

The sandwich with the two aluminum foils facing outward is placed between two ferro-type plates and placed on the bed of a heated press. The electrode is then sintered at 375° C for 2 min under pressures of 1800 to 3000 psi.

Following that treatment, the aluminum foil is dissolved in warm 20% NaOH. The electrode is rinsed with water and air dried.

This technique may be used to study and evaluate various catalysts, the effect of amount Teflon and catalyst, and the ratio of binder to catalyst, sintering temperature and pressure, and so on.

A simple modification (74) of this technique involves combining the catalyst with a support, for example, active carbon. The carbon may be catalyzed before or after electrode construction. The slurry of Teflon-30, support, and catalyst is worked into a dough-like consistency by warming slightly, or with an admixture of isopropyl alcohol or other similar agent, to coagulate the Teflon suspension. A typical basic mix consists of approximately

Carbon	11 g
Ethyl alcohol	10 to 60 ml
2% Polyvinyl alcohol or distilled water	50 ml
Teflon 30-B	15 ml

Combine sufficient ethyl alcohol to the carbon to form a wet but not runny paste. Then add and mix thoroughly the 50 ml of 2% PVA solution. For a low-activity carbon, substitute distilled water for the PVA solution. Mix in 15 ml of Teflon 30-B emulsion and continue stirring rigorously. Heat if necessary to coagulate the Teflon emulsion. Coagulation is complete when the carbon-bearing phase pulls free from the container walls leaving a clear liquid. Pour off the excess liquid and reduce the liquid content by working the plastic mix into paper towels to the total weight 50 to 80 g depending on the fluffiness of the carbon. Mixes that are too dry tend to crumble and do not knit well. The doughy mixture is then uniformly worked into the screen or expanded metal with a spatula and pressed at about 4000 psi between the plates of a press. The electrode, which is then sintered at 375° C fo 2 min to sinter the Teflon, possesses good strength. The Teflon content controls the wetproofness and wetting characteristics.

Vogel et al. (75) describe a similar method wherein the catalyst and/or support was sprayed rather than molded. In most preparations, care should be exercised to control the particle size of the catalyst and support.

C. Gas Electrode Characteristics

The techniques to determine surface area, and so on described in the preceeding section b are all applicable to gas electrodes. In addition, characteristics such as permeability, porosity, and pore size, are especially important in defining the nature of the structure of a gas electrode. The permeability of a gas electrode constitutes an essential characteristic of the electrode and determines the upper limit for electrode operations with respect to gas transport. As gas electrodes become thinner, the need to determine the gas permeability lessens.

The flow rate of a gas through a porous medium is given by

$$V = K\Delta P \tag{26}$$

where V is the mass flow rate, ΔP is the pressure drop through the medium, and K is the permeability constant. Assuming nonturbulent flow and a nonadsorbing gas, the constant K may be defined

$$K = \frac{\beta}{\eta}\bar{p} + k\bar{v} \tag{27}$$

The first term represents viscous flow, and the second term represents the contribution of slip flow or Knudsen flow, where β and K are constants, η is the gas viscosity, $\bar{p}$ is the average pressure, and $\bar{v}$ is the mean molecular velocity. Equation 27 may be rewritten as

$$K = a\,\bar{p} + b \tag{28}$$

Plots of K against average pressure should yield a straight line whose y intercept represents the portion resulting from slip or Knudsen flow. While values of a and b may be calculated, comparisons of two different gases show the applicability of the equations. For oxygen and helium flow,

$$a_{O_2} = a_{He}\left(\frac{\eta_{He}}{\eta_{O_2}}\right) \tag{29}$$

$$b_{O_2} = b_{He}\left(\frac{M_{He}}{M_{O_2}}\right)^{1/2} \tag{30}$$

Figure 41 shows results from flow measurements. If an adsorbable gas is present, surface diffusion may contribute. If it is, the experimental value of K is greater than those calculated from a and b of a nonadsorbing gas. The good agreement between H_e and O_2 in Fig. 41 precludes a surface adsorption as an important contributor. Knudsen flow constitutes a sizable portion of the total flow through the particular porous medium.

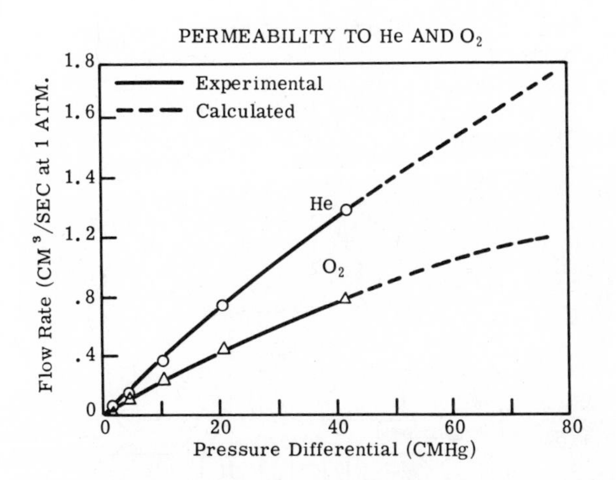

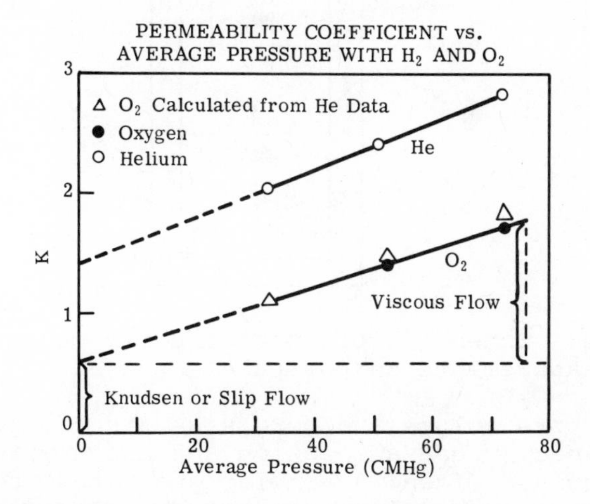

Fig. 41. Typical results of permeability versus gas pressure from flow measurements on porous electrode materials (76).

The apparatus for measuring permeability is shown in Fig. 42. The sample is clamped in place and gas flowed from one side to the other by means of a diaphragm or other suitable gas pump for reduced pressures and clean compressed gases for higher pressures. A monometer measures the pressure gradient across the sample. Mercury or dibutylphthalate may be used, depending on the magnitude of the pressure gradient. The gas volumes are measured with burettes or gas meters. The flow rates may also be determined by calibrated flow meters.

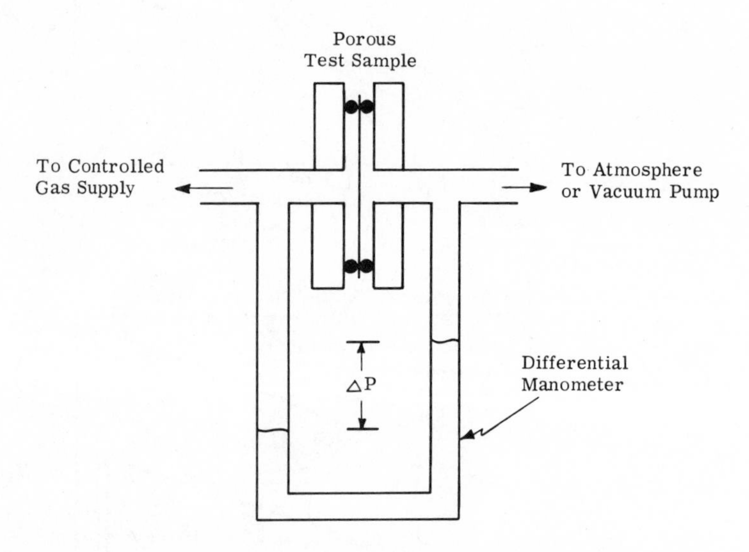

Fig. 42. Schematic of apparatus for measuring gas flow measurements on porous electrodes.

The pore volume and pore size distribution may be determined by gas adsorption and mercury intrusion techniques. Often, a simpler quick test to determine the average pore size is required. The average and maximum pore size may be conveniently determined using the ASTM standard method, E-128-61 (38) described in the preceding section. This method relies on the wetting and capillary forces of the porous matrix. Modification of the method finds use to measure gas-bubble penetration of a separator matrix. Here, the term "bubble pressure" is used to define the minimum pressure to force gas through the matrix.

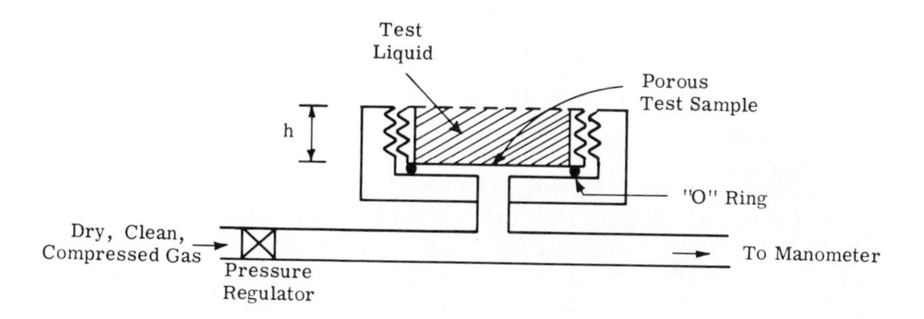

Fig. 43. Schematic of apparatus for measuring porosity and pore size of porous materials.

Figure 43 depicts a simple convenient means to determine the maximum and average pore diameter as well as permeability (38). The porous material is immersed in a test liquid that wets the materials, for example, isopropyl alcohol, and air pressure is applied until the first bubble appears through a fixed height of the liquid. The maximum pore diameter is calculated from the surface tension of the test:

$$D + 30\ \gamma/p \qquad (31)$$

where D is the pore diameter in micrometers, γ is the surface tension of the test liquid in dynes/cm, and p is the air pressure in mmHg.

Continue to increase the air pressure until bubbles appear uniformly over the surface. This pressure value determines the average pore diameter, which is also calculated using Eq. 31. With a little practice the method yields reproducible results that compare favorably with the results from the more complex and time-consuming mercury intrusion technique.

This test may also be used to determine the bubble pressure for the separator in matrix fuel cells or in sealed batteries.

D. Electrochemical Characterization of Gas Electrodes

The testing of fuel cell electrodes follows closely the technique for testing for activity of battery material (described in Section III). A simple diagram of a test cell is found in Fig. 44. The setup can be used singly as shown in the diagram or with many cells in combination for multiple testing. The cell finds application for both short-term polarization studies and for long-term life tests. In the multiple-cell configuration with a circulating electrolyte, some provision must be made to restrict intercell electrolyte connection (such as a dripping or long narrow path).

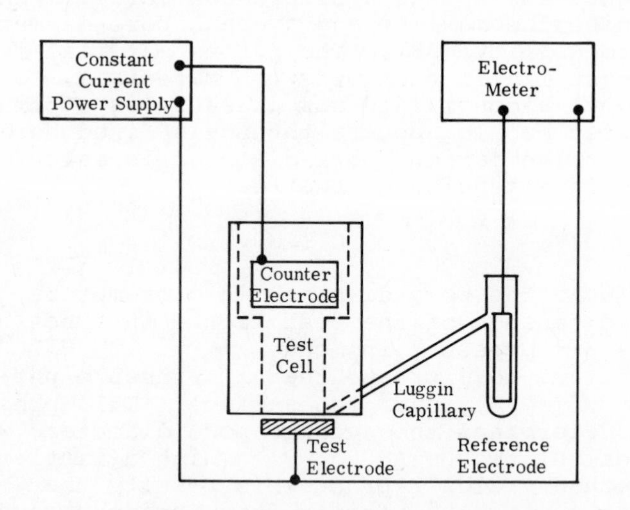

Fig. 44. Block diagram of a single electrode test setup (77).

Figures 45 and 46 show an expanded version of the electrode holders for the test cell. The test electrode is mounted horizontally so as to minimize the pressure difference along the electrode that arises with vertically oriented gas electrodes. The gas pressure is adjusted so that there is bubbling through the electrode. The electrode holder is held in place with a jaw clamp.

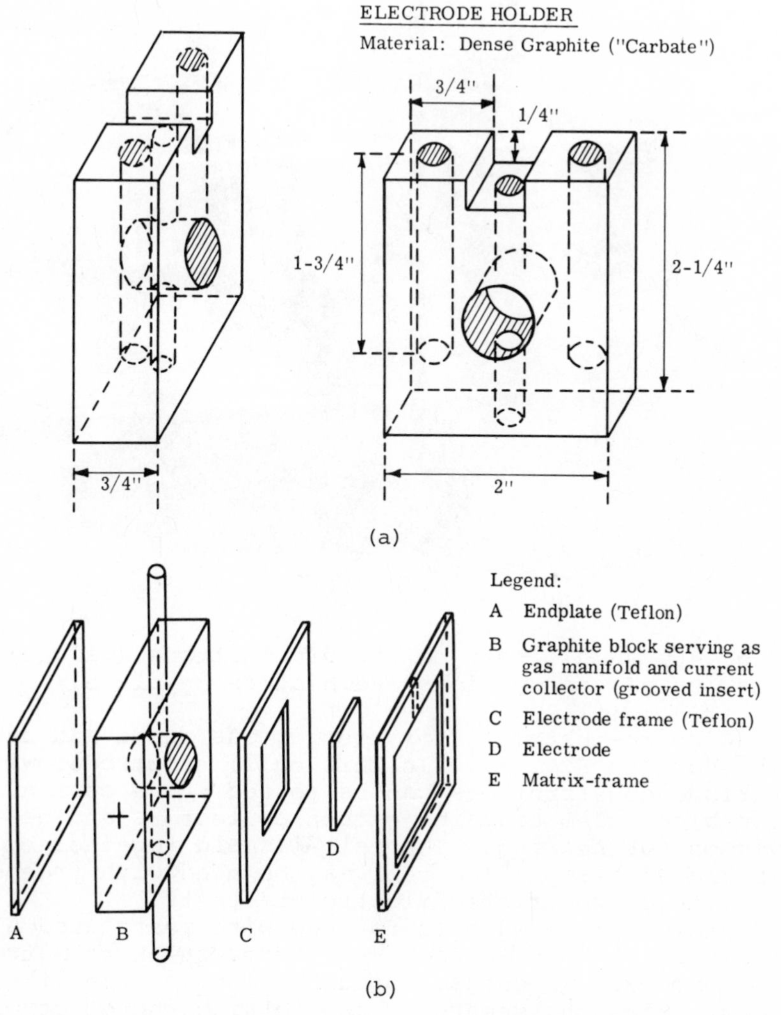

Fig. 45 (a) Quick-test cell. (b) Arrangement of a quick-test half cell unit for studying gas electrodes. (may be used with matrix or flowing electrolyte.).

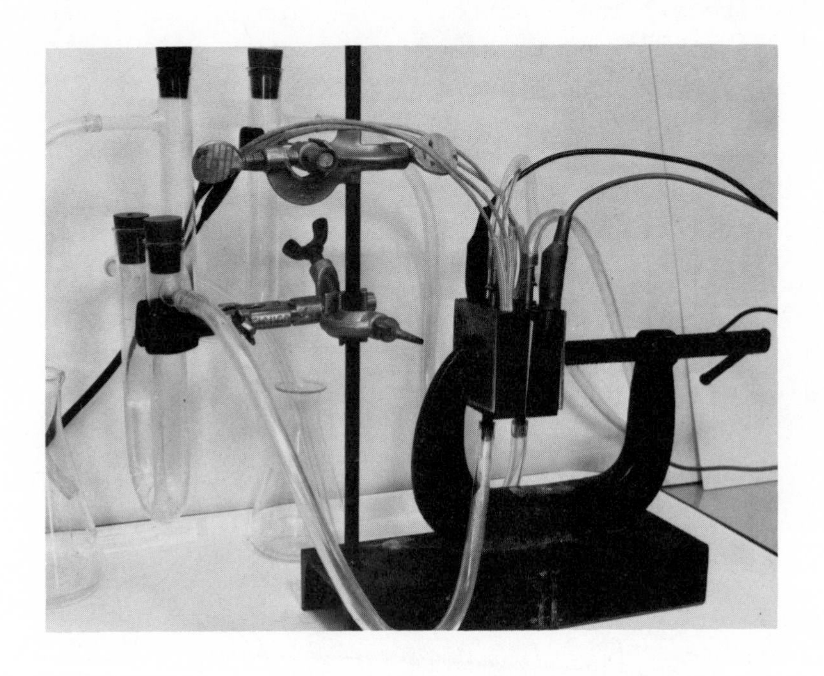

Fig. 46. Small test cells for electrode evaluation showing gas connections and heaters in place.

A quick-test cell for testing gas anode and cathode is shown in Fig. 47. In this cell, electrodes with various constructions can be tested. The central Teflon block also contains carbon collectors and gas provision for heating. The cell is held together by an insulated clamp. The cell may be used with free electrolyte or restricted electrolyte tests.

Another type of cell for use with restricted electrolyte is shown in Fig. 48. Here, an inert matrix, for example, asbestos, porous Teflon, or the like, c constitutes the separator and retains the electrolyte in its pores. The cell may be constructed of graphite, gold-plated aluminum, or other corrosion-resistant materials, depending on the cell environment.

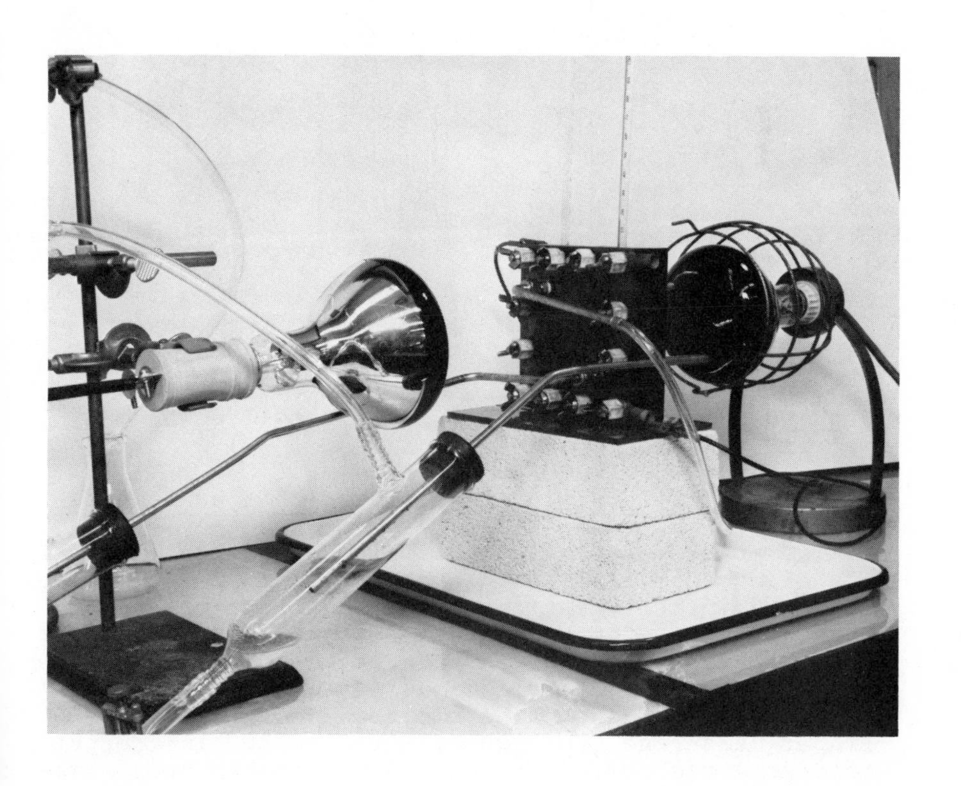

Fig. 47. Cells (3 x 3 in.) as used for life testing. Similar in construction to Fig. 45, but larger in size.

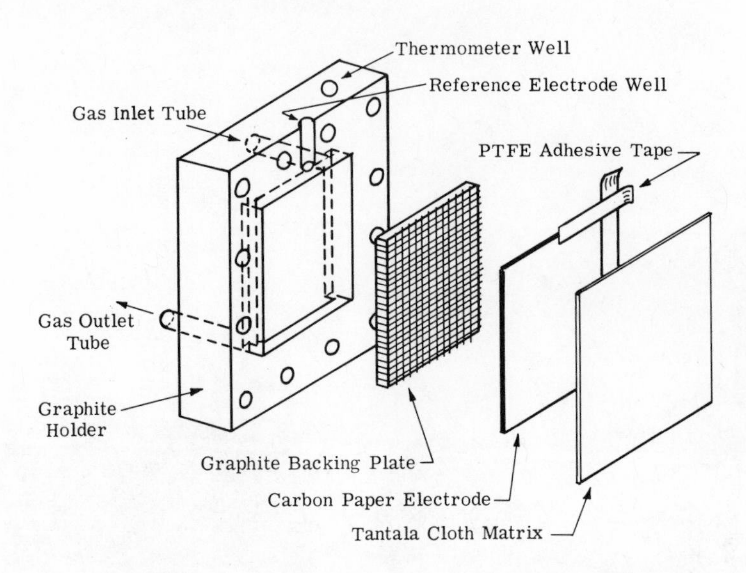

Fig. 48. Arrangement of large test cell using a matrix separator.

E. Interpretation of Polarization Measurements

The measurement of polarization characteristics of gas electrodes follows that described in Section III. As noted in the Equation 23 the resistive component governs to a large extent the current distribution, local heating, and uneven polarization over the electrode surface. Good electrode design minimizes the ohmic contribution to the electrode polarization. Proper measuring techniques, for example, interruption to remove the ohmic component from experimental measurements, should also be used to ensure that the gas electrode is also corrected for gas polarization.

To determine the real gas pressure at the electrode electrolyte interface requires a knowledge of the pressure drop through the electrode structure. Kordesch (78) reported a method for evaluating electrode performance independent of the gas concentration polarization and in terms of the relative value of a characteristic constant. Equation 32 gives the limiting current, i_{lm}, for an electrode:

$$i_{lm} = -K \log (1-p) \quad (32)$$

where p is the inert gas fraction, and K is a constant. Equation 32 permits the direct evaluation of the characteristic constant K, the limiting current for pure gas. The larger values of the constant reflect better electrode performance. Figure 49 and 50 depict typical results for the hydrogen electrode, although the equation applied to the performance of all gas electrodes. The evaluation of the characteristic constant K becomes easier at low percentage gas mixtures.

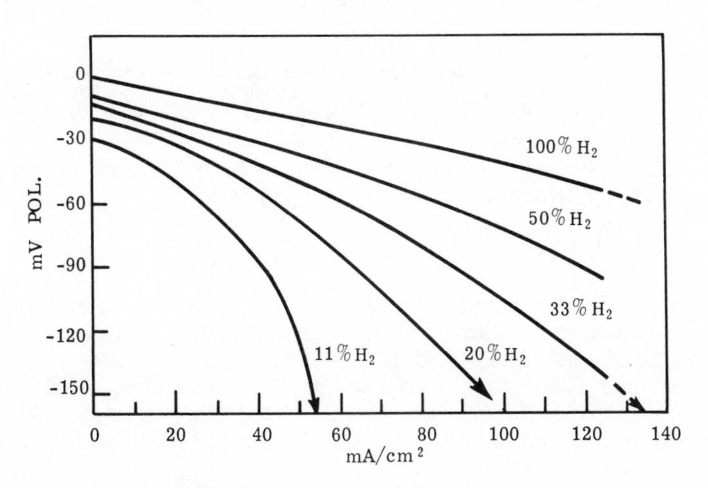

Fig. 49. Experimental data of hydrogen anode polarization in various H_2-N_2 mixtures (78).

Qualitatively, the performance of a gas electrode at set current, say, 100 mA/cm^2, with varying partial pressures indicates the overall performance of the electrode. Good gas electrodes follow approximately the Nernst equation for pressure changes. Poor electrodes exhibit excessive polarization, especially at low partial pressures.

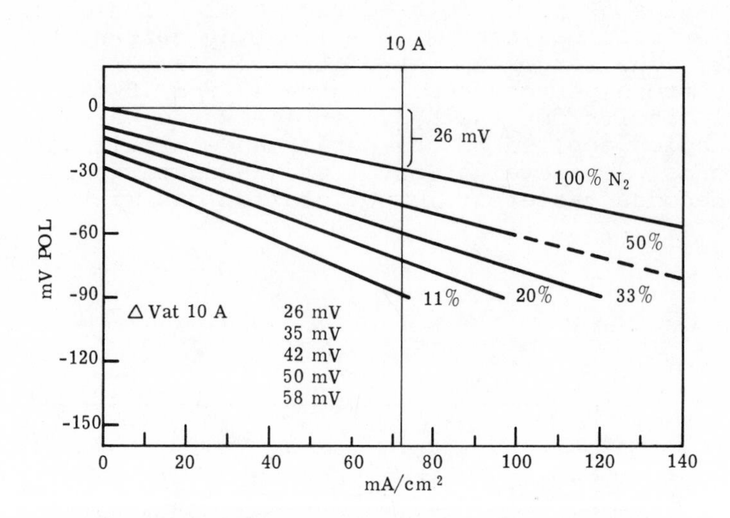

Fig. 50. Polarization of hydrogen electrode as a function of current density (78).

REFERENCES

1. U. S. National Bureau of Standards, Selected Values of Chemical Thermodynamic Properties, Circular 500, 1952.
2. W. M. Latimer, Oxidation Potentials, 2nd ed., Prentice-Hall, Englewood Cliffs, N.J., 1952.
3. S. B. Greene and N. D. Greene, Electrochem. Tech. 1, 276 (1963).
4. M. Bonnemay, G. Bronoël, E. Levart, and G. Peslerbe J. Electrochem. Soc. 111, 265 (1964).
5. M. Bonnemay, G. Bronoël, and E. Levart, Electrochim. Acta 9, 727 (1964).
6. Radiography in Modern Industry, 2nd ed., Eastman Kodak, Rochester, New York, 1957.
7. H. Berger, Neutron Radiography Methods, Capability and Applications, Elsevier, New York, 1965.

8. N. Hackerman and R. A. Powers, J. Phys. Chem. 57, 139 (1953); N. Hackerman and S. J. Stevens, J. Phys. Chem. 58, 904 (1954).
9. J. O'M Bockris and D. A. J. Swinkels, J. Electrochem. Soc. 111, 736 (1964).
10. G. A. Boyd, Autoradiography in Biology and Medicine, Academic Press, New York, 1955.
11. W. A. Roiter, E. S. Polujan, and W. A. Juza, Acta Physiochim. USSR 10, 845 (1939).
12. S. Tudor, A. Weisstuch, and S. H. Davang, Electrochem. Tech. 3, 90 (1965).
13. D. L. Kantro, S. Brunauer, and C. H. Weise, Adv. Chem. Ser. 33, 199 (1961).
14. J. Hagymassy, Jr., and S. Brunauer, J. Colloid Interface Sci. 33, 317 (1970).
15. A. Kozawa, J. Electrochem. Soc. 106, 552 (1959).
16. A. Kozawa, J. Inorg. Nucl. Chem. 21, 315 (1961).
17. V. A. Sljaka and W. H. Pie, Jr., U. S. Pat. 3,229,507.
18. C. Wagner, J. Electrochem. Soc. 97, 71 (1950).
19. R. J. Brodd and N. Hackerman, J. Electrochem. Soc. 104, 704 (1957).
20. D. Berndt, Electrochim. Acta 10, 1067 (1965).
21. R. Osteryoung, G. Laura, and F. Anson, J. Electrochem. Soc. 110, 926 (1963).
22. M. A. Short and P. L. Walker Jr., Carbon 1, 3 (1963).
23. M. L. Kronenberg, J. Electroanal. Chem. 24, 357 (1970).
24. H. B. Morley and F. E. Wetmore, Can. J. Chem. 34, 359 (1956).
25. N. C. Cahoon, Trans. Electrochem. Soc. 92, 159 (1948).
26. H. F. McMurdie, D. N. Craig, and G. W. Vinal, Trans. Electrochem. Soc. 90, 509 (1946).
27. P. Benson, W. B. Price, and T. L. Tye, Electrochem. Tech. 5, 517 (1967).
28. A. K. Covington, T. Cressey, B. G. Lever, and H. R. Thirsk, Trans. Faraday Soc. 58, 1975 (1962).
29. N. C. Cahoon, Electrochem. Tech. 3, 3 (1965).
30. D. J. G. Ives and G. J. Janz, Reference Electrodes, Academic Press, New York, 1961.
31. J. P. Hoare and S. Schuldiner, J. Phys. Chem. 61, 399 (1957).
32. J. P. Hoare, G. M. Eng. J. 9 (1), 14 (1962).

33. R. Piontelli, G. Bianchi, and R. Aletti, Z. Elektrochem. 56, 86 (1952).
34. S. Barnartt, J. Electrochem. Soc. 99, 549 (1952); 108, 102 (1961).
35. M. Eisenberg, C. W. Tobias, and C. R. Wilke, J. Electrochem. Soc. 102, 415 (1955).
36. R. Glicksman and C. K. Morehouse, J. Electrochem. Soc. 103, 149 (1956).
37. K. J. Euler and L. Horn, Elektrotech. Z. 16, 566 (1960).
38. ASTM Test E-128-61, 1971 Annual Book of ASTM Standards, Part 30, American Society of Testing and Materials, Philadelphia, Pa., 1971.
39. J. C. Cessna, Corrosion 27, 244 (1971).
40. R. J. Brodd, J. Electrochem. Soc. 106, 471 (1959).
41. A. Tvarusko, J. Electrochem. Soc. 109, 557 (1962).
42. J. Geard, G. Berbier, and J. P. Gabano, in Batteries 2, D. Collins, Ed., Pergamon Press, Oxford, 1965, pp. 219-232.
43. K. J. Euler and K. Dehmelt, Z. Elektrochem. 61, 1200 (1957).
44. R. J. Brodd and H. J. DeWane, J. Electrochem. Soc. 110, 1091 (1963).
45. R. J. Brodd and H. J. DeWane, Nat. Bur. Std. (U. S.) Technical Note 190, 1963.
46. K. J. Euler, Mettaloberflache angew. Elektrochem. 26, 257 (1972).
47. K. J. Euler, Electrochim. Acta 4, 27 (1961).
48. K. J. Euler, Z. angew. Phys. 31, 62 (1971).
49. T. Hirai, M. Fukuda, and Y. Amano, Nat. Tech. Report 8, 160 (1962).
50. K. J. Euler and W. Nonnemacher, Electrochim. Acta 2, 268 (1960).
51. K. V. Kordesch and A. Marko, J. Electrochem. Soc. 107, 480 (1960).
52. L. W. Niedrach and M. Tochner, Electrochem. Tech. 5, 220 (1967).
53. G. Schreiber and W. Schwarz, Electrochim. Acta 11, 211 (1966).
54. A. Kozawa and R. A. Powers, Electrochem. Tech. 5, 535 (1967).
54a. A. Kozawa and J. F. Yeager, J. Electrochem. Soc. 112, 959 (1965). For RuO_2, see Extended Abstract of ECS Boston Meeting, 1973.
55. T. Hirai and M. Fukuda, in Electrochemistry of

Manganese Dioxide, S. Yoshiyawa, K. Takahashi, and A. Kozawa, Eds., U. S. Branch Office of The Electrochemical Society of Japan, Cleveland, 1971, Vol. 1, Chap. 7.
56. N. C. Cahoon, J. Electrochem. Soc. 99, 343 (1952).
57. K. Kornfeil, J. Electrochem. Soc. 106, 1062 (1959).
58. K. Appelt and E. Forecki, Electrochim. Acta 8, 639 (1963).
59. H. Schweigart, South African CSIR Report, Dec. 1969.
60. G. S. Bell and J. Bauer, Electrochim. Acta 14, 453 (1969).
61. R. Huber and J. Kandler, Electrochim. Acta 8 265 (1963).
62. L. Balewski and J. P. Brenet, Electrochem. Tech. 5, 527 (1967).
62a. G. S. Bell, Electrochem. Tech. 5, 512 (1967).
63. N. Cahoon and G. W. Heise, The Primary Battery, Wiley-Interscience, New York, 1976.
63a. K. V. Kordesch, Ed., Batteries, Marcel Decker, New York, 1974, Vol. 1.
64. A. Ralston and H. S. Will, Mathematical Methods for Digital Computers, John Wiley, New York, 1960.
65. American National Consumer Standard for Dry Cell Batteries, Publication C18.2 - 1971, American National Standards Institute, Inc., 1971.
66. Primary Cells and Batteries, International Electrochemical Commission, Publication 86-1, 3d ed., 1971.
67. C. M. Shephard, J. Electrochem. Soc. 112, 657 (1965).
68. K. J. Euler, R. Ludwig, and K. N. Muller, Electrochim. Acta 9, 495 (1964).
69. E. A. Grens II and C. W. Tobias, Bu. Bunseges. Phys. Chem. 68, 236 (1964).
70. R. Selim and P. Bro, J. Electrochem. Soc. 118, 829 (1971).
71. C. Berger, Handbook of Fuel Cell Technology, Prentice-Hall, Inc. Englewood Cliffs, J.J., 1968.
72. H. A. Liebhafsky and E. J. Cairns, Fuel Cells and Fuel Batteries, John Wiley, New York, 1968.
73. J. L. Niedrach and H. A. Alford, J. Electrochem. Soc. 112, 117 (1965).
74. A. Kozawa, V. E. Zilionis, and R. J. Brodd, NIH Report.

References

75. W. Bogel, J. Lundquist, and A. Bradford, Electrochim. Acta 17, 1735 (1972).
76. M. B. Clark, unpublished results.
77. M. W. Reed and R. J. Brodd, Carbon 3, 241 (1965.
78. K. V. Kordesch, Electrochim. Acta 16, 597 (1971)
79. A. J. Salkind, The Measurement of Surface Area and Porosity, in Techniques of Electrochemistry, Vol. 1.
80. R. Bates, Measurement of Reversible Electrode Potentials in Techniques of Electrochemistry, Vol. 1, E. Yeager and A. Salkind, editors, Wiley-Interscience, N.Y. 1972.

V. A REVIEW OF SECONDARY BATTERIES AND EVALUATION TECHNIQUES*

Gerald Halpert

Goddard Space Flight Center
Greenbelt, Maryland

*Special additional material supplied by Mr. James Doe and Dr. A. J. Salkind, ESB Technology Center.

I. INTRODUCTION

The objective of this chapter is to provide the reader with a working knowledge of the properties, characteristics, and terminology of the most frequently used secondary (rechargeable) electrochemical batteries as well as the electrochemical techniques used in their testing and characterization. Included in the discussion are the alkaline electrolyte systems; nickel-iron, nickel-cadmium, nickel-zinc, silver-cadmium, as well as silver-zinc; and the acid-electrolyte lead-lead dioxide system (commonly called the lead-acid battery), the most used of all battery systems. In addition, there is a brief discussion of some new secondary battery systems, including the metal-hydrogen and metal-air systems and those that use molten-salt electrolytes. To simplify the discussion, each of these systems is described in terms of a single electrochemical cell, recognizing that a multiplicity of electrically connected cells constitutes a battery.

Many of the chemical, electrochemical, and physical properties associated with secondary electrochemical system operations are related to those utilized in the primary battery system described by Brodd and Kozawa (1). A major difference between primary and secondary battery systems is in the nature of the electrochemically active material. Primary cells are discarded when they can no longer deliver the required energy, whereas secondary cells can be recharged from an external source to their original fully charged condition as many as several thousand times, depending on use and conditions. In order to accomplish this, the secondary cell must utilize reactions that approach 100% of electrochemical reversibility; the more reversible the reactions, the longer the operating life of the cell. As a cell is operated, its efficiency gradually decreases and ultimately it is no longer capable of being recharged. The cell is then said to have "failed." Loss of porosity in the active mass, creation of electrically nonconductive species, corrosion of conductive grids and connectors, and reduction of the particle size of the active material are mechanisms that lead to such failure.

The characteristics of each of the electrochemical cells previously mentioned is discussed from a system point of view. In addition, the composition and role of each of the four major components, positive electrode, negative electrode, separator, and electrolyte, is considered.

II. ELECTRICAL OPERATION AND THERMODYNAMICS

A. Charge and Discharge

The secondary cell takes advantage of substantially reversible chemical/electrochemical reactions to provide electrical energy from stored chemical energy. During the discharge phase the cell acts as a voltaic cell in which the difference in electrochemical potential serves as the driving force to supply electrons through the load. The electrons exit from the anode where oxidation occurs through the load to the cathode where reduction takes place. When the potential difference reaches zero volts, the source of electrons is exhausted and the cell needs to be recharged.

During the charge process the secondary cell behaves as an electrolysis cell in which an external power source provides the necessary electrons to convert the electrical energy into stored chemical energy. The anode that served as a source of electrons during discharge becomes the cathode during charge and accepts the electrons. The cathode during discharge becomes the anode during charge. Although the roles of the electrodes depend on whether the process is charge or discharge, the positive electrode is identified as always being the electrode connected to the positive lead of the voltmeter. Similarly, the negative electrode is connected to the negative lead. The combined charge/discharge sequence is referred to as a cycle. The electrical arrangement used for charge and discharge is given in Fig. 1.

B. Free Energy

The calculations of energy density require the determination of the free energy, given by

$$\Delta G = -nFE \tag{1}$$

where n is the number of electrons transferred in the reaction, F is the Faraday constant, given as 23,060 calories per volt equivalent, and E is the potential

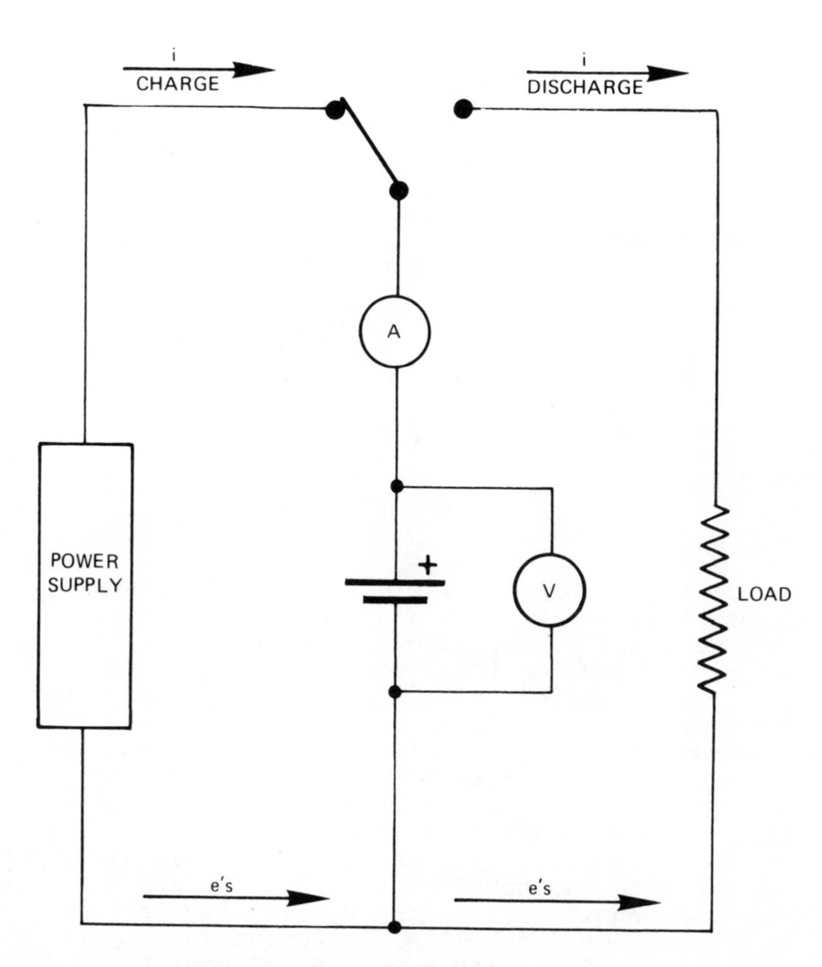

Fig. 1. Electrical arrangement for charge and discharge.

difference between the two electrode reactions. The discharge of a cell produces the desired positive value for E and thus a negative free energy representing a spontaneous reaction. The potential difference between the electrodes when written as charge reactions is negative. This results in a positive free energy indicative of a nonspontaneous reaction. Therefore, an external power supply is required to return the discharged active material to the charged condition.

C. Efficiency

Although electrochemical efficiency is given as

$$E_{eff} = \frac{\int Ei\ dt}{n\Delta G(4.186)} \tag{2}$$

storage battery efficiency is generally expressed in terms of ampere-hour or watt-hour efficiency. The ampere-hour efficiency is given as

$$C_{eff} = \frac{\int_{DIS} i\ dt}{\int_{CHG} i\ dt} \tag{3}$$

and watt-hour efficiency is expressed as

$$P_{eff} = \frac{\int_{DIS} Ei\ dt}{\int_{CHG} Ei\ dt} \tag{4}$$

where E is the potential difference (voltage), i the current, t the time, n the number of electrons in the reaction and ΔG the free energy from Eq. 1 for the specific reaction involved.

If the chemical reaction were ideal and reversible, the efficiency would be equal to 1. Actually, the value of P_{eff} is generally greater than 0.7 and that for C_{eff} greater than 0.9 at room temperature. P_{eff} is lower because of the lower voltage on discharge compared with charge.

The values for P_{eff} and C_{eff} are strongly dependent on the rate and temperature of charge and discharge and must be considered for the cycling regime employed. For most secondary cells, operation at temperatures between 0 and 25°C with a minimum of voercharge and low rates of discharge will maximize efficiency. The value for P_{eff} is also dependent on the internal resistance of the cell. For secondary batteries with strong alkaline or acid electrolytes, the solution resistance (R_{soln}) is less than 0.1 ohms. Therefore the potential drop due to the iR_{soln} loss is minimized. However, most of the internal electrical resistance of a cell is in the active material. Therefore, conductive material in the form of grids, plaques, or additives must be included in the plate structure. In addition, these elec trical supports also provide mechanical support. They help maintain uniform current distributions in the electrodes, which reduces the tendency of electrodes to warp and or shed during cycling.

D. Energy Density

Watt hours per unit weight (or per unit volume) is referred to as energy density. The most widely used watt hours per pound or per kilogram can be expressed as

$$\text{Energy density} = \frac{n\ F\ E}{\text{Weight}} \qquad (5)$$

where n is the number of electrons or equivalents involved in the reaction, F is the Faraday constant, which is given as 26.8 A hr per equivalent, and E the average voltage of the reaction on discharge.

III. THE ELECTROCHEMICAL SYSTEMS

Six secondary cell systems in large-scale commercial use are described below. The description is given in terms of the chemical and electrochemical processes occuring in each cell relating to the storage of energy. Table 1 provides a comparison of the characteristics of the six conventional systems. Although the lead-acid system constitutes the mjaority of batteries sold in the world, the use of the alkaline systems in more critical applications, such as space probes and satellites, has resulted in more sophisticated analytical and test procedures for the latter. A few of the more unfamiliar aqueous and molten salt secondary systems are also discussed below. Table 2 provides a listing of the characteristics of these systems. The approximate operating voltages and other parametric data for the more conventional electrodes are given in Table 3.

A. Lead-Acid Cell (14, 15)

The positive active material of a lead-acid cell is lead dioxide and the negative active material is sponge lead. The electrolyte ia an aqueous solution of sulfuric acid. The basic material for fabrication of both electrodes is a lead oxide paste mixed with a dilute sulfuric acid solution. The past, applied to a lead

TABLE 1

	Pb/PbO_2	Ni-Fe	Ni-Cd	Ni-Zn	Ag-Cd	Ag-Zn	Refs.
Open Circuit Potential	2.095	1.370	1.299	1.735	1.380	1.856	
Nominal* Discharge Voltage	1.95	1.20	1.25	1.65	1.18** 1.04	1.65** 1.50	
Theoretical *** Energy Density (WH/Kg at 25°C)	246	267	235	293	267 171	458 273	2, 4
Practical* range Energy Density (WH/Kg at 25°C)	17–25	17–30	20–35	44–60	50–80 =	120–130	3, 58
Practical* Energy Volume WH/dm^3 at 25°C	40-90	60-90	30-110	70-130	60-180	110-270	4
Electrolyte	30% H_2SO_4	20% KOH	25.31 KOH	40% KOH	42% KOH	40% KOH	
Discharged Positive Active Material	$PbSO_4$	$Ni(OH)_2$	$Ni(OH)_2$	$Ni(OH)_2$	Ag_2O Ag	Ag_2O Ag	
Discharged Negative Active Material	$PbSO_4$	$Fe(OH)_2$	$Cd(OH)_2$	$Zn(OH)_2$	$Cd(OH)_2$	$Zn(OH)_2$	
Estimated $/Kg	20	31	48	3	190	187	3
Estimated Cycle Life	1600	~2000	~2000 ~20000 (sealed)	300	1000	300	3, 5

*at the 2 hour rate, ambient temperature. *** without electrolyte.
**two steps.

Footnote: References to the technical literature on the subject of secondary Alkaline Batteries and Battery Materials have been compiled in two bibliographies – NASA Special Publication SP7027 and SP7044.

alloy grid and formed during the initial charge, is converted to the active material, PbO_2, on the positiv plate, and sponge lead on the negative plate. The electrochemical reaction is given as

$$Pb+PbO_2+2H_2SO_4 \underset{\text{Discharge}}{\overset{\text{Charge}}{\rightleftharpoons}} 2H_2O+2PbSO_4 \qquad (6)$$

TABLE 2

	Ni/H_2	Ag/H_2	Fe/Air	Zn/Air	Zn/Cl_2	Na/S	Refs.
Open Circuit Potential	1.358	1.398	1.278	1.645	2.12	2.08	6, 7, 11
Nominal Discharge* Voltage (2 hr rate)	1.25	1.10	0.5-0.8	1.30	1.70	1.75	8, 10, 12
Theoretical Energy Density WH/kg at 25°C	378	523	756	275	825	758	6, 8, 12, 13
Practical Energy* Density WH/kg at 25°C	60	110	40	245	66	95	7, 9, 12, 13
Electrolyte Composition	30% KOH	25% KOH	25% KOH	20% KOH	10% $ZnCl_2$	Beta alumina (sodium aluminate)	
Discharged Positive Active Material	$Ni(OH)_2$	Ag	H_2O	H_2O	$ZnCl_2$	Na_2S_3	
Discharged Negative Active Material	H_2O	H_2O	$Te(OH)_2$	$Zn(OH)_2$	$ZnCl_2$	Na_2S_3	

*at the 2 hour rate.

At both electrodes $PbSO_4$ is produced on discharge. A key limiting feature of this system is that the electrolyte, sulfuric acid, is involved in the reaction and must be present in stoichiometric quantities. The system is thermodynamically reversible and has high efficiency.

The grids on which the active material is pasted include lead-antimony alloys, lead-calcium alloys, and some pure lead variations. The advantage of the lead-calcium and pure lead grids is that they result in an increase in hydrogen overpotential and reduction of float current, which reduces water consumption. Grids are usually cast, but in some new designs they are made from expanded metal or are forged.

In another construction for lead-acid positive electrodes, the active material is enclosed in tubes of perforated plastic or fiberglass mat. In this "tubular" electrode, there is also a central conductive spine of lead. Multiple tubes are assembled with a lead cross-bar top and a lead or plastic bottom bar, to form a single electrode. Separator materials are usually made of microporous rubber, plastic impregnated paper, or sintered porous polyvinyl chloride (PVC).

TABLE 3

	Ni-Fe	Ni-Cd	Ni-Zn	Ag-Cd	Ag-Zn	Pb/PbO_2
Open Circuit Potential	1.37	1.24	1.735	1.40	1.86	2.10
Nominal* Discharge Voltage	1.20	1.25	1.65	1.18 ** 1.04	1.65 ** 1.50	1.95
Theoretical Energy Density (WH/kg at 25°C)	265	235 209 (5)	300 (3) 326 (5)	440 270 (3)	458 440	175 (5) 170 (3)
Practical Energy Density (WH/kg at 25°C)	28 17 (3)	35 20 (3)	44 59 (4)	80 50 (3)	130 120 (3)	24 17 (3)
Practical Energy Density WH/dm³ at 25°C	60-90 (5)	30-110 (5)	70-130 (5)	60-180 (5)	110-270 (5)	40-90 (5)
Electrolyte	20% KOH	25.31% KOH	35% KOH	42% KOH	40% KOH	50% $Hz SO_4$
Discharged Positive Active Material	$N_1(OH)_2$	$N_1(OH)_2$	$N_1(OH)_2$	Ag_2O Ag	Ag_2O Ag	$PbSO_4$
Discharged Negative Active Material	$Fe(OH)_2$	$Cd(OH)_2$	$Zn(OH)_2$	$Cd(OH)_2$	$Zn(OH)_2$	$PbSO_4$
Estimated $/KG	31 (3)	48 (3)	3 (4)	190	187 (3)	20 (3)
Estimated Cycle Life	~2000	~2000 ~20000 (sealed)	300 (4)	1000	300	1600 (6)

* at the 2 hour rate

B. The Alkaline Systems

The alkaline electrolyte systems have the advantage that the electrolyte is not part of the electrochemical reactions and does not have to be present in stoichiometric, quantities. Another benefit is that the electrodes can be spaced closer together, this decreases internal electrical resistance.

1. The Nickel-Iron Alkaline Cell (4, 15)

The active materials in many alkaline cell systems are fabricated into electrode structures by mechanically containing the materials in punched or porous steel tubes or pockets. The positive electrode in this cell is usually of tubular or pocket construction containing nickel hydroxide, $Ni(OH)_2$, as the uncharged active material. The negative electrode utilizes ferrous hydroxide active material and usually has a pocket type of construction. The overall cell reaction is given as

$$2NiOOH + Fe + 2H_2O \underset{\text{Charge}}{\overset{\text{Discharge}}{\rightleftharpoons}} 2\,Ni(OH)_2 + Fe(OH)_2 \quad (7)$$

The positive plate tubular construction, which originated with Edison and is still in use today, consists of alternate layers of nickel hydroxide and nickel flake tamped into nickel-plated perforated tubes which, when mounted in a conductive frame, becomes the positive plate. The nickel hydroxide is prepared by dissolving nickel sulfate also contains cobalt sulfate, which is coprecipitated with the hydroxide. The cobalt tends to increase efficiency of the positive electrode.

The ferrous hydroxide serving as the negative electrode is produced from pure iron pellets through a series of steps to a purified oxide. Recent crystallographic data suggests that it is a mixture of Fe_3O_4 and Fe. Additives include cadmium oxide, mercuric oxide, or ferrous sulfide. They tend to reduce self-discharge during open-circuit stand, increase conductivity, or act as depassivating agents. The iron oxide is converted to the hydroxide by continuous cycling in potassium hydroxide electrolyte.

The recommended electrolyte depends on the condition of the cell. For the new cells, 20% aqueous potassium hydroxide is diluted with lithium hydroxide to bring the total alkalinity to approximately 30%. After the initial cycles are performed, the electrolyte is change to 20% KOH with enough LiOH to a total alkalinity of 25%. The lithium has been shown to enhance the iron electrode.

Separation of the plates is maintained by hard rubbe hairpins between plates and by channels of hard rubber along the edge of the cell.

2. The Nickel-Cadmium Alkaline Cell (4, 15, 16, 17)

As in the nickel-iron cell, the active material of the positive electrode is nickel hydroxide. The negati electrode active material is cadmium hydroxide. The overall cell reaction is given as

$$2\ Ni\ OOH + Cd + 2H_2O \underset{\text{Charge}}{\overset{\text{Discharge}}{\rightleftharpoons}} 2Ni(OH)_2 + Cd(OH)_2 \quad (8)$$

The electrodes of the nickel-cadmium alkaline cell are, depending on application, of three different types of construction, the pocket type, the sintered plate type, and the pasted-on plastic-bound type.

In the pocket plate construction, the active materia preparation and electrode fabrication are similar to th those described previously for nickel-iron cells. However, graphite is usually used instead of nickel flake as an aid to conductivity.

In the negative pocket plate, cadmium hydroxide is the active material. It is prepared by coprecipitation of cadmium, spongy masses of iron, and sometimes nickel from a sulfate electrolyte, or comixing of cadmium as a dry powder with iron oxide of the type used in iron electrodes. Nickel and graphite are also used for improvement of conductivity, and solid rubber or latex particles to reduce agglomeration of the cadmium; agglomeration results in a loss of surface area and effective capacity.

The sintered-plate electrode construction is also based on nickel hydroxide and cadmium hydroxide active materials. However, in this type, the active materials are precipitated within the pores of a highly porous sintered nickel matrix (plaque) usually from their respective nitrates. The plaque is produced from special type of pure nickel powder sintered in a reducing furnace, which results in an approximately 80% porous sturcture. A sequence of (a) vacuum impregnation of nitrate solutions into the pores of the plaque, (b) conversion with NaOH, and (c) washing and drying is repeated several times until the required quantity of active material is realized. There are also new methods for depositing active materials in a continuous one-step process. Positive and negative electrodes also differ in thickness and pore volume depending on application.

The electrolyte used for the sintered plate cell is 31% to 34% potassium hydroxide. The pocket plate cells utilize approximately 20% aqueous KOH electrolyte. In these cells the plates are flooded in the electrolyte. The sintered plate cells can be operated flooded for high-drain-rate capability or in the semiwet (starved) condition for sealed-cell applications. In this condition the entire volume of electrolyte is immobilized in the plate and separator structures.

The sintered Ni-Cd cell utilizes a nonwoven (felt-like) nylon or polypropylene separator material. It serves two purposes in the hermetically sealed cell. In addition to physically separating the plates, it serves as a sponge that immobilizes the electrolyte so that it is not free flowing. The electrolytic path between plates is ensured because the cell pack consisting of plates and separators is maintained under compression.

A cheaper construction, which has the same electrochemistry, is one in which the active materials are contained in a plastic matrix that has a carbon additive to enhance conductivity. This construction does not result in the same long life or high performance characteristics of the sintered plate cells. Although the active material tends to shed, reasonable life is ensured because the cell pack consisting of plates and separators is maintained under compression in the cell case.

The separators in the flooded or pocket cells are hard rubber or polystyrene hairpins of the type used in the nickel-iron cells. Sheets of ribbed polystyrene with perforations to allow electrolyte access is also used.

3. The Nickel-Zinc Alkaline Cell (3, 18, 19)

This system is not at present in large-scale commercial production but has been studied for many years.

Since it offers the advantage of higher energy density, it may be important for electric vehicle applications. Early cells utilized the pocket plate construction containing $Ni(OH)_2$ as the active material. Later, sintere plates were used in their place. The negative plate was an electrolytically deposited or compressed zinc/zinc hydroxide powder. The overall cell reaction is given as

$$2NiOOH + Zn + 2H_2O \underset{\text{Charge}}{\overset{\text{Discharge}}{\rightleftharpoons}} 2Ni(OH)_2 + Zn(OH)_2 \qquad (9)$$

Recent changes in construction to the sintered plate and pressed electrodes have significantly improved cell performance. The major problems are attributed to zinc dendritic growth and shape change occurring in the negative plate.

The electrolyte used in this system is aqueous 40% KOH solution. The cells are usually flooded, but some attempts have been made to produce sealed units.

The separators for this system are combinations or layers of the type used in silver zinc or silver cadmium cells. A new approach is the use of inorganic separator materials, which appear to improve the electrolyte wetability through loosely bound water, which facilitates proton transfer.

A new construction under study vibrates the zinc electrodes during charge in order to create a moving film of the electrolyte-plate interface. This reduces shape change and dendrite formation. Such cells do not use membrane-type separators, and the zinc active material dissolves in the electrolyte during each discharge.

4. The Silver-Cadmium Alkaline Cell (4, 20)

The positive active material of this cell type consists of Ag_2O_2 and the negative active material $Cd(OH)_2$. The overall cell reaction is written as a two step reaction because the silver electrode reaction takes place in two steps:

$$Ag_2O_2 + Cd + H_2O \underset{\text{Charge}}{\overset{\text{Discharge}}{\rightleftharpoons}} Ag_2O + Cd(OH)_2 \qquad (10)$$

$$Ag_2O + Cd + H_2O \underset{\text{Charge}}{\overset{\text{Discharge}}{\rightleftharpoons}} 2Ag + Cd(OH)_2 \qquad (11)$$

The silver electrode has been prepared by slurry pasting or dry pressing of Ag_2O or Ag_2O_2 into a conductive base, followed by conversion in KOH solution. Another method consists of sintering of silver powder with pore formers on a grid, followed by electrochemical oxidation in potassium hydroxide. For secondary cells a dense layer of Ag_2O_2 is required for stronger structure and greater capacity. It is produced from a water paste of Ag_2O slurry, drying, and decomposition to metallic silver. The silver is then compressed to the grid and electrochemically oxidized to Ag_2O_2 in KOH.

The negative electrode can be a sintered plate type as in the Ni-Cd cell or it can be prepared by pasting cadmium as the oxide or hydroxide on an expanded metal conductive grid or electrodeposited on cadmium directly on a nickel screen.

The electrolyte in the silver cadmium cell is 30% to 40% KOH solution in the flooded cells and 42% KOH in the sealed-cell types.

The separator must be a compromise between retarding silver migration and maximizing the conductivity by increasing the absorption of electrolyte. Among the complex combinations of materials used are four to five layers of silver-treated cellophane wrapped around a woven nylon fabric. The fabric is held against the positive plate and keeps the cellophane from adhering to the plate. Around the negative plate is a wrap of Aldox, a treated cellulosic film that is used to maintain the structure of the plate. Irridiated membranes have also been utilized as has the new inorganic Zirconia separator.

This cell has the unusual property of being nonmagnetic and has been utilized in specific aerospace applications for that very purpose.

5. The Silver-Zinc Alkaline Cell (4, 21)

The silver positive electrode used in this cell is of the same type described above in the silver-cadmium cell. The active material of the negative electrode is zinc hydroxide. The cell reaction as with the silver cadmium cell is considered to be in two steps:

$$Ag_2O_2+Zn+H_2O \underset{\text{Charge}}{\overset{\text{Discharge}}{\rightleftharpoons}} Ag_2O+Zn(OH)_2 \tag{12}$$

$$Ag_2O+Zn+H_2O \underset{\text{Charge}}{\overset{\text{Discharge}}{\rightleftharpoons}} 2Ag+Zn(OH)_2 \tag{13}$$

The negative electrode can be produced by a compressed dry powder process, slurry or pasting method, and by electroforming. The majority of the secondary zinc electrodes are the dry powder type with zinc oxide as the active ingredient. Mercury as the oxide is used as an additive to raise the hydrogen overvoltage, improving the utilization of the electrode. The electrodes are placed in potassium hydroxide and electrolyzed as dummy electrodes to produce the active zinc mass.

The electrolyte is generally an aqueous 40% solution of KOH as in the silver cadmium electrochemical system. Usually zinc oxide is added to the electrolyte to reduce the solubility of the zinc electrode. The problem of silver migration and zinc dendrite growth within the cell during cycling limits use of this system to mainly short-term applications. It has, however, a significantly higher energy density than any of the other five conventional electrochemical systems considered in this chapter. The separator consists of several layers of membrane material such as cellophane or irradiated polyethylene and nonwoven nylon. A layer of Aldox described above is also used. Potassium titonate inorganic separators have also been used with this system.

C. Other Aqueous Electrolyte Systems

Some pertinent characteristics for these systems are tabulated in Table 2.

1. Nickel-Hydrogen (12, 22)

The nickel-hydrogen cell is a combination created by the secondary nickel hydroxide electrode commonly used in a nickel-cadmium cell and a hydrogen gas fuel cell type, diffusion electrode. The overally electrochemical reaction is

$$NiOOH+\frac{1}{2}H_2 \underset{\text{Charge}}{\overset{\text{Discharge}}{\rightleftharpoons}} Ni(OH)_2 \tag{14}$$

The positive electrode is the standard nickel-cadmium cell sintered nickel anode, which is impregnated with active nickel hydroxide. The hydrogen electrode (negative) is a sintered nickel or carbon structure, Teflon impregnated for waterproofing and catalyzed with platinum black.

The electrolyte is a 7N potassium hydroxide solution, and the separator is nonwoven nylon or polypropylene. Electrolyte distribution affected by H_2 evolution during charge and recombination during discharge is the most difficult problem with this system.

Each cell is stacked with alternating nickel hydroxide and hydrogen fuel cell electrodes insulated from each other by the separator which maintains the electrolyte between the electrodes. This stack is clamped to the desired pressure and inserted into a cylinder made of Inconel or steel. The cylinder is evacuated, electrolyte introduced, and finally hydrogen gas is used to fill the cylinder under 500 to 600 psi pressure. The capacity of the cathode is equivalent to the partial pressure of hydrogen in the cylinder. The pressure in the cell at any time is also a direct measure of the state of charge.

2. Silver-Hydrogen (10, 22, 23)

This system, analogously to the nickel-hydrogen system, utilizes the rechargeable silver electrode from the silver-zinc system as the positive and a hydrogen fuel cell electrode as the negative. The reaction is written.

$$Ag_2O_2 + 2H_2 \underset{\text{Charge}}{\overset{\text{Discharge}}{\rightleftharpoons}} 2Ag + 2H_2O \qquad (15)$$

The electrode construction has been described earlier. The important factor here, as in the nickel-hydrogen system, is the problem of electrolyte management within each cell. Water produced during charge and removed during discharge must be moved through capillary forces to maintain uniform wetness throughout the stack. The chemically inert separator must permit good conductivity and electrolyte movement and it must retard soluble silver migration. The inorganic material, potassium titanate, whose composition has been considered proprietary by the manufacturers has been used in combination with other materials.

The system has a higher energy density than its nickel counterpart but the silver electrode has not exhibited the long-lived use as the nickel electrode of the nickel-cadmium cell.

3. Iron-Air (9)

One way to achieve higher energy densities is to use oxygen in the air as one of the reactants in a cell. The iron-air cell is one such type of metal-air system that is undergoing study for use in electric vehicles. The overall electrochemical reaction is

$$2Fe+O_2+2H_2O \underset{\text{Charge}}{\overset{\text{Discharge}}{\rightleftharpoons}} 2Fe(OH)_2 \quad (16)$$

The negative iron electrode described by Lindström (9) is a sintered porous iron structure fabricated by pressing electrolytic iron powder into a nickel supporting mesh and heating to 650°C in hydrogen. Up to 65% active material utilization is possible and the electrode porosity is 60% to 70%. The air electrode consists of three layers: (a) a fine-pore nickel layer which, when flooded with electrolyte, acts as the current collector, gas-bubble barrier, and reacharging electrode; (b) a partially wetproofed middle layer containing silverized graphite as an oxygen discharge catalyst which does not participate in the charging; and (c) a PTFE barrier layer on the gas side for final wetproofing, which is highly permeable to gas and increases cathode life.

The air electrode loses its ability to reduce the oxygen from the air and its ability to anodically liberate oxygen as it is put through repeated charge-discharge cycles. The oxygen liberated from the air electrode during charging can oxidize the negative substrate and catalysts; it alters the diameters of the discrete metal particles in the negative, thus changing the negative porosity and damaging the wetproofing film. Also, dissolution of the catalyst can result. The voltage efficiency is relatively poor with this type of negative; this is essentially beacuse of the electrode mechanism that is occurring. Thus the inefficiency is translated into heat that must be removed from the system.

The electrolyte is about 25 wt% potassium hydroxide and includes a small percentage of lithium hydroxide to enhance the anode discharge performance.

Each iron-air cell consists of one iron electrode surrounded by two air electrodes in plastic frames. Each negative shares a common air compartment with negative of neighboring cells, and the cells are assembled in a modular fashion for form a pile. A number of these piles are put into one common battery box, which also serves as a tank for the electrolyte. The electrolyte is circulated between each cell during operation.

The auxiliary system for this iron-air battery contains two subsystems, the electrolyte subsystem and the air subsystem. The electrolyte subsystem circulates the electrolyte, facilitates waste heat removal, and simplifies the battery stack maintenance, i.e., draining, filling, and maintaining electrolyte balance between cells. This sybsystem is designed to minimize shunt current losses and electrolyte pumping power. The air supply subsystem takes care of the following operations: air supply to the air electrodes (CO_2 free), pressure difference over air electrodes, and venting during charge.

4. Zinc-Air (24)

The zinc-air system has seen service as a primary electrochemical system. Recently, it has been designed into a rechargeable system and is somewhat analogous to the iron-air system. Correspondingly, the reaction is written

$$2Zn + O_2 + 2H_2O \underset{\text{Charge}}{\overset{\text{Discharge}}{\rightleftharpoons}} 2Zn(OH)_2 \tag{17}$$

The system has the built-in complexity of having to separate the zinc hydroxide produced on discharge from the electrolyte (7N aqueous KOH) by means of a pump. A pump is required to provide the atmospheric air in a compressed state. The used air is bubbled through the electrodes into the electrolyte where it is alternately separated from the electrolyte and expelled.

Among the problems associated with this system are water loss, temperature control, and zinc dendrite formation in addition to those problems associated with the circulation and separation system.

5. Zinc-Chlorine Hydrate (11, 25)

A unique battery has been described that utilizes chlorine gas in a manageable for (chlorine hydrate, $Cl_2 \cdot 6H_2O$) at normal temperatures. The zinc-chlorine hydrate battery is currently under extensive development for use in utility load-leveling and electric vehicles. The overall electrochemical reaction is

$$Zn + Cl_2 \underset{\text{Charge}}{\overset{\text{Discharge}}{\rightleftharpoons}} ZnCl_2 \qquad (18)$$

The negative electrode consists of zinc, electroplated on a suitable conductive, inactive substrate (such as carbon). This feature is critical for the secondary characteristics of this battery.

The positive chlorine electrode also uses a conductive inactive substrate (such as carbon). It has been reported that noble-metal-coated titanium electrodes are also being considered in the zinc-chlorine hydrate battery.

The electrolyte is about 10% aqueous zinc chloride and includes a proprietary compound to enhance the electrodeposition of zinc.

No separator is used in the zinc-chlorine hydrate battery.

The cells are bipolar configurations stacked togethe in a pile fashion to form the desired battery unit. During operation, the electrolyte is circulated between each cell and stored externally from the battery unit.

Outside of the battery unit, a mixture of chlorine gas and zinc chloride electrolyte is cooled to about 0°C (32°F). Under these conditions, chlorine hydrate ($Cl_2 \cdot 6H_2O$), a pale yellow solid, is formed. Although stored separately from the battery unit, it is in the electrolyte flow loop system. As more chlorine gas is required for the battery, electrolyte passes into the chlorine hydrate storage area. The chlorine concentration increases in the electrolyte as the chlorine hydrate decomposes. A water balance is achieved in the system by the chlorine hydrate and the zinc chloride electrochemical reactions. Thus the electrolyte concentration remains relatively constant during discharge

The charge process is merely the replating of zinc back on its substrate material, and the evolution of chlorine gas to reuse in the formation of chlorine hydrate.

The auxiliary system for this zinc-chlorine hydrate battery contains two sybsystems, the electrolyte subsystem and the chlorine hydrate production subsystem. The electrolyte system circulates the electrolyte, the chlorine, and the chlorine hydrate, and faciliates waste heat removal and simplifies the battery unit maintenance, including, draining, filling, and electrolyte balance between cells. The system design minimizes shunt current losses and electrolyte pumping power. The chlorine hydrate production system cools the electrolyte to produce chlorine hydrate and then stores it for future use.

D. Molten-Salt Systems

Chemicals such as sodium chloride, which are solid and poor electrolytic conductors at room temperature, have sufficient conductivity at 200 to 500°C to be used as electrolytes in some special cells. Such batteries are generally referred to as molten-salt systems and are important new systems for the future since they provide the possibility of higher energy density than aqueous electrolyte systems as well as moderate cost. This is important for such emerging uses as utility load levelling and electric cars and vans.

The alkali metals (sodium, lithium, potassium) are among the cheapest, lightest, and most reactive metals, and are logical electrode materials for use in molten-salt electrolytes. Many cathode materials are under study, but sulfur, metal halides, or halides directly absorbed on carbon have been studies most extensively. Four molten-salt systems representative of those under development as secondary batteries are described below. The techniques used for evaluation are discussed in Section IX.

1. Sodium-Sulfur (7, 8, 26)

This high-temperature battery utilizes molten sodium as the negative electrode and molten sulfur as the positive electrode. Graphite fibers are added to the positive to enhance conductivity and β-alumina, which is a sodium-ion conducting solid ceramic membrane, is used both to separate the molten reactants and to serve as electrolyte.

The overall electrochemical reaction is

$$2Na + 3S \underset{\text{Charge}}{\overset{\text{Discharge}}{\rightleftharpoons}} Na_2S_3 \quad (19)$$

Since the temperature of this system is 300 to 350°C, it is imperative that it be sealed in a suitable container such as stainless steel, titanium, or aluminum. The cell construction is as follows:

Conductive sulfur containing graphite is added to the container, a tubular ceramic separator-electrolyte (β-alumina) with sodium inserted, and the system sealed with a combination of β-slumina/glass/metal seals. The negative connection is a stainless steel rod inserted into the sodium, and the positive connection is the container.

Additional information on this system is in Table 2.

2. Sodium-Antimony Trichloride (27)

This is a unique molten-salt battery that operates at lower temperatures (180 to 200°C) than any other high-temperature system. The molten reactants are sodium, as the negative electrode, and sodium chloroaluminate with antimony trichloride, a Faradaic additive as the positive electrode. Carbon powder is added to the positive to enhance conductivity, and β-alumina is used as both a separator between molten reactants and as electrolyte.

The overall electrochemical reaction is

$$3Na+SbCl_3(NaAlCl_4) \underset{\text{Charge}}{\overset{\text{Discharge}}{\rightleftharpoons}} 3NaCl+Sb \quad (20)$$

The cell construction includes a conductive positive material placed inside a β-alumina tube, inserted into the negative stainless steel container, and then sealed with silicone rubber, Teflon, or glass-to-ceramic seals. The negative terminal is the steel container and the positive terminal is either tungsten or molybdenum.

This system is currently in the research stage and much effort is being undertaken to achieve as much of the theoretical specific energy (825 W hr/kg) as possible. The open-circuit potential is approximately 3 V.

3. Lithium-Tellurium Tetrachloride (8, 28, 29)

This molten-salt battery system has an open circuit potential of 3.1 V per cell and is operated at 400°C. The negative electrode is a lithium-aluminum alloy and the positive electrode is carbon treated with tellurium tetrachloride. These electrodes are separated by a boron nitride separator, and the electrolyte is a molten eutectic of lithium-chloride and potassium chloride operating at 400°C.

The electrochemical reaction is as follows:

$$XLi+TeCl_x \rightarrow XLiCl+XTe \quad (21)$$

The battery is a hermetically sealed steel container. Steel serves as the connector to the negative, and a graphite header serves as the connector for the positive.

4. Lithium-Iron Sulfide (8, 30)

This battery is another of the molten-salt types being developed for electric vehicle propulsion. The cell has a lithium-aluminum alloy as the negative electrode and either FeS or FeS_2 as the positive. The molten-salt electrolyte is a eutectic of lithium chloride and potassium chloride and the operating temperature of this system is 400 to 500°C. The reactants are separated by a ceramic material (e.g., BN, Y_2O_3, $CaZnO_3$) and the battery container is hermetically sealed stainless steel.

The overall electrochemical reactions are as follows:

$$2LiAl+FeS_2 \rightarrow FeS+Li_2S+2Al \quad (22)$$

The theoretical specific energy projected for the FeS_2 system varies between 650 and 458 W hr/kg based on the FeS system.

IV. SECONDARY CELL COMPONENTS AND DESIGN

The secondary electrochemical cells described in this chapter all contain the following components; one or more plates in parallel comprising the cathode, one or more plates in parallel comprising the anode, the electrolyte, and a separator. The component construction depends on the application for which it is intended. For the most part, the discussion below is directed at the aqueous systems.

A. Plate Configuration

Although plates are available in a variety of configurations as described above, all have basically the same general construction. Each plate has two distinct structures that provide the essential properties for storing and converting the chemical energy into electrical energy and vice versa. The inert electrical conductor or substrate serves as the supporting structure and electron collector while the active material serves as the source of the chemical/electrical energy. Both parts have specific criteria that must be met to ensure long term reliability.

B. Conductor/Support/Sintered Plaque

Usually, the conductor is an elemental metal, for example, nickel, iron, silver, lead, or in some cases graphite. Each must be inert to the electrolyte and the atmosphere in the cell. The inert metallic conductor must have sufficiently high electrical conductivity to collect and transfer the electrons from one electrode to the other during charge and discharge. It must also have the mechanical properties necessary to support the active material without distortion, separation, or breaking.

On the other hand, it is important that it be light weight so as not to reduce the energy density, yet ductile enough to be formable into shapes such as screen perforated sheet, pockets, tubes, and frames. The substrate also is perforated with slits or holes to allow the electrolyte to pass through and enter into direct contact with the active material. From an electrical viewpoint it is also desirable for the inert conductor to be an integral part of the cell terminal, therby eliminating another source of iR loss.

The sintered plate construction used in some nickel cadmium and silver cell systems has an additional feature that must be considered part of the conductor assembly. On both sides of a perforated sheet, screen, or expanded metal layer serving as a substrate is sintered a layer of pure metal powder, for example, nickel or silver. This process provides a highly porous (∿80%) structure into which the active material can be deposited. This configuration is analogous to the storage of active material in a tubular container.

C. Active Material

The second constituent of a plate is the active material. The various types used in secondary cells were discussed in the previous section. The criteria for successful electrode active material include high ampere hour per gram energy density, thermodynamic or reaction reversibility, and minimum complexity in the fabrication of electrode structures. For the most part the electrochemical systems discussed in this chapter are relatively low in energy density compared with the lithium system or similar high-energy systems. However, they operate in aqueous electrolyte solutions at environmental conditions that are convenient for day-to-day applications. The electrode reactions of the more conventional electrochemical systems are listed in Table 3 with the theoretical energy densities.

D. Separator

The separator material in a cell, which must be inert to the electrolyte and permeable to ions, serves to mechanically and electrically isolate positive and negative plates from one another. Separator materials are usually of a fibrous sheet nature. Materials such as regenerated cellulose (sausage casing) are continuous membranes. Polyethylene, Teflon, PVC, Dynel, and Nylon are some of the fiber materials that have been used. Paper-based materials with protective resin impregnants are also common. The separator is either wrapped around one of the plates or sandwiched between them. On occasion, a bag is made from the separator material and slipped over one of the plates. In the silver-zinc and silver-cadmium cells where migration and dendritic growth are of concern, several layers of material are used. In sealed nickel-cadmium and lead-acid cells where the quantity of electrolyte must be minimized, the spongy nature of the separator also serves to immobilize the electrolyte. In molten-salt cells, the separator is Beta Alumina, a rigid material through which sodium ion can diffuse at high temperatures.

E. Electrolyte

The electrolyte used in secondary cells must be of high ionic conductivity and must not chemically attack the other components of the cell. For the most part, 30 to 40% potassium hydroxide solutions are used in the alkaline cells while 20 to 30% sulfuric acid solutions is the electrolyte in the lead-acid cell. The concentration depends on cell construction and intended application.

V. PHYSICAL/CHEMICAL METHODS OF ANALYSIS

The positive and negative electrode assemblies in secondary cells usually consist of one to several plates connected in parallel. The positive plates are alternately arranged between the negative plates. The more plates that are used for each electrode, the higher the cell capacity. Also, the more plates there are in parallel, the greater is the averaging effect of the electrode properties. There are several requirements that a cell plate must meet in order for it to be judged adequate. Among the significant properties are surface area, porosity and pore size distribution, electrochemical activity, and stability.

In the tubular or pocket type of plate the active material is contained within the inert metallic conductor, which must have the mechanical properties to restrict expansion. In an analogous manner, the sintered plate not only serves to contain the active material but must have specific properties for optimizing electrical conductivity and active material utilization. Because the plaque structure plays such an important role in the plate performance, the discussion starts with evaluation and analytical methods utilized in selection and control of the powders, follows with the methods used in plaque characterization, and finally describes analytical methods relative to the plate and its active materials.

A. Powders

The three powder properties that contribute most to high plaque porosity and electrical conductivity are apparent or bulk density, average particle, and grain size. These parameters are measured on the nickel powder used in the manufacture of plaque for Ni-Cd cells and for silver in the Ag-Cd and Ag-Zn cells. Other powders of interest when considering secondary cell plates include CdO for the negative plate in a Ag-Cd cell, ZnO and HgO for the negative plates of the Ag-Zn cell, and PbO for both electrodes in the lead-acid cell.

1. Apparent or Bulk Density

The production of an 80% porous nickel plaque structure of the type utilized in a sintered nickel plate requires a very special type of powder. Only two grades of powder, Inco 255 and Inco 287, both of which are produced with the same process, have been found to be generally acceptable for this application. A scanning electron micrographic photo (31) Fig. 2) provides a visual impression of this irregular, fragile, filamentary shaped particle. The difference between the two powders is the apparent or bulk density. The 255 powder is 0.55 g/cc and the 287 powder is 0.87 g/cc nominal. Both of these are significantly different from the density of pure nickel of 8.9 g/cc.

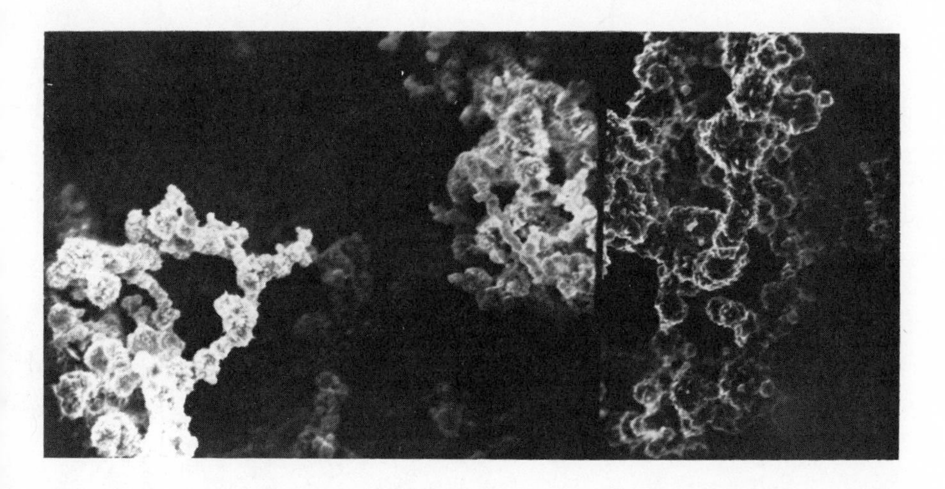

Fig. 2. SEM photomicrographs of INCO 255 and 287 powders.

The generally accepted method for characterizing this type of powder is the measurement of apparent density Two ASTM (32) methods that have been utilized are the B329-61 test using the Scott volumeter and the older B202-48 method. Both utilize the technique of flowing the powder through a funnel to separate the particles into a cup of known volume. The Scott Volumeter (Fig. 3) utilizes a series of baffles between the funnel and cup to further individualize the particles. The weight of powder in the cup of known volume

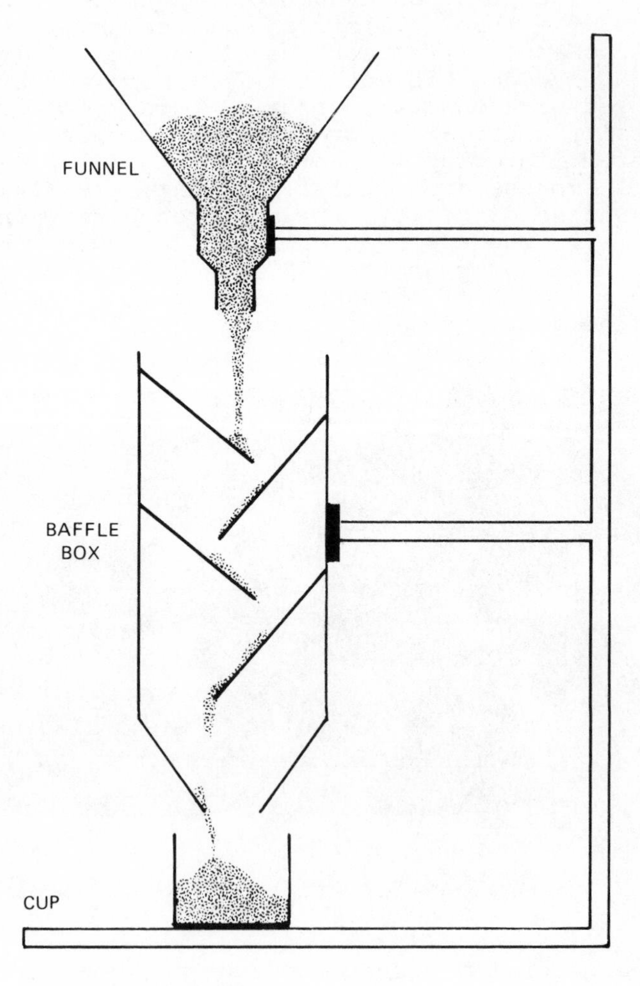

Fig. 3. Scott Volumeter.

is the apparent or bulk density. In addition to the nickel powders, the method is also applicable to silver and cadmium oxide powders used in pressed electrodes for silver cadmium and silver zinc cells. Table 4 provides sample data resulting from measurements on several samples (33) of nickel powder and Table 5 the expected ranges for representative powders used in secondary cell plates.

TABLE 4

Sample No.	Powder Type	Bulk Density g/cm^3	Grain Size Å	Fisher Number (μ)	Surface Area m^2/g
704-1	287	0.91	449	3.22	0.386
704-2	255	0.68	369	2.82	0.535
704-3	255	0.61	392	2.63	0.540
704-4	257	1.02	458	3.27	0.394
704-5	257	1.07	474	3.21	0.420
704-6	257	1.07	450	3.23	0.454
704-7	257	1.01	447	3.39	0.386
704-8	287	1.04	449	3.18	0.446
704-9	287	0.99	456	3.15	0.471
704-10	287	0.97	494	3.18	0.457
704-11	287	1.02	436	3.18	0.429
704-12	287	0.98	457	3.16	0.437
704-13	287	0.97	452	3.23	0.405
704-14	255	0.68	384	2.78	0.507
704-15	287	0.99	424	3.14	0.446
704-16	S.G.*	0.95	406	2.17	0.756
704-17	S.G.*	0.99	390	3.58	0.496
704-18	287	0.99	464	3.51	0.448

*Sherritt Gordon.

TABLE 5

	Type	Pure Density g/ml	Apparent Density g/ml	Use in Cell
Ni	Inco 255	8.9	2.6-3.4	Ni-Cd
Ni	Inco 287	8.9	2.9-3.6	Ni-Cd
Ag	H&H* 130	10.5	1.4-1.9	Ag-Cd
Ag	–	10.5	.73-.98	Ag-Zn
CdO	–	8.15	,30-.90	Ag-Cd

*Handy and Harmon.

2. Average Particle Size and Distribution

The average particle size can be determined in two ways. The simplest method is by using a screen/sifting procedure either with or without vibration wherein the mesh has the desired size characteristic. This method is satisfactory for some types of powders, that is, silver and zinc. However, in the case of the fragile nickel powders described above, the filamentary powders will ultimately break and all pass through the screen mesh. The Fisher Subsieve Sizer has been utilized as a comparative method in determining average particle size. This instrument utilizes the principle of air flowing with a constant force through a powder bed in a tube of known cross section. The change in regulated air flow is a function of particle size. The measurement provides a means of determining the average particle size for irregularly shaped particles. The fundamental equation of Gordon and Smith (34) is

$$d_m = \frac{60,000}{14}\left(\frac{\eta CFDL^2M^2}{(VD - M)^3 \ (P - F)}\right)^{1/2} \qquad (23)$$

where d_m is the average particle diameter in microns, η the viscosity, C the conductance of the manometer, F the pressure differential, D the density, L the length of the compacted sample, M the mass, V the apparent volume, and P the overall pressure head.

By using a sample weight equal to the density, packing the sample to a known dimension and fixing the other variables, the particle diameter becomes a function of the flow. Fisher particle diameters for several nickel powders are compared with Sherritt Gordon samples in Table 4.

Particle size distribution can be determined using a cumulative sedimentation technique. Essentially, the procedure involves distributing the powder in a sedimentation fluid whose viscosity and density is large enough to restrict settling of the larger particles, while allowing the smaller particles to settle in a short time. A 35% glycerol in water solution was found to be satisfactory for nickel powders. The increase in weight is measured on a Cahn recording electrobalance.

The weight at a specific time includes those particles whose time to fall is less than that time and those particles whose time of fall is greater but in an intermediate position in the column. Expressed mathematically, the weight settled is

$$W = \int_{M_1}^{M_{max}} f(M)\ dM + \int_{M_{min}}^{M_1} \frac{vt}{h} f(M)\ dM \qquad (24a)$$

where f(M) is the frequency of occurrence of a particle of size M and f(M)dM is the fraction between M and M + dM. M_1 is the size that would fall the full height, h, of the column; v is the velocity of fall of a particle of size $M<M_1$; M_{min} and M_{max} are the smallest and largest sizes, respectively.

Differentiating with respect to time,

$$\frac{dW}{dt} = \int_{M_{min}}^{M_1} \frac{v}{h} f(M)\ dM \qquad (24b)$$

or

$$t\frac{dW}{dt} = \int_{M_{min}}^{M_1} \frac{vt}{h}\ f(M)\ dM \qquad (24c)$$

We may then write the original equation as

$$W = \int_{M_1}^{M_{max}} f(M)\,dM + t\frac{dW}{dt} \qquad (24d)$$

Since the weight fraction, w, of particles greater than M_1 is given by

$$w = \int_{M_1}^{M_{max}} f(M)\ dM \qquad (24e)$$

we can write

$$w = W - t\frac{dW}{dt} \qquad (24f)$$

or

$$w = W - \frac{dW}{d \ln t} \tag{24g}$$

We can thus analyze the weight-time curve plotted on a logarithmic time scale to define a particle size distribution. Samples of several nickel powder analyzed by this sedimentation technique is given in Table 6. Comparing the Inco 255 and 287 powders it can be seen there is little difference in weight distribution. However, recalling the data in Table 5 it is clear that the only difference is in particle diameter.

Other techniques used in particle evaluation include surface area described in detail in Volume 1 of this series (35) and grain size using X-ray techniques.

B. Porous Plaques

The plaque serves as the basic support structure and contains the active materials of the electrodes. Composed of pure nickel powder in the nickel and cadmium electrodes of the Ni-Cd cell and sintered in a reducing atmosphere, the structure produced is approximately 80% porous. The physical properties of importance to proper plaque performance include porosity or void volume, mean pore size and distribution, and sinter conductivit[y]. An example of the plaque structure after sintering the powders described earlier is given in the scanning elec[-]tron photomicrograph in Fig. 4.

1. Apparent Porosity

Several methods have been utilized to determine porosity. The least complex is to measure the physical dimension of the sample and then determine its weight. The volume of the sample is multiplied times the density of the pure metal and compared with the measured weight:

$$\%\text{Porosity} = \frac{\text{Sample weight X 100}}{\text{Sample volume X density}} \tag{25}$$

This method is subject to error and because the plaque is not a flat structure andy may have part of the conductor (screen or perforated sheet) protruding from one side or the other. The metal conductor also provides

TABLE 6

	Median Diameter μ	Weight Percentage Distribution							
		5-7 μ	7-9 μ	9-11 μ	11-13 μ	13-15 μ	15-17 μ	17-19 μ	19-21 μ	Total % $< 21\ \mu$
704-1	15.8	4	7.5	9	10.5	13	12.5	11	11	88.5
704-2*	12.8	7.5	10	15	16.5	15	12	9	7	93
704-3*	13.4	7	9	12	15	20	15	11	8	97
704-4	12.0	8	13.5	17	18	18.5	14.5	8	–	97.5
704-5	12.8	7.5	10	15	16	17	13	11	9	98.5
704-6	12.2	5	12	17	19.5	18	16.5	9	–	97.0
704-7	12.2	10.5	12.5	13.5	17.5	20	15	8	–	97
704-8	11.5	8	15	17	19	20	13	3	–	95
704-9	11.7	8.5	13	17.5	18	18	15	5	–	95
704-10	12	5	12	16	24	22	11	4	–	94
704-11	12.4	8	13	13	16	15.5	15	11	4	95.5
704-12	12.2	7.5	12	15.5	18	17.5	16	10	–	96.5
704-13	11.7	9.5	13	18.5	18.5	18.5	14.5	3	–	95.5
704-14*	12.0	9	12.5	15	17	19	20	3.5	–	96.0
704-15	12.7	8	11	12.5	14.5	15	13	10	10	94.0
704-16+	11.7	11.5	13	17.5	19	17	14	7.5	–	99.5
704-17+	15.2	7.5	8	10	11	12	12.5	11	13	83.0
704-18	11.4	12.5	13.5	16	19.5	20	14	–	–	95.0

*INCO Type 255 – +Sherritt Gordon powders.

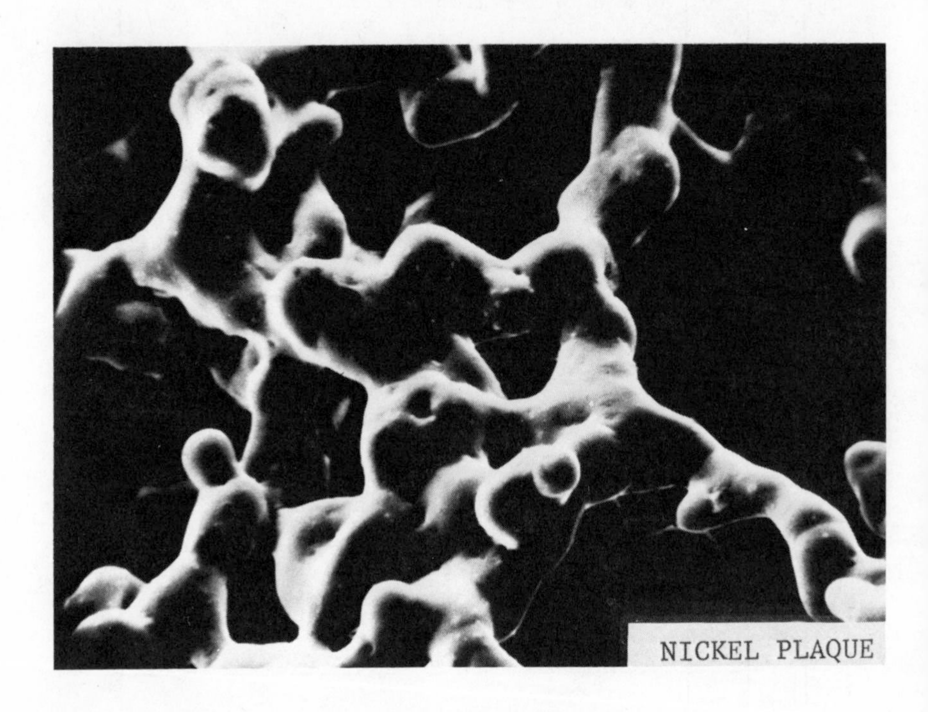

Fig. 4. SEM photomicrograph of sintered nickel plaque.

an error because its density may be different from the pure metal of the plaque. The conductor takes the form of mesh screen or perforated sheet. Subtracting its weight and volume from the values required in Eq. 14 before the calculation is made, will remove some of the error but at best is only acceptable as a qualitative measure of apparent porosity.

A second method for determining porosity is to immerse a preweighed sample of plaque or pressed powder or impregnated plaque in a solution that will neither wet nor react with the material. The sample is weighed dry then sealed in a closed flask, and a vacuum is applied to remove all entrapped gases. The liquid that has been subjected to vacuum in a separate container is allowed to run into the flask until the sample is submerged. Water, methanol, toluene, and other liquids merged. Water, methanol, toluene, and other liquids have been used for this measurement. The sample is then removed and reweighed. There is a basic problem with this technique also. When the sample is removed, an attempt is made to remove all excess liquid from the surface. Wiping the sample against Teflon or a plate of glass has not been found to be very reproducible especially when comparing results from different groups of samples. The method can be used, however, for relative measurements. The difference between wet and dry weights divided by the density of the liquid is equal to the volume adsorbed in the pores. Comparing this to the total sample volume results in a measure of porosity as given in Eq. 15:

$$\%\text{Porosity} = \frac{\text{Sample weight(wet)} - \text{Sample weight(dry)}}{\text{Density of liquid} \times \text{sample volume}} \times 100 \qquad (26)$$

The third method for determining porosity while at the same time providing mean pore diameter utilizes mercury intrusion. This method is a special case of the volumetric method described in the preceding paragraph. Commercially available instruments have been utilized for this purpose; one such example is the Micromeritics Corp. Model MIC 901. This method is described in detail in Volume 1 of this series (35). A computer program written for use with this instrument at Goddard Space Flight Center provided the sample line-printed curve given in Fig. 5.

2. Plaque Continuity

The measurement of particle continuity in the sintered plaque has two purposes: to ensure that the sintering operation has been complete enough to maintain structural integrity even after the corrosive process of impregnation and to determine the inherent iR loss through the structure. Two methods have been used for this purpose. The mechanical strength method measures the force necessary to crack the sinter under compression. The electrical method provides either a conductivity value or its inverse (resistivity).

PORE SIZE DISTRIBUTION BY MERCURY INTRUSION - MICROMERITICS DATE 03 MAR 1970 PAGE 0025

PORE SIZE DISTRIBUTION BY MERCURY INTRUSION

CLEVITE .020 PLAQUE 6 NOV 69 MPD 10 MICRONS

SAMPLE HOLDER NUMBER IS 13 SAMPLE WEIGHT = .59310

PRESSURE APPLIED (PSIA)	PORE RADIUS (MICRON)	PENETRATION COUNTER	CORRECTED COUNTER	VOLUME OF PORES LARGER THAN RADIUS (CC/GM)	D(R)	R AVG MICRON	LOG R AVG ANGSTROMS
2.00	44.09800	3.00	1.00	.00260	.000	66.14700	5.82051
4.00	22.04900	6.00	4.00	.01039	.026	33.07350	5.51948
6.00	14.69933	9.00	7.00	.01818	.044	18.37417	5.26421
8.00	11.02450	12.00	10.00	.02597	.062	12.86192	5.10931
10.00	8.81960	15.00	13.00	.03375	.080	9.92205	4.99660
12.00	7.34967	19.00	17.00	.04414	.131	8.08463	4.90766
14.00	6.29971	24.00	22.00	.05712	.194	6.82469	4.83408
14.35	6.14606	25.00	23.00	.05972	.242	6.22289	4.79399
15.85	5.56442	74.00	71.96	.18685	2.944	5.85524	4.76754
20.35	4.33396	105.00	102.84	.26701	.739	4.94919	4.69453
35.35	2.49494	107.00	104.41	.27111	.017	3.41445	4.53332
45.35	1.94479	107.00	104.13	.27038	-.007	2.21986	4.34633
75.35	1.17048	107.00	104.00	.27004	-.002	1.55763	4.19247
114.35	.77128	107.00	104.00	.27004	.000	.97088	3.98717
165.35	.53339	107.00	104.00	.27004	.000	.65234	3.81447
1100.00	.08018	107.00	103.90	.26978	-.000	.30678	3.48683
2000.00	.04410	108.00	104.00	.27004	.001	.06214	2.79336
4000.00	.02205	109.00	103.00	.26744	-.009	.03307	2.51948
6000.00	.01470	111.00	103.00	.26744	-.000	.01837	2.26421
8100.00	.01089	112.00	101.95	.26472	-.021	.01279	2.10700
10000.00	.00882	113.00	102.00	.26485	.001	.00985	1.99361
14000.00	.00630	116.00	101.00	.26225	-.018	.00756	1.87850
16100.00	.00548	118.00	100.87	.26192	-.005	.00589	1.77003
18000.00	.00490	119.00	99.50	.25835	-.074	.00519	1.71507
20100.00	.00439	121.00	98.90	.25680	-.033	.00464	1.66688
24000.00	.00367	124.00	98.00	.25446	-.030	.00403	1.60545
28000.00	.00315	127.00	98.00	.25446	.000	.00341	1.53305
32000.00	.00276	131.00	99.00	.25706	.045	.00295	1.47026
36000.00	.00245	135.00	99.00	.25706	.000	.00260	1.41548
40000.00	.00220	137.00	98.00	.25446	-.057	.00233	1.36687
44000.00	.00200	141.00	97.00	.25186	-.063	.00210	1.32319
46600.00	.00189	146.00	98.24	.25508	.129	.00195	1.28971

PORE SIZE DISTRIBUTION BY MERCURY INTRUSION - MICROMERITICS   DATE 03 MAR 1970   PAGE 0026

VOLUME OF PORES G.T. RADIUS

.500
.400
.300
.200
.100
.000

.000 .500 1.000 1.500 2.000 2.500 3.000 3.500 4.000 4.500 5.000

LOG PRESSURE APPLIED (PSIA)

1.0 3.16 10.0 31.6 100. 316. 1.0K 3.16K 10.0K 31.6K 100.K

PRESSURE (PSIA)

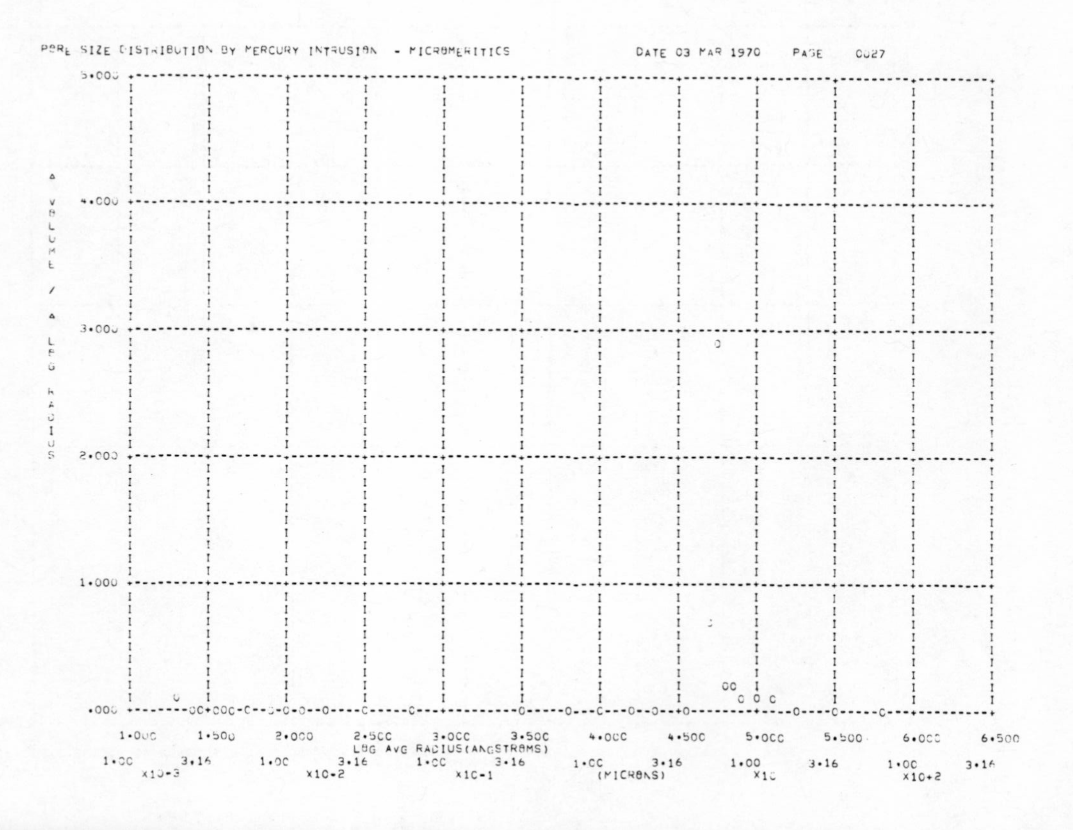
PORE SIZE DISTRIBUTION BY MERCURY INTRUSION - MICROMERITICS
DATE 03 MAR 1970
PAGE 0027
A VOLUME / A LOG RADIUS
5.000
4.000
3.000
2.000
1.000
.000
1.000 1.500 2.000 2.500 3.000 3.500 4.000 4.500 5.000 5.500 6.000 6.500
LOG AVG RADIUS(ANGSTROMS)
1.00 3.16 1.00 3.16 1.00 3.16 1.00 3.16 1.00 3.16 1.00 3.16
X10-3 X10-2 X10-1 (MICRONS) X10 X10+2

3. Mechanical Strength

The mechanical strength of the sinter is a measure of the completeness of the sintering operation. A four-point band test was adapted for this purpose by Parry (36). A constant-driving crosshead-motion compression force was applied to a piece of plaque. The displacement caused by the force of the load was recorded as a function of time using the device given in Fig. 6. The force in kg/cm^2 required to permanently deform the 1 x 2-in. sample is a measure of its strength. The curve generated during this measurement is given in Fig. 7.

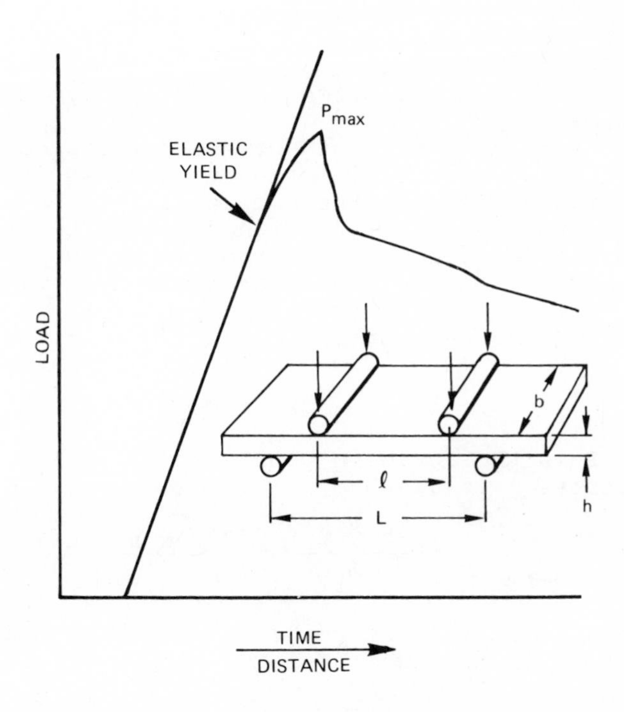

Fig. 6. Stress curve for Mechanical strength measurement.

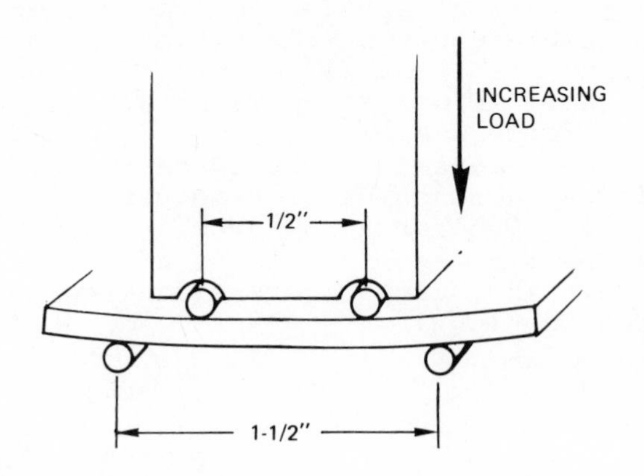

Fig. 7. Mechanical strength test apparatus.

The effect on strength of the perforated sheet or screen conductor is minimized because it is centered (relatively) between the compressive force at the top of the sample and the tension force at the bottom. The center position is referred to as the point of zero moment. An Instron testor was used with a light load cell of 0 to 60 psi.

Examinging the bending moment (M_b) in the figure,

$$M_b = \frac{P}{2} x \ a \tag{27}$$

where P is the force in pounds and a is 1/2 ($L - \ell$). The stress σ is given by

$$\sigma = \frac{M_b \ x \ h/2}{I_{yy}} = \frac{P/2 \ x \ a \ x \ h/2}{I_{yy}} \tag{28}$$

where h/2 defines the point of zero moment and I_{yy} the moment of inertia which, for a sample of width b, is

$$I_{yy} = \frac{bh^3}{12} \tag{29}$$

Thus, the equation for the stress becomes

$$\sigma = \frac{3Pa}{bh^2} = \frac{3P}{2bh^2} \tag{30}$$

for a situation in which L = 3/2 and ℓ = 1/2. Converting to units of kg/cm^2 from the usual pounds and inches used by the metallurgists results in

$$\sigma = \frac{0.109P}{bh^2} \tag{31}$$

4. Electrical Resistivity

Plaque resistivity or conductivity is also a measure of plaque continuity. This electrical method requires the attachment of one copper bar on each side of a sample of plaque. Four current leads are attached to each bar to allow current to pass uniformly across the plaque. A pair of spring-loaded voltage probes are placed in the center of the sample. A fixed current is imposed across the sample. The voltage drop measured is a function of the resistivity (ρ) according to Eq. 21:

$$\sigma = \frac{EA}{i\ell} \tag{32}$$

where E is the voltage drop, A the cross sectional area, i the current in amperes, and ℓ the distance between probes. The method adapted by Kantner (37) utilizes the experimental setup of Fig. 8.

Sample data utilizing both techniques are given in Figs. 9a and 9b. The expected inverse relationship between the two exists as shown in Fig. 10. As one might expect, there is also a relationship between porosity and resistivity.

C. Electrode Plates

Both the positive and negative plates contain the active material used in the storage and exchange of chemical and electrochemical energy. In order to satisfy the requirements of secondary cell electrodes, they must have specific properties that will optimize the reversible electrode processes. Among the physical properties of significance are dimensional stability, surface area, and chemical composition and active material structure.

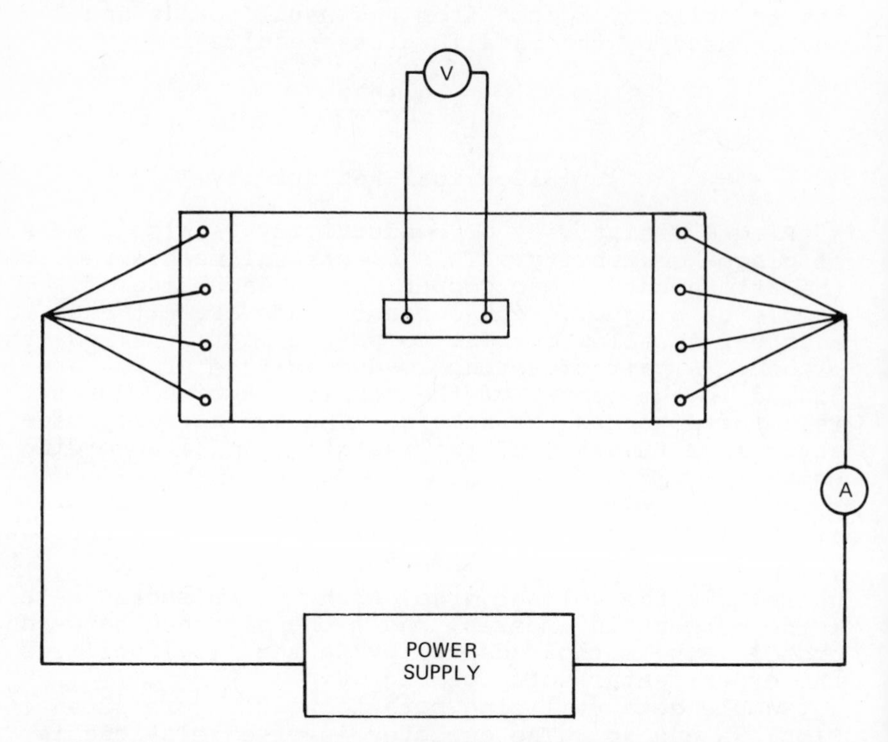

Fig. 8. Apparatus for resistivity measurement.

1. Dimensional Stability

One of the most predominant causes of cell failure is change in the physical structure of the plates. This problem leads to shorts between opposing electrodes or shedding of the active material, during which there is loss of active material from the electrode and permanent loss of energy. The change in plate structure can occur during the charge/discharge process or during open-circuit stand. The change in plate shape is the result of chemical changes in the active material. Other than manufacturing defects,

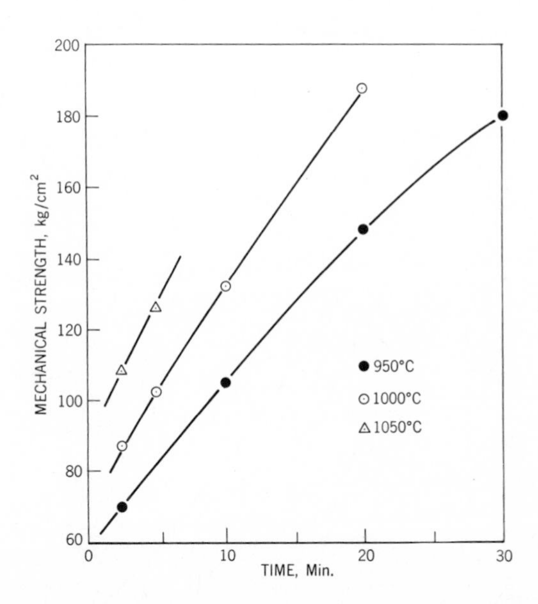

Fig. 9a. Mechanical strength versus sintering time.

the tubular and pocket electrodes are least likely to experience a dimensional problem because the active material is held fixed within the mechanical structure. Sintered or pressed electrodes can experience warping in flooded electrolyte if not held firmly in place by means of compression. In the electrolyte-starved nickel-cadmium cell, sintered plates experience thickness expansion during cycling particularly in the positive plate as indicated in Table 7. For example, these measurements were made on ten plates taken from a 20-A hr cells used in the Orbiting Astronomical Observatory (OAO) Satellite. Plates as received from the manufacturer are compared ith plates standing on pen circuit for 2 yr shorted and cells operated in a simulated satellite cycling regime of 90 min (60 min/ charge, 30 min/discharge) for more than 2000 orbits. The rather large 17.9% increase in positive plate thickness has been related to changes in the structure of the active material, although the chemical basis for this is not clear. Excessive structural changes result in plate blistering, which occurs on plates with high loading of active material per cubic centimeter of voids and at very high rates of overcharge.

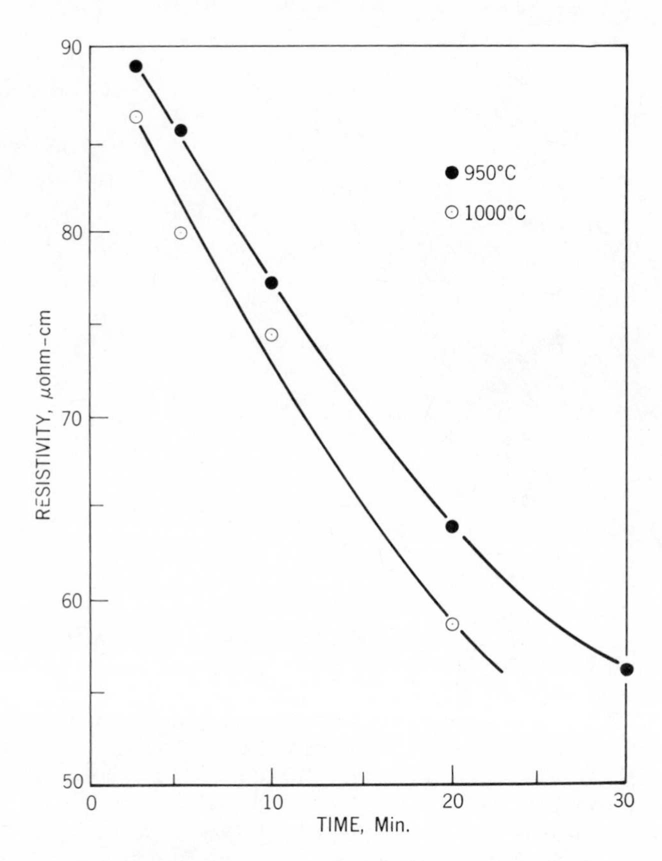

Fig. 9b. Resistivity versus sintering time.

Physical changes in the zinc electrode are accompanied by the loss of geometric plate area, which result from active material shape change. The finely divided zinc grows into a dense nodular structure. Utilization of the active material decreases. Larger crystals increase at the expense of the smaller, and ultimately dendritic growth of the zinc will cause a short between a positive and negative plate. Silver and cadmium, to a lesser extent, experience recrystallization described as migration, which ultimately leads to failure. In cells with this active material, several layers of separator are used to delay the shorting of the cell.

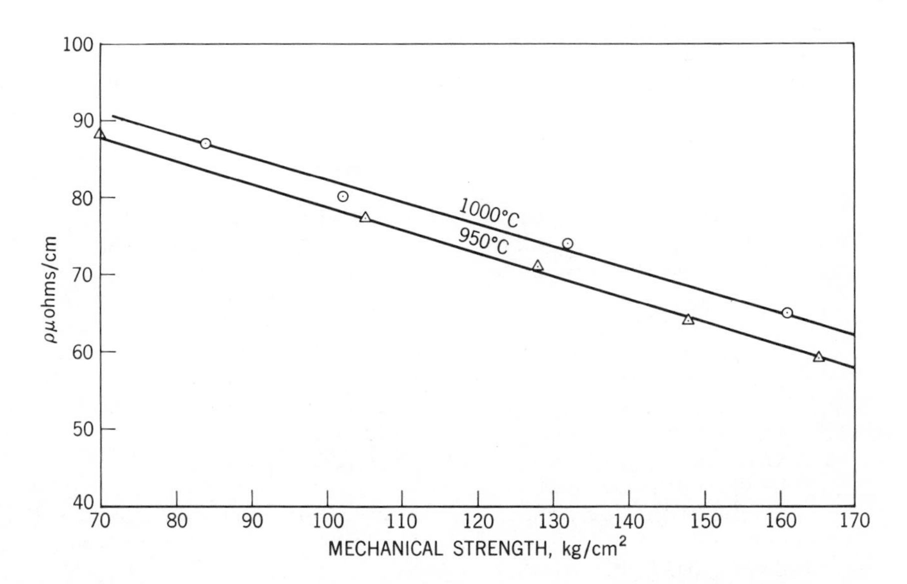

Fig. 10. Relationship between mechanical strength and Resistivity.

TABLE 7

	Dry Thickness (mils)	Wet Storage (mils)	% Change	After 2 Year Cycling (mils)	% Change
Positive	35.7 $^{+.6}_{-.1}$	37.4 $^{+1.4}_{-1.0}$	4.7	42.1 $^{+3.4}_{-3.0}$	17.9
Negative	32.9±1.0	33.6 $^{+.5}_{-.9}$	2.1	35.4 $^{+1.4}_{-2.1}$	7.6

The positive plate in the lead acid cell experiences shedding of the active material and disintegration of the plate caused by grid corrosion. The corrosion of the grid is related to the effects of continuous overcharging. The new calcium grid alloy plates, which require significantly less overcharge, have alleviated this problem. Plates from the lead-acid cell can also become warped if charged after being open-circuited for extended periods. The process responsible is sulfonation, which results in a glaze on the plate surface In recharging the cell, the active material recrystallizes and the stress on the surface from within causes the plate to buckle or warp.

2. Surface Area

One of the more important properties related to polarization and active material utilization is the surface area of a plate. The more the surface of the active material is exposed and in contact with the electrolyte, the faster and more efficient will be the ion/electron or electron/ion transfer process. Several methods have been utilized for measurement of surface area. The theoretical discussion of these methods has been given by Salkind (35). These methods can be applied in a practical way to evaluate secondary battery plate materials.

The BET gas adsorption method has been utilized to characterize sintered nickel hydroxide and cadmium hydroxide plates used in a Ni-Cd cell. Milner and Thomas (16) in their 1967 review article summarize the results of others as being 70 m^2/g for positive plates and 6 m^2/g for negatives. Since that time, measurements have been made at Goddard Space Flight Center, in their evaluation of plate materials for high-reliability aerospace applications. The instrument utilized was a Micromeritics (Numinco) Model No. MIC 103 Surface Area Analyzer. An improvement in the technique was made by Campbell (38), who adapted a flat stainless sample holder that was easily assembled and disassembled while holding the required vacuum under liquid-nitrogen conditions (Fig. 11). The data were recorded directly on 80-column computer paper and the results calculated using a computer program developed specifically for the purpose. In the calculation, the value for the equilibrium pressure (P_2) divided by the saturation pressure (P_S) is plotted against the calculated value for

$$\frac{P_2}{V(P_S - P_2)} \qquad (33)$$

where V is the empty volume of the chamber.

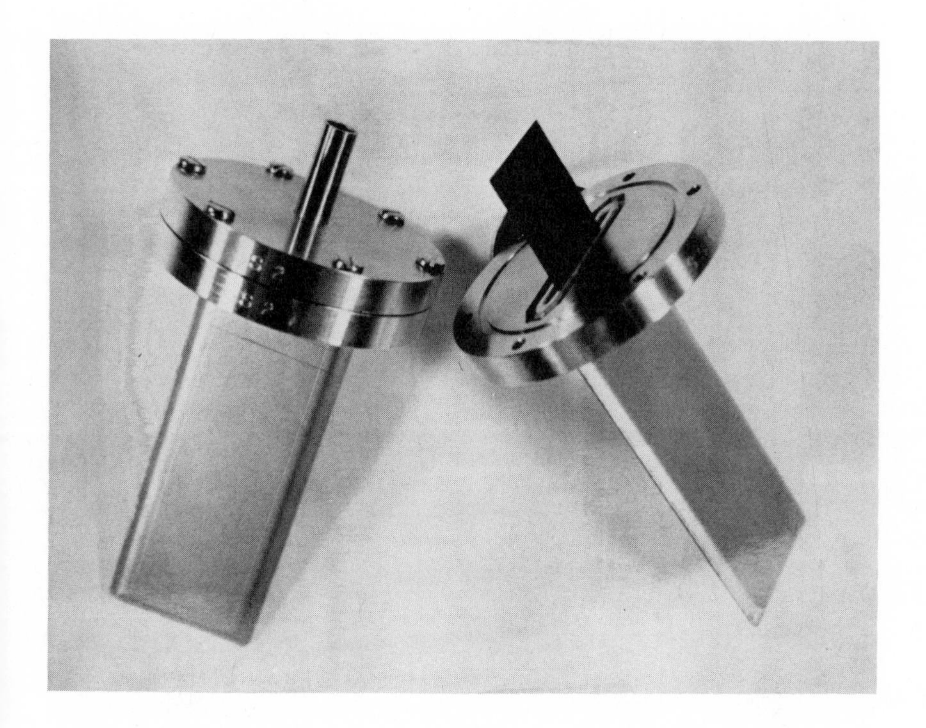

Fig. 11. Plate sample holders for BET analysis.

The relationship is linear between P_2/P_s values of 0.1 and 0.35. Utilizing the slope and intercept determined from the least-squares method, the surface area (S_W) for the nitrogen BET analysis in units of m^2/g is determined from

$$S_W = \frac{0.269}{\text{slope + intercept}} \tag{34}$$

Surface area data from the positive and negative plates of 20-A hr nickel-cadmium cells is given in Tables 8 and 9. Measurements of uncycled plates are compared with data from plates of the same group after six deep discharge cycles in flooded 31% KOH solution in part A

TABLE 8

	Surface Area Measurements					
	m^2/gm	m^2/cm^2	m^2/cc	Thick (mils)	Density gm/cc	Porosity %
A. Uncycled Plates						
As Received	13.36	4.25	46.1	36.5	3.44	35.22
After Flooded Cycling	28.50	9.13	93.4	38.5	3.28	37.90
B. Plates from Sealed Cells (Opened, Washed & Dried)						
Cell 743 (Stored 2 yrs.)	23.85	7.29	76.8	37.4	3.22	36.03
Cell 666 (2000 cycles)	27.23	8.84	82.5	42.1	3.04	43.85
C. Cycled Flooded, Washed & Dried – Charged Condition						
Cell 743 Plates	26.19	8.61	81.6	41.5	3.12	44.13
Cell 666 Plates	24.09	7.80	75.6	40.6	2.51	41.33
Cycled Flooded, Washed & Dried – Discharged Condition						
Cell 743 Plates	26.74	9.00	84.4	42.0	3.16	36.43
Cell 666 Plates	26.46	8.78	79.6	43.3	3.01	44.34

TABLE 9

	m^2/gm	m^2/cm^2	m^2/cc	Thick (mils)	Density gm/cc	Porosity %
A. Uncycled Plates						
As Received	2.53	0.79	9.26	33.7	3.66	35.72
After Flooded Cycling	1.93	0.57	6.83	32.8	3.55	39.83
B. Plates from Sealed Cells Opened Washed, Dried						
Cell 743 (Stored 2 yrs.)	1.38	.42	4.93	33.4	3.60	39.81
Cell 666 (2000 cycles)	1.52	.44	4.97	35.4	3.27	41.30
C. Cycled Flooded, Washed, Dried Charged						
Cell 743	2.88	.77	9.22	32.8	3.21	49.37
Cell 666	2.46	.65	7.52	34.0	3.07	53.53
Cycled Flooded, Washed, Dried Discharged						
Cell 743	1.54	.45	5.44	32.8	3.54	43.64
Cell 666	1.63	.45	5.40	33.1	3.33	45.76

CONDENSER
WATER COOLED
EXTRACTOR
BOILER
INERT GAS
IN
HEATER
CONTROL

of each of these tables. The measurements are performed on three 1-in-diameter disks punched from plate samples after washing in deionized water in a specially constructed nickel soxlet extractor modified by Ungar and Stemmle (39) (Fig. 12). The samples are then dried at 35°C in a vacuum oven. It appears that the electrical testing has provided additional surface area on the positive plates, due in part to the expansion of the plates, while the negative plates have decreased in surface area with an accompanying slight decrease in thickness. In part B of these tables, two cells (666 and 743) from the same manufacturing lot are compared. Cell 666 was cycled in a simulated satellite orbit at 25% depth of discharge for 2000 cycles. Cell 743 was given the same initial cycles then allowed to stand on open circuit for a period of a year. The cells were disassembled and the plates were subjected to an electrolyte extraction as described above. The results indicate that the positive plates of 666 that were continually operated for an extended period expanded but maintained a significantly high surface area, while in the cell 743 positive plate area decreased on standing. In part C, the surface areas of charged plates are compared with discharged plates. The most significant point is the large increase in surface area of the charged negative plates versus the discharged negative plates.

The surface area of the PbO_2 electrode in a lead acid storage battery plate has been reported (14) to be 7 to 7.5 m^2/g using a similar gas absorption method.

3. Pore Volume/Porosity

This subject was described earlier under the measurements and characterization of plaque materials. Porosity also plays an important part in plate performance in that it provides access for the electrolyte to the active material. The methods of evaluation were described earlier.

4. Active Material Analysis

The quantity of active material in a plate is a measure of its stored energy. However, the utilization of the chemically active material as electrical energy can vary over a wide range. Therefore, to evaluate the operation of a battery plate, it is desirable to chemically analyze the active materials contained therein.

Wet chemical analytical techniques have been utilized in the evaluation of Ni-Cd cell plates. The stepwise procedure for the analysis has been documented by Halpert et al (40). A similar procedure was described by Kroger and Catotti (41).

The principle used in the analysis of the positive plate is somewhat inaccurate when a significant quantity of charged active material is present, but is acceptable when the plate is completely discharged. The total discharged material in the positive plate is acknowledged to be $Ni(OH)_2$ and the charged material NiOOH. To determine the amount of charged material, the mixture of sintered nickel and active material is treated with excess of ferrous ammonium sulfate in acetic acid solution. This results in a reduction of higher valence (both trivalent and tetravalent) nickel to divalent nickel by the simultaneous oxidation of ferrous ions to ferric ions:

$$NiO\ OH + Fe^{2+} + 3H^{+} \rightarrow Ni^{2+} + Fe^{+++} + 2H_2O \quad (35)$$

The excess of ferrous ions in the solution is then titrated with a standard potassium permanganate solution:

$$5Fe^{2+} + MnO_4^{-} + 8H^{+} \rightarrow 5Fe^{3+} + Mn^{2+} + 4H_2O \quad (36)$$

To determine the total amount of active material, the mixture of sintered nickel and active material is leached in acetic acid; hydrazine sulfate is added as an inhibitor for metallic nickel dissolution. The nickel in the solution is then titrated with a standard EDTA solution. The amount of discharged material plus metallic nickel is determined by subtracting the charged material from the total. To determine the amount of metallic nickel, nitric acid is used to dissolve the metallic residue from the extraction that was described in the preceding paragraph. The nickel in the solution is then titrated with a standard EDTA solution.

Cadmium hydroxide ($Cd(OH)_2$) is recognized as the discharged active material and cadmium metal (Cd) as the charged active material in the negative plate of nickel-cadmium cells. The analysis procedure involves the separation of the cadmium ion of the cadmium hydroxide from the cadmium metal in the neckel plaque comprising a nickel or steel substrate. The first step of the separation is accomplished by extraction of the plate with ammonia solution to form the $[Cd(NH_3)_6]^{++}$ complex. The complex is broken with the addition of formaldehyde and the free Cd^{++} is then titrated with standardized disodium ethylene-diamine-tetraacetate (EDTA). The second step of analyzing for the charged cadmium involves dissolving the remainder of the sample from above with nitric acid to form the nitrate. The cadmium ion (Cd^{++}) thus formed is complexed in basic ammonia solution to form the cadmium-ammonia complex as above. However, consideration must be given to the nickel and iron ions also present in solution. After standing overnight, iron hydroxide, which appears as a brown precipitate, is removed by filtration. The nickel and cadmium ions are converted from the basic solution to a cyanide complex by adding sodium cyanide. Then formaldehyde is added to break the cadmium complex without affecting the nickel complex. The free cadmium ion is then titrated with the standardized EDTA solution.

The analysis of the silver electrode from silver cadmium battery is described by Thirsk and Lax (42). The procedure requires dissolving the oxide in dilute ammonia solution. The silver is estimated volumetrically using Volhard's method. A gravimetric method using benzimidazole was found to be a suitable alternative procedure to the volumetric method. The zinc electrode can be analyzed using the procedure described by Keralla (43). This method involves adjusting the pH of solutions containing zinc to 4.6 to 5 with acetic acid/sodium acetate buffer. Two drops of ($CuSO_4$)-EDTA solution and 4 to 5 drops of 0.1% ethanolic PAN(1-2 pyridyl-AZO-2 Napthol) indicator solution are added. The solution is titrated with standardized 0.1N EDTA solution until the color changes from purple to yellow.

The analysis of materials used in lead-acid battery manufacture has been delineated by Denby (44). Appropriate analytical procedures for determination of lead (Pb), lead monoxide (PbO), lead dioxide (PbO_2), and lead sulfate ($PbSO_4$) in battery pasts are described.

The determination of free lead in the battery paste requires the dissolving of the lead sulfates and oxides in a solution of sodium hydroxide (dissolves PbO), mannitol (complexing agent for $PbSO_4$), and hydrazine dihydrochloride (reducing agent for PbO_2). After the lead metal has precipitated, the solution is filtered; the lead is then dried and its amount determined by direct weighing.

For lead monoxide (PbO) determination, the battery paste is leached of its PbO by a solution of dilute acetic acid. This solution is titrated with 0.05 M EDTA at a pH of 5.5 to 6.5 using methyl thymol blue as the indicator. Some error is introduced by the dissolution of some free lead, and there is a possible redox reaction between Pb and PbO_2 in acid solution to form PbO.

In the case of the determination of PbO_2 in battery paste and in the presence of Pb, and PbO_2 in the paste is dissolved in a boiling solution of hydrazine sulfate and ammonium acetate. Next, these lead ions are precipitated as lead carbonate by a solution of sodium bicarbonate. The solution is then filtered to remove the precipitate. Finally, the filtrate is titrated with 0.1N iodine solution to a persistant yellow endpoint. This determines the unoxidized hydrazine and then one can calculate the amount of PbO_2. Because of the boiling procedure, it is imperative that a blank that contains an equal quantity of a divalent lead compound be treated in an equivalent manner to compensate for some hydrazine decomposition.

For the $PbSO_4$ determination, the battery paste is boiled in a solution of sodium carbonate to dissolve the $PbSO_4$. The lead ions then react with the carbonate ions to form $PbCO_3$ which precipitates from the solution. The Na_2SO_4 formed remains in solution, and this solution is filtered to remove the precipitate. Next, $BaCl_2$ solution, which has been acidified with HCl, is added. The resulting precipitate is $BaSO_4$; and after normal gravimetric finish procedures, one can determine the $PbSO_4$ originally present in the sample. If a positive plate paste is to be analyzed, the $PbSO_4$ can be decomposed by dilute HCl, and the lead removed by reduction with zinc. The sulfate can be determined as before.

Another procedure by Gabrielson (45) for determining sulfate in the filtrate is to prepare the sulfate solution as before but treat the filtrate with a cation exchange resin. Then, after the CO_2 has been removed from filtrate by boiling, the sulfuric acid can be titrated with standard sodium hydroxide. All of the above procedures are capable of giving good results in the presence of all the usual battery paste constituents.

Dodson's (46) X-ray diffraction technique is used to differentiate between α-PbO_2 (tetragonal) and β-PbO_2 (orthorhombic). After the total PbO_2 has been determined by the above technique, the relative amounts of α-PbO_2 and β-PbO_2 are determined by X-ray diffraction. First, a standard plot is prepared from known mixtures of α-PbO_2 and β-PbO_2. Three diffraction lines are chosen for analytical purposes, the 3.49 and the 2.79 line of β-PbO_2 and the 2.61 line of α-PbO. A plot of the ratio of the average intensities of the first two lines to the intensity of the third,

$$\log \left| \frac{\text{Ave. intensity of 3.49 and 2.79 lines of } \beta\text{-PbO}_2}{\text{Intensity of 2.61 ines of } \alpha\text{-PbO}_2} \right|$$

versus the composition of standard mixtures of pure α-PbO_2 and β-PbO_2 (percentage of α-PbO_2 in mixtures of α- and β-PbO_2). The plot approximates a straight line on a semi-log scale and is used to determine the composition of the mixture of the two lead dioxides. This analysis was used on positive plate material and gave very reproducible results in repeated analyses of the standard mixtures and of various samples of positive plate material.

Caulder and Simon (47) have reported that studies of the positive active material (PbO_2) have revealed that there is an electrochemically inactive form of PbO_2 in the charged positive plate, in addition to α- and β-PbO_2. Nuclear magnetic resonance (NMR), differential thermal analysis (DTA), optical and scanning electron microscopy, and mass spectrographic studies revealed that, as the $PbO_2/PbSO_4$ plate was cycled, the electrochemically active PbO_2, which may be an amorphous compound, underwent a structural reordering with the loss of a hydrogen species. This structural reordering led to an electrochemically inactive PbO_2 compound, which gave NMR, DTA, and mass spectroscopy results similar to those obtained on reagent PbO_2.

This material does not take part in the discharge process, and seems to be responsible for the appearance of the reticulate structure that accompanies the presence of tetrabasic lead sulfate crystals in the paste. The continual conversion of electrochemically active PbO_2 to the electrochemically inactive PbO_2 is one of the major factors that causes battery capacity loss and ultimate failure.

VI. ELECTRICAL TESTS

Electrical performance and evaluation testing is a significant consideration in the characterization of secondary cells. In the case of primary cells, testing must be performed on a sample basis because each test is destructive. However, most secondary cells undergo a series of electrical tests both by manufacturer or user to determine the capability of meeting performance requirements. For the most part, each user has a preferred test procedure that simulates the specific application. However, there are some basic tests that a cell must pass in order to be considered acceptable. Among those of interest are measurements of capacity, usually at two or more temperatures, charge retention, and charge efficiency. Internal impedance as an indicator of quality of a cell was described by Brodd and Kozawa in Chapter 4 of this book. These tests are usually conducted after the cells have been filled with the proper electrolyte and then "activated" by imposing a series of conditioning or formation cycles. The activation process adjusts the active material into the desired composition, morphology, and structure.

A. Capacity Tests

The capacity of a cell is the number of ampere hours of electrical energy produced until failure of the firs of two electrodes (approximate cell voltage of 0.5 V) after a prescribed charge. The discharge rate and temperature must be considered in performing this test for the specific application.

B. Ampere Hour Efficiency

The charge efficiency (or A hr efficiency) is a measure of how efficiently the energy utilized in charging the cell is converted to usable discharge energy. Several alternative procedures can be utilized; however, it is recognized that all are rate and temperature dependent. In the nickel-iron system a 125% recharge at the 5-hr rate is expected to return the required cell capacity at the 5-hr rate.

A special low-rate efficiency test is performed on the nickel-cadmium sealed sintered electrode cells. This test is performed along with several other evaluation tests at the Naval Ammunition Depot, Crane, Indiana. Each cell is charged at the C/40 rate at 20° C, after being shorted for an extended period. The discharge is performed at the 2-hr rate to an end voltage of 0.5 V per cell. In this test, the cells are charged to only 50% of their capacity, then discharged, thereby minimizing the inefficient, nonreversible gassing process occurring at the end of charge. A minimum efficiency of 55% is required. This test is considered to be a measure of the low-rate charge acceptance of a sealed sintered plate cell.

C. Charge Retention

Charge retention is a measure of the capability of a cell to maintain its charge over prolonged periods of inactivity. The two most widely known causes of loss of available capacity are electrode instability and electrolyte impurities. The retention of charge also referred to as self-discharge is strongly dependent on temperature and to some extent dependent on the history during a cell's previous operations.

Most charge-retention tests involve bringing a cell to its full state of charge and then allowing it to stand on open circuit at an ambient temperature for periods up to 30 days. The cell capacity is then determined in the same manner as that performed during the capacity tests.

For the lead-acid cell, the sequence involves discharge at the 8-hr rate to 1.75 V per cell. The battery is then recharged to 110% of the previous discharge. After a 12-hr open-circuit stand, the cells are again discharged at the 8-hr rate to 1.75 V per cell. The capacity on the last discharge must meet 85% of the rated cell capacity. In effect, this test is a measure of the loss of charge experienced on open-circuit stand.

Electrical Tests

An unusual charge-retention test referred to as the internal short test is performed on the sealed nickel-cadmium cell. Each cell after a normal capacity measurement is shorted for a 24-hr period. The shorts are removed and each cell is allowed to stand on open circuit for a period of 16 hr. (Sometimes a 5-min charge is given the cells after removal of the shorting bar.) The requirement for acceptance is for each cell to reach a minimum value of 1.15 V per cell within that period. This test is performed in order to ascertain whether there are any high-resistance shorts between the cell plates that would not have been indicated by the high-rate impedance test.

VII. SEPARATORS

The separator serves to maintain a physical separation between positive and negative plates to prohibit shorting diffusion and migration of the plate materials In some cases, notably in the nickel-cadmium sealed cell, the separator also is required to store the electrolyte and maintain it immobilized, analogously to a sponge. Separators of the film or membrane type are used for separation. Nonwoven types are used for separation and electrolyte adsorption as well. The physical methods employed to evaluate and characterize the separators include mechanical/dimensional, adsorption, retention and wet-out, tensile strength, porosity and permeability, chemical stability and analysis. Some of these are described in detail and others only briefly.

A. Mechanical/Dimensional Stability

The two most important criteria under this subject are the thickness and density. For films and membranes the measurements are simple. However, if these separators change their properties by swelling in the electrolyte, the procedures are more complex. In addition, the thickness of the nonwoven separator materials that are easily compressed requires some uniform method to ensure that the measurement has some meaning. One such method is with the Ames Type 240 gauge, which utilizes a 0.5-in. anvil and spring tension to compress the separator in a uniform manner. There is a 50-g load applied during measurement and the gauge can be read to 0.0001 in. In the NASA/GSFC specification (48), nonwoven separators for Ni-Cd cells requires 0.030 ± 0.005 mm thickness. Once thickness is known, the density for the specific separator material of interest can easily be determined.

B. Adsorption, Retention, and Wet-Out

This characteristic is determined for both film/membrane and nonwoven types of separators. These measurements provide data on the level of ionic conductivity or internal impedance that a cell may experience. These tests include the measurement of the quantity of electrolyte adsorbed in the separator, the rate of pickup, and the ability to retain it during acceleration.

In the procedure described by Shair and Seiger (49), a standard sample of separator (3/4 x 1-1/4 in.) is weighed dry, then immersed in electrolyte for 5 min or until constant weight is achieved. The sample of separator is wiped across a glass or Plexiglass plate until no droplets of electrolyte remain on the glass. Then the separator is reweighed. To determine how well the separator retains the electrolyte, the sample is centrifuged at 25 g for 2 min and reweighed. Maximum adsorption is desired. Those of the nylon nonwoven type absorb 12 to 13 g of 30% KOH electrolyte per gram of separator. The limit on retention of electrolyte desired is greater than 90%.

TABLE 10

Sample	Time		Circle Range	
	Min	Sec		
A. Pellon 2505-K4 (Calendered)	3	0	3-1/2-5	Nylon
B. Pellon 2505-K4 (Calendered) acid treated	21	54	3-4	Nylon
C. Pellon FT 2140	25	6	1-1/2-2	Polypropylene
E. Kendall E 1451	4	8	5	Polypropylene
H. GAF WEX 1242-304-PO	6	51	5	Polypropylene
J. Pellon 2505 ML	3	36	2-1/2-3	Polypropylene

A. Average of 5 specimen

B. Drop of KOH not completely absorbed by the time test concluded.

Wet-out time is a measure of the speed with which a separator sample adsorbs the electrolyte. The method (50) utilizes a standard 2.5-cm-diameter specimen on a piece of nylon netting. A volume of 0.1 ml of 30% KOH containing phenol purple is placed in the center. The time to reach the outer circle is considered to be the wet-out time. Some data on nonwoven materials appear in Table 10.

C. Porosity and Permeability

The technique described in the section on porous plaques is applicable also for separator materials. The weight of electrolyte picked up during the adsorption test is divided by the weight of electrolyte assuming the void volume of the separator was filled with electrolyte. This procedure is better for the nonwoven separators.

A procedure recommended by Cooke and Lander (51) based on laminar flow throu fibers is useful in the determination of por sizes in the 10 to 100 Å range. It utilizes the permeation of water through a membrane. However, it requires a specially designed cell and osmometer. Pore radius is equal to

$$\text{Pore radius} = (2)\left(\text{wet film thickness}\right)\left(\frac{(2)\ (\text{viscosity of water})\left(\dfrac{\text{volume of water passed}}{\text{area of film}}\right)}{(980.6)\left(\begin{array}{c}\text{Avg. height of}\\ \text{water column}\end{array}\right)\dfrac{(\text{wet-dry weight})}{(\text{density of liquid}}\left(\begin{array}{c}\text{penetra-}\\ \text{tion time}\end{array}\right)}\right)$$

$$r = 2\ell\left(\frac{2\ \eta\ \dfrac{V}{a}}{980h\ \dfrac{W_w - Wd}{\rho}\ t}\right)^{1/2}$$

The method is somewhat inaccurate for materials that are very different, but it is useful in comparing the same material. Salkind and Kelley (52) have described a similar measurement technique that is specifically used with alkaline electrolytes. Utilizing the same Poisseuille's law as the basis, their equation becomes

$$r = 3.18 \times 10^{-3}\ L\left(\frac{1}{B_t}\ \log\frac{H}{h}\right)\ 1/2 \qquad (37)$$

for a solution of 31% KOH whose density is 1.3 g/ml and viscosity 0.022 g/cm sec and for a 0.025-cm^2 cross sectional tube area with 0.785 cm^2 of separator exposed. L is the separator thickness; B is the pore volume of film; H is the original height of the liquid; h is the height of the liquid at time t. This apparatus utilizes a capillary tube with Teflon stopcock and polyethylene holder all inert to the alkali. Comparison of this with mercury porosimetry data was good.

D. Air Permeability

The air permeability method has been utilized to determine differences in structured samples. The Gurley Densometer Model 4110 with 5-oz inner cylinder is an instrument that provides a constant air pressure equivalent to 1.22 in. H_2O across a fixed sample area (0.1 in^2). An example of air permeability measurements on several separator materials before and after extraction with water was reported by Baker, Tower, and Cuthrell (Table 11) (50) as the time required for 300 ml of air to pass through the 0.1-in^2 sample.

E. Tensile Strength

This measurement to ascertain how well the fibers of the separator are held together. It requires a standard tensile strength sample that has the appearance of a bow tie. Prior conditioning of the samples is required to reduce spurious errors. The preconditioning should include placing the sample in a constant relative humidity environment at 24°C for 24 hr prior to pull. A controlled cross-head pully speed of 2 in/min is desired for uniformity of results. The results should be compared with similar samples that have been exposed to the electrolyte of interest in order to ascertain whether there has been some effect on the fiber bonds. The samples must be labeled as to which of the two directions (machine or cross) the sample represents. Signficant differences will exist between the two directions.

F. Chemical Stability and Analysis

The chemical stability of a separator in the electrolyte, whether it be 30% KOH or H_2SO_4, is important

TABLE 11

	Type **	Time to Pass 300 ml of Air(s) As Received	After Extraction with Methanol
A. Pellon – 2505-K4	N	5.9	4.2
B. Pellon – 2505-K4 – Calendered. Acid treated in 0.7% HCl. Washed in deionized water until neutral. Dried in hot air oven at 50°C.	N	3.9	3.9
C. Pellon – FT-2140	P	8.5	7.1
D. Micro polypropylene fiber RT-37-2665-15	P	84.6	36.7
E. E-1451	P	223.7	196.1
F. E-1451 methanol washed	P	154.9	185.3
G. E-1451 methanol washed – air dried – acid washed (conc. H SO) approx. 1/2 hour. Washed in deionized water until neutral. Dried in hot air oven overnight 85-90°C.	P	148.2	177.2
H. WEX-1242-304 PO	P	3.7	4.6
I. WEX-1242-304 PO methanol washed and air dried.	P	2.7	2.7
J. Pellon nylon – 2505 ML	N	2.2	2.2
K. Kendall XM1253.2 (Florida)	P	4.6	4.7
L. Celgard 2400 wettable	P	0	0
M. Celgard 2400 non-wettable	P	0	0
N. 10SA6.4 + Triton X	T	2446.4	2674.8
P. 13684-105 wettable	T	525.2	603.1
R. 71-156-3	T	2702.6	3359.7
S. 13684-104	T	1362.4	1166.9
T. Micro polypropylene fiber No. 2711-54, 1 oz./sq. yd.	P	4.6	3.4
U. Micro polypropylene fiber No. 2711-55, 1.5 oz./sq. yd.	P	12.2	9.4
V. Micro polypropylene fiber No. 2711-56, 2 oz./sq. yd.	P	7.3	7.5

*Average of 10 specimens

**N - nylon; P - polypropylene; T - polyterrafluoroethylene

to the long-term operation of a secondary cell. If the electrolyte reacts with the separator and changes its properties, the result will have an effect on the properties of the cell. The changes in separator properties may be somewhat subtle and require a long period of time to be recognized. One such example is the nylon in the nonwoven separator of a nickel-cadmium cell. Nylon, a polyamide, is attacked by concentration alkali. A test program for chemical stability should include exposure of the separator to the electrolyte at the temperatures of interest. In addition, exposure to the oxygen gas normally existing during charge of the cell. Comparison should be made of the separator properties before and after the exposure to similar cell conditions. Even though the nylon material does not have the long-term chemical stability desired, it has the best electrolyte wetting and retention properties of the materials available.

Other sources of separator problems are the inorganic and organic extractables that are present in the separator. During the course of oxidation and reduction in the cell, foreign ions and organics can produce deleterious effects on cell performance. The foreign materials are introduced during the separator manufacture either as antistatic agents, process agents, or in the wash water. Another source of impurities is from the fiber itself or solvents used in the fiber manufacture. Organic extractables are removed by extraction with soxlet extractor. Methanol has been found to be a suitable solvent. Samples of 10-cm^2 size are washed for 24 hr in the extractor. The extractable can be identified by evaporating the solvent and using standard infrared analysis techniques. Inorganic content is determined by ignition of the separator sample containing the foreign materials. The residue is a measure of the inorganic composition. Identification is made through emission spectroscopy or atomic absorption techniques.

G. Impedance Measurement

The impedance or electrical resistance of a separator is important in that it provides a measure of tortuosity or how easily the electrolyte or ions pass through the membrane or fibers of the separator. Several methods have been devised to measure this characteristic. The method of Salkind and Kelley (53), originally devised for membranes, has been adapted to nonwoven fiber types also by NBS (50).

Making use of the theory that if a current flows through an electrolyte path of uniform crosssection the voltage drop measured between two identical electrodes on opposing sides of the separator is related to the resistance by Ohms law, one obtains

$$R=\frac{E}{I} \tag{38}$$

The specific resistance ρ is given by

$$\rho=\frac{AR}{\ell} \tag{39}$$

where A is the cross-sectional area, R is the resistanc and ℓ is the distance between electrode tips.

A potential measurement (E') is made between the electrodes in the electrolyte of interest without the separator. Then the separator is spaced between the electrodes and a second potential measurement (E) is made. The first measurement is the iR drop across the solution and the second, iR', includes the separator:

$$E=IR + IR' \tag{40}$$

$$R'=\frac{E' - E}{I} \tag{41}$$

$$R''=AR'=\frac{(E' - E)A}{I} \tag{42}$$

where R" is in ohm cm^2 and A is the area in cm^2.

A bridge circuit can be used for measurement; a.c. resistance measurements are preferred over d.c. methods because the polarization effects are minimized. If R_C is the resistance of the cell without separator and R_S with separator, then

$$R'=R_S - R_C \tag{43}$$

Electrodes should be fabricated from inert materials that is, platinum or nickel. Vents are needed to allow

TABLE 12

Sample	Specific Resistivity, ohm-cm					Material
	Original Extracted	Soxhlet Extracted After Cycling				
		Number of Cycles				
A. Pellon 2505 K4	2.88	2.86	4.07	3.14	3.25	
B. Pellon 2505 K4 Acid Treated	3.49	2.92	3.11	2.92	2.87	Nylon
C. Pellon FT 2140 Herclues	5.67	5.70	5.10	e	7.16	Nylon
D. Microporous Fiber RT 37-2665-15	b	64.07[c]	13.69[d]	e	b	Polyprop
H. GAF WEX 1242-304 Po	3.67	5.95	3.57	3.92	6.59	Polyprop
I. GAF WEX 1242-304 Po Methanol Washed	3.22	3.22	3.44	e	9.72	Polyprop
J. Pellon 2505 ml	2.06	3.88	2.80	2.87	3.00	Nylon

a/ Average of 10 specimens.
b/ Sample would not wet.
c/ Value for one specimen after 24 hours, but resistance was still dropping very slowly.
d/ Average value obtained after 24 hours of test.
e/ Sample has not been received for testing.

gas bubbles to excape. A 31% KOH electrolyte in a cell of the type designed for this purpose produced a cell resistance of 0.3 ohms. Measurements from the Baker, Tower, and Cuthrell (50) on separator materials are given in Table 12.

VIII. ELECTROLYTE

Secondary cell electrolytes are solutions of high ionic conductivity that must be inert to the electrodes separator, and internal cell materials. The electrolyt for the lead/lead dioxide cell, sulfuric acid, and that for the alkaline systems, potassium hydroxide, provide the desired conductive and stable media necessary for long-life rechargeable systems.

Both electrolytes, even though on the opposite end of the pH scale, have similar physical characteristics with regard to conductivity/resistivity, viscosity, and specific gravity. They differ in chemical properties insofar as the impurities or additive cause side reactions that affect the overall electrode reaction.

A. Conductivity/Resistivity

Electrical conductivity or resistance within an electrolyte depends on the nature and number of ions present. The resistance of any conductor is given by

$$R=\frac{\rho \ell}{A} \tag{44}$$

where ρ is the specific resistivity (characteristic of the conductor), ℓ the length of the path, and A the cross-sectional area. The inverse of specific resistance is the specific conductivity κ given by

$$\kappa = \frac{1}{\rho} = \frac{\ell}{AR} \tag{45}$$

The value of κ reaches a maximum of 7.4 mho/cm for 31.5 wt % H_2SO_4 and 5.4 mho/cm for 27.5 wt % KOH at 18°C as noted on the curves plotted in Fig. 13. The maxima change slightly with temperature.

Also plotted in the fiture is the more comparative measure of electrolyte conductivity, namely the equivalence conductant, Λ_c, which is based on the conductivit

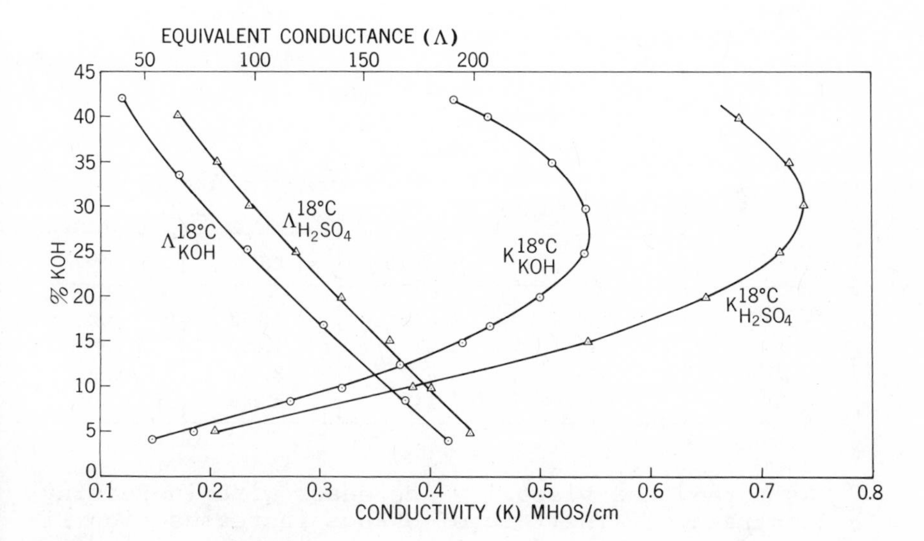

Fig. 13. Specific and equivalent conductant for H_2SO_4 and KOH at 18°C.

per gram equivalent weight. The equation for equivalent conductance is

$$\Lambda_c = \frac{1000\kappa_c}{cn_e} = \frac{1000\ell}{ARcn_c} \tag{46}$$

where c is the concentration in gram moles/liter and n_e is the number of equivalents per mole of anion plus cation. The equivalent conductance increases as concentration decreases and also varies with temperature.

B. Viscosity

Another of the electrolyte parameters of interest is the viscosity, which has an effect on ion mobility and electrolyte conductance. The equation relating viscosity η with concentration is

$$\log \frac{\eta}{\eta_0} = \frac{A_c}{1-B_c} \tag{47}$$

where η_0 is the viscosity (cp) of pure water c is the concentration in moles/liter and A and B are constants for the specific solutions involved. For KOH A_C = 0.0476, B_C = 0.0199 at 25°C. The values for η for the two electrolytes are given in Table 13.

TABLE 13

% By Wt.	H_2SO_4		KOH	
	20°C	30°C	20°C	40°C
10	1.228	0.976	1.232	.831
20	1.545	1.225	1.629	1.105
30	2.006	1.596	2.421	1.607
40	2.700	2.163	4.159	2.595

As noted, the viscosity increases with increasing concentration. The rate of change increases significantly faster above 30 wt%. Of greater significance is the increase in viscosity as temperature falls, especially below -20°C. The viscosity change with temperature is 0.05 cp/°C from 20 to 40°C, 0.1 cp/°C from 0 to 20°C, and as much as 1.5 cp/°C from -20 to -40°C. This has a profound effect on the ion migration and therefore cell impedance at the lower temperatures.

C. Specific Gravity/Density/Concentration

The parameter of most practical value is the specific gravity of the electrolyte measured with a hydrometer. The definition of specific gravity is given as

$$S.G. = \frac{\text{Density of Solution at } T_1}{\text{Density of pure water at } T_1}$$

The density of water is approximately 1.0 at ambient temperatures and therefore the specific gravity of an electrolyte can be given as a unitless value of density Density is another means of describing concentration, thus, relating the three characteristics. Both the lead acid cell and the alkaline cells exhibit a decrease in electrolyte and an increase in water during a discharge with the reverse occurring during charge. Therefore, starting with a known concentration and in absence of spillage, the specific gravity of the electrolyte provides a measure of the state of charge the battery as exemplified by, the following lead acid cell data:

	Specific Gravity
Fully charged	1.260-1.280
75% charged	1.230-1.215
50% charged	1.180-1.170
25% charged	1.130-1.120
Discharged	1.080-1.070

The temperature coefficient for sulfuric acid electrolyte is approximately 0.001 unit/°F. Generally, the specific gravity measurements are corrected to 77°F for comparative measurements.

Another interesting feature of electrolyte dependent properties is the change in volume during cell operation. During cell overcharge, while water is converted to oxygen and hydrogen, there is an obvious decrease in volume. There is also a decrease in volume when discharging a cell. It occurs in the lead/lead dioxide cell by the conversion of 2 mol of sulfuric acid to 2 mol of water. The net result is a loss of 1 ml of volume of a cell is low, the full capacity of the cell may not be realized during the discharge.

The concentration of potassium hydroxide in the nickel-cadmium sealed cell changes from 28.5% in the discharged state to 22.9% in the charged condition (54). This is caused by the cell reaction in which 2 moles of water are formed for every mole of Cd and 2 moles of NiOOH formed during the charging process.

The open-circuit potential of each electrode is also affected by electrolyte concentration as reported by May (55). He reported that the cell potential was related to

$$E_{cell} = E^{\circ}_{cell} + 1.985 \times 10^{-4}\ T \log a_w \qquad (48)$$

where T is the absolute temperature and a_w the activity of water at the given concentration of KOH. The values of a_w are found from the vapor pressure lowering of the hydroxide solutions and therefore the potential is

$$E_{cell} = E^\circ_{cell} + 8.619 \times 10^{-5}\ T \ln \frac{p}{p^0} \quad (49)$$

where p is vapor pressure of water over hydroxide solution and p^0 is the vapor pressure of water at the same temperature. The net effect of concentration on potential is given in Fig. 14.

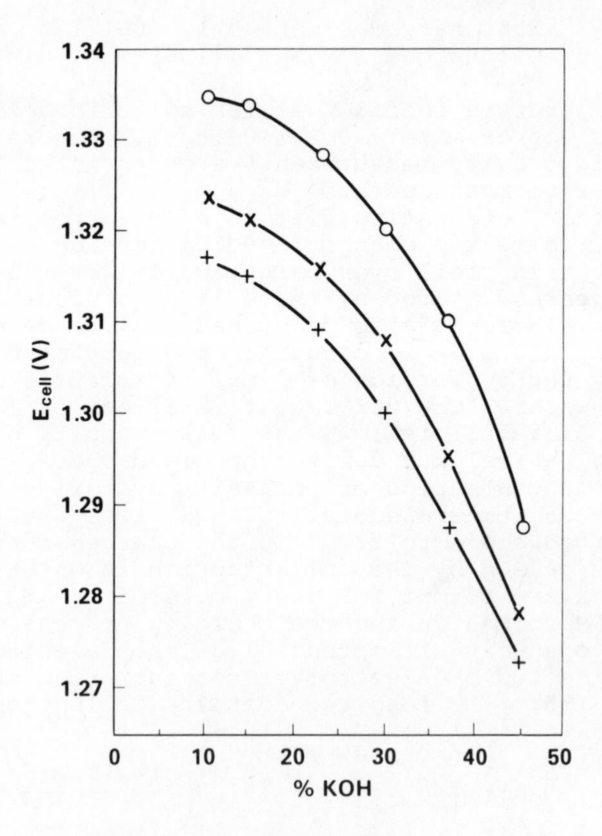

Fig. 14. Variation of cell potential with KOH concentrations at 0, 25, and 40°C.

D. Impurities

Impurities in the electrolyte can have an effect on cell performance particularly for the longer-life applications. The general effect of metal ions is to plate out on the cathode and affect the cell efficiency. In the lead-acid cell, bismuth, antimony and arsenic form sulfates that interfere with the production of lead sulfate. Iron is the most common impurity. Oxidized at the anode and reduced at the negative, the net effect is the discharge of both. Manganese in the form of MnO_4^- oxidizes the separator material, rubber ribs, or any organic parts internal to the cell. Chlorides in the form of HCl and nitrates as the acid tend to dissolve lead and lead dioxide and produce the discharged lead sulfate. This process is known as sulfation; it is an irreversible process that leads to cell degradation.

In the alkaline cells, impurities are also of concern. The one most often cited is carbonate. Its effect on potential was reported by May (55), and its effect on cadmium migration was reported most recently by Mayer (56) in which cadmium migration increased by a factor of 3 in concentrations of 0.8 to 6.2% K_2CO_3 in 34% KOH. Others report that carbonate clogs pores in the electrodes, causing agglomeration. Casey, et al. (57) refer to it as the carbonate choke.

Nitrate functions in the nickel-cadmium cell similarly to iron in the lead-acid cell. Referred to as the nitrate shuttle (57), cadmium reduces the nitrate to nitrite and the NiOOH oxidizes nitrite to nitrate thereby causing a net effect of cell discharge. It was reported that 7×10^{-3} M KNO_3 in a cell at 55°C was equivalent to a 100-ohm resistor connected directly across the terminals.

The major additive to the alkaline electrolytes is lithium hydroxide to increase the efficiency of the positive plates in the nickel-iron cell and to a lesser extent the nickel-cadmium cell.

Cellulose (~1%) in various forms was added to electrolyte of nickel-cadmium cells to reduce the agglomeration of the negative plate, thereby increasing its efficiency. Usually it is added during plate activation tests, not to the cell electrolyte.

Zinc oxide is added to the electrolyte of silver-zinc and nickel-zinc cells. The electrolyte is saturated with the zinc oxide to prevent the dissolution of zinc from the plates. This electrode generally limits the cell life by reducing the solubility and dendrite formation. The addition results in decreased conductivity and increased charge voltage.

Other alkaline electrolyte impurities include calcium, magnesium, thallium, and aluminum, which have a deleterious effect on the negative plate, and silicon and iron, which affect the positive plate. The effect of these ions are discussed in other chapters.

IX. MOLTEN-SALT CELL EVALUATION TECHNIQUES

Quality assurance tests can be directed at the particular requirements for each application and will include component as well as cell tests. Battery life in tube-type β-alumina type cells is primarily a functi of the characteristics of the β-alumina. Recommended tests are as follows for β-alumina:

Test	Minimum Requirement
Sodium-sodium test	
Charge sodium metal through β-alumina tube at a current density of 0.2 A/cm^2 of β-alumina at recommended all operating temperature in range 200° to 300°C. Test can be accelerated to 1.2 A/cm^2 for excellent β-alumina.	1000 A hr/cm^2 to failure by sodium penetration and cracking of β-alumina, or 5000 hr 1500 A hr/cm^2 to failure or 1250 hr operation.
β-alumina surface morphology	
Microscopic examination	Average particle size 5 μ. Absence of Duplex structure (large aggregate crystals to 100 μ).
Diametral break strength	
Apply pressure to ring of β-alumina along diameter to breakpoint.	20,000 psi

d.c. resistance test

β-alumina in Na at 200°C	8 to 12 ohm cm
β-alumina in Na at 300°C	4 to 6 ohm cm

Visual check of β-alumina

Tube with 6 W light bulb inside tube (on each tube).	No cracks. No foreign particles. No dark spots. Even luminosity all over.

Helium leak check

On each tube	Leakage no greater than 10^{-8} ml/(sec/cm^2)

Dimensional check

On each tube. Diameter.	Pass

Chemical composition

By analysis of major components. Absence of critical contaminants K and Si, which cause premature failure of β-alumina.	Desired ± 0.1%. K^+, 500 ppm max.; Si, 500 ppm max

The sodium-sodium test and the chemical analysis should be performed on lots, not on each tube; however, for strict quality control, samples from each tube are removed from the excess tube length at machining to the length dimension and submitted for the break strength test, d.c. resistance test, and morphology test.

Chemical analysis should be performed on each active material to ensure conformance to material specification requirements. Normally, ACS reagent-grade chemical purity is required for all chemical components, but particular analysis varies with the electrochemical system.

During assembly, precautions are taken to eliminate water and oxygen by using dry-box helium atmospheres less that 1 50 2 ppm in water and oxygen. Also, oil vapors or film precursors must be eliminated in purification to prevent contamination of β-alumina surfaces and subsequent failure of sodium to "wet" beta at 200 to 300°C.

Preproduction qualifications and lot acceptance cell tests are designed around requirements for a particular application and include:

1. Formation charge as specified by manufacturer.
2. Performance characterization tests at operating temperature: discharge at the 1-, 3-, 5-, 7-, and 10-hr rates; charge at the 5-, 7-, and 10-hr rates.
3. Life cycle tests in a routine equivalent to expected service, or accelerated by increasing the current density and the number of cycles per day, or both.
4. Environmental shock and vibration tests.

Design performance parameters measured by the above tests would be calculated from the data and would include: (a) discharge capacity, energy, and average voltage at each discharge rate; (b) charge capacity, energy, and average voltage at each charge rate; (c) turn-around energy efficiency at each charge/discharge condition; (d) cycle life to 50% of initial or maximum capacity; (e) modes of failure exhibited in performance and environmental tests.

Molten salt batteries are being developed to achieve higher delivered energy per unit battery weight and volume coupled with longer cycle life than existing commercial systems. Cycle lives of 10 to 20 yr are goals. For this service expectation, a cost-effective quality assurance program is a very realistic requirement.

ACKNOWLEDGMENTS

The author wishes to acknowledge the assistance and understanding of D. A. J. Salkind, the special additional material supplied by Mr. James Doe and Dr. Salkind of the ESM Technology Center, and the advice of Dr. Leopold May of Catholic University.

REFERENCES

1. R. J. Brodd and A. Kozawa, in Techniques of Electrochemistry, Vol. III, E. Yeager and A. J. Salkin

Eds. John Wiley, New York, 1976.
2. M. Salomon, Batteries for the electric automobile, private communication.
3. E. Luksha, Long-Life Rechargeable Nickel Zinc Battery, NASA Contract NASA 3-16809, final report, Sept. 1974.
4. S. U. Falk and A. J. Salkind, Alkaline Batteries, John Wiley, New York, 1969.
5. C. L. Mantell, Batteries and Energy Systems, McGraw-Hill, New York, 1970.
6. M. Klein, 1972 IECEC (Intersociety Energy Conversion Engineering Conference Proceedings), pp. 79-85.
7. S. Gratch and J. F. Petrocelli et al., 1972 IECEC Proc., p. 38-41.
8. E. J. Cairns and J. S. Dunning, Proceedings of the Symposium and Workshop on Advanced Battery Research and Design March 22-24, 1976; Argonne National Laboratory and Chicago Section of the Electrochemical Society, Report No. ANL-76-8.
9. O. Lindström, in Power Sources-5, D. H. Collins, Ed., Academic Press, London, 1975, pp. 283-302.
10. R. J. Haas and D. C. Briggs, 1973 IECEC Proc., pp. 116-120.
11. C. J. Amato, Paper 730248, and P. C. Symons, Paper 730253, presented at the SAE International Automotive Engineering Congress and Exposition, Detroit, 1974.
12. M. Klein and M. George, Proceedings of 26th Annual Power Sources Conference, 1974, pp. 18-20.
13. L. E. Miller, Proceedings of the 26th Annual Power Sources Conference, 1974, pp. 21-24.
14. J. Burbank, A. C. Simon, and E. Willinghanz, The Lead Acid Cell, in Advances in Electrochemistry and Electrochemical Engineering, Vol. 8, John Wiley, New York, 1972, p. 157.
15. G. Vinal, Storage Batteries, John Wiley, New York, 1955.
16. P. C. Milner and G. B. Thomas, The Nickel-Cadmium Cell, in Advances in Electrochemistry and Electrochemical Engineering, Vol. 5, John Wiley, New York, 1967, p. 1.
17. P. Bauer, Batteries for space power systems, NASA Report SP-172 (1968).
18. M. Sulkes, Proceedings of 23rd Annual Power Sources Conference, 1969, p. 112.

References

19. A. Drarkey, Proceedings of 25th Annual Power Sources Conference, 1972, p. 64.
20. T. P. Jackson and E. F. Colston, Goddard Space Flight Center X-Document No. X-716-70-113, March 1970.
21. A. Fleischer and J. Lander, Ed., Zinc-Silver Oxide Batteries, John Wiley, New York, 1971.
22. J. Dunlop, J. Stockel, and G. Van Ommering, in Power Sources 5, Academic Press, New York, 1975, pp. 315-329.
23. M. Klein, in Power Source 5, Academic Press, New York, 1975, pp. 347-359.
24. D. S. Adams, in Power Sources 4, Oriel Press, Newcastle upon Tyne, England, 1973, pp. 347-361.
25. P. C. Symons and P. Carr, 1973 IECEC Proc., p. 72-77.
26. L. J. Miles and I. Wynn Jones, in Power Sources 3, O. H. Collins, Ed., Oriel Press Newcastle upon Type, England, 1971, p. 245.
27. J. J. Werth, U. S. Patent No. 3,877,84, April 15, 1975.
28. J. E. Metcalf, E. J. Chaney, and R. A. Rightmeyer, 1971 IECEC Proc., p. 685.
29. J. C. Schaefer, T. M. Noveske, J. S. Thompson, and B. Profeta, 1975 IECEC Proc., p. 649.
30. P. A. Nelson, E. C. Gay, and W. J. Walsh, 26th Annual Power Sources Conference, 1974, p. 65.
31. G. Halpert, Nickel Powder and Plaques for Nickel Cadmium Cell Plates, GSFC X-735-68-400, Oct. 1968.
32. American Society for Testing and Materials (ASTM) Part 7, p, 442, (1970).
33. J. Parry, Development of uniform and predictable materials for nickel cadmium aerospace cells, NASA/GSFC Contract No. NAS-5-11561, 1st Quarterly Report, 1968.
34. E. L. Gordon and C. M. Smith, Ind. Eng. Chem. Anal. Ed. 12, 479 (1940).
35. A. J. Salkind, in Techniques of Electrochemistry, E. Yeager and A. J. Salkind, Ed., Vol. I, Chapter 4, John Wiley, New York, 1972.
36. J. Parry, Development of uniform and predictable materials for nickel cadmium aerospace cells, NASA/GSFC Contract No. NAS-5-11561, 2nd Quarterly Report.

37. E. Kantner, Production of Uniform Nickel Cadmium Battery Plate Materials, 1st Quarterly Report NASA/GSFC Contract NAS-5-21045.
38. W. Campbell, Holder for Pure Size Distribution Measurements, NASA SP 5926(02) 635.
39. J. F. Unger and J. T. Stemmle, A soxhlet extractor system for cleaning electrochemical cell components, X-761-71-157, April 1971.
40. G. Halpert, et al., Procedure for analysis of nickel cadmium cell materials, GSFC X-711-74-279.
41. H. Kroger and A. J. Cattoti, 24th Annual Power Sources Conference, Vol. 24, 1970, p. 4.
42. H. R. Thirsk and D. Lax, in Zinc-Silver Oxide Batteries, A. Fleischer and J. J. Lander, Eds., Chapter 12, John Wiley, New York, 1971.
43. J. A. Keralla, in Zinc-Silver Oxide Batteries, A. Fleischer and J. J. Lander, Eds., Chapter 13, AD 447301, 1964.
44. M. Denby, in Batteries, K. H. Collins, Ed., MacMillan, New York, 1963, p. 439.
45. G. Gabrielson, Anal. Chim. Acta 15, 426 (1956).
46. W. H. Dodson, J. Electrochem. Soc., 108, 401 (1961).
47. S. M. Caulder and A. C. Simon, J. Electrochem. Soc. 121, 1546 (1974).
48. Specification for the Manufacturing of Aerospace Nickel Cadmium Storage Cells, NASA GSFC Specification 74-15000 with Amendments, March 1975.
49. R. C. Shair and H. N. Seiger, in Characteristics of Separators for Alkaline Silver Oxide Zinc Secondary Batteries - Screening Methods, J. E. Cooper and A. Fleisher, Eds., Chapter 4, A. F. Aero Propulsion Labs, Dayton, Ohio.
50. H. A. Baker, S. D. Tower, and W. F. Cuthrell; Evaluation of separator materials used in nickel cadmium satellite batteries, NBS Report 10956, Nov. 1972.
51. L. M. Cooke and J. J. Lander, in Characteristics of Separators for Alkaline Silver Oxide Zinc Secondary Batteries $\frac{1}{M}$ Screening Methods, J. E. Cooper and A. Fleischer, Eds., Chapter 5A.
52. A. J. Salkind and J. J. Kelley, in Characteristics of Separators for Alkaline Silver Oxide Zinc Secondary Batteries $\frac{1}{M}$ Screening Methods, J. E. Cooper and A. Fleischer, Eds., AF Aero Propulsion Labs, Dayton, Ohio, AD 447301, 1964, Chapter 5B.

References

53. A. J. Salkind and J. J. Kelley, in Characteristics of Separators for Alkaline Silver Oxide Zinc Secondary Batteries 1/M Screening Methods, J.E. Cooper and A. Fleischer, Eds., AF Aero Propulsion Labs, Dayton, Ohio, AD 447301, 1964, Chapter 6B.
54. G. Halpert, Electrolyte concentration changes during operation of the nickel cadmium cell, GSFC X-Document X-711-75-135, May 1975.
55. L. May. The effect of the electrolyte on the open circuit potentials in the nickel cadmium cell, GSFC X-Document X-711-75-271.
56. S. W. Mayer, J. Electrochem. Soc. 123, 159 (1976).
57. E. J. Casey, A. R. Dubois, P. E. Lake, and N. J. Moroz, J. Electrochem. Soc. 112, 371 (1965).
58. F. J. Port, Proceeding of the 4th Electric Vehicle Symposium, Dusseldorf, W. Germany, 1976.

VI. ELECTRODEPOSITION

Dodd S. Carr

International Lead Zinc
Research Organization, Inc.
New York, New York

I. INTRODUCTION

For many years, the process of electrodepositing metals was considered a closely held secret. From the time that the first metals were plated, around 1800, until after World War I there was relatively little exchange of information in this field. However, since 1920 the subject of electrodeposition has formed the basis of many technical articles, books, and scientific symposia. Thus the field of electroplating was transformed from an art into a worldwide subject of scientific investigation and commercial exploitation. This chapter describes some of the more important techniques used today in electroplating. These are divided into four major categories: (a) preparation of plating solutions, (b) preparation of surfaces to be plated, (c) evaluation of electrodeposits, and (d) special plating techniques.

II. PREPARATION OF PLATING SOLUTIONS

One of the most important prerequisites for successful electroplating is the proper preparation of the plating solutions to be used. This involves the selection, purification, evaluation, and control of each of the solutions used in the plating cycle.

Preparation of Plating Solutions

A. Selection

Quite often, a plating cycle will specify the use of two or more plating solutions to meet the functional requirements of the finished product. For example, a zinc die casting will normally receive three successive electrodeposits (copper, nickel, and chromium) to provide the requisite corrosion resistance and decorative appearance for exterior automotive hardware applications. A mildly alkaline copper plating bath is selected to prevent chemical attack of the zinc during plating, to provide complete coverage in recessed areas, and to permit buffing when needed. Next a duplex, acid-nickel plating system is used (consisting of a sulfur-free, leveling bath followed by a high-sulfur, bright nickel bath) to provide a smooth, corrosion-resistant, decorative finish. Finally, an acid-chromium plating solution is selected to give the part a hard, scratch-resistant, and tarnish-resistant finish. Thus, each plating solution in the cycle is selected to meet the processing or performance requirements of the zinc die casting. Of course, the thicknesses of the plated coatings must meet or exceed the minimum standards established for outdoor weathering exposures. Such standards have been published by the American Society for Testing and Materials as well as by other standards-setting organizations around the world.

The selection of plating solutions to satisfy functional specifications is also illustrated by the following examples of electrodeposited coatings:

1. Protective: hard chromium coatings on steel dies to reduce wear, maintain close tolerances, and provide longer life.
2. Reflective: gold-plated infrared reflectors for heat-seeking rocket; rhodium-plated reflectors in searchlights.
3. Electrical: platinum-plated titanium anodes for precious metal plating; gold-plated contacts on printed circuits for low contact resistance.
4. Structural: thick copper electroformed waveguides; heavy electroformed nickel cavities for molding plastics.
5. Masking: copper plating of selective areas on steel parts to prevent carburizing; gold plating for

selective etching of printed circuits.

6. Joining: tin-lead alloy plating of copper wire for improved solderability; gold plating of semiconductors for diffusion bonding silicon chips.

7. Lubrication: copper plating of steel to permit improved wire drawing; indium alloy plating of steel bearing surfaces.

8. Magnetic: cobalt alloy plating to provide computer memory drums with a magnetic surface.

In addition to the selection of plating solutions for a given job, it is necessary to select suitable pretreatment solutions, such as cleaners and acid activators. Data on the preferred chemical composition for cleaners, acid activators, metal plating solutions, and recommended operating conditions (temperature, time, current, density, pH) are given in the standard plating handbooks.

B. Purification

Before a new plating solution can be expected to give satisfactory electrodeposits, it must be given a thorough purification. The presence of inorganic, organic, and gaseous impurities in plating baths causes the majority of plating problems, such as roughness, brittleness, pitting, cracking, peeling, and high stress in the electrodeposited metal. Therefore, it is also important to remove harmful impurities at regular intervals during the operation of plating baths.

Inorganic impurities in plating solutions may be introduced from one or more of the following sources: (a) contaminated water containing rust, dirt, or excessive amounts of calcium or magnesium salts ("hard" water); (b) impure plating salts containing harmful quantities of foreign metals; (c) corrosion products resulting from chemical or electrochemical attack of the plating solution on exposed metallic tank walls, anode and cathode rods, plating racks, overhead ducts and cranes, and fallen work in the bath; (d) accidental addition of wrong chemicals during make up or adjustment of plating baths and through dragover of alkaline cleaners, acid pickles, or various strike solutions; (e) decomposition, precipitation, or supersaturation of plating chemicals during electrolysis or on standing; and (f) metallic solids or dirt from broken anode bags, dragover of polishing and buffing compounds, and air-borne solids landing in the plating bath.

Organic impurities may enter plating solutions from many sources, including the following: (a) oil, grease, or paint accidentally dropped into the plating bath from overhead mechanical equipment; (b) anode bags that have not been laundered to remove the sizing from the cotton; (c) new tank linings, rack coatings, or stop-off lacquers, which release contaminants into the plating bath; (d) make-up water containing algae, fungi, or oil; (e) entrained oil in compressed air wrongly used to agitate the plating bath; (f) impure or decomposed brighteners and wetting agents or an overdose of such plating bath additives; and (g) other sources, such as buffing compounds on unclean work and fallen objects like wood or insects.

Gaseous impurities in plating solutions, which tend to produce pitted, brittle, or cracked electrodeposits, may arise from the following sources: (a) electrolysis of the plating bath and resultant formation of hydrogen bubbles on the plated surface (cathode); (b) release of dissolved air when a cold plating solution is heated; (c) aspirated air, as finely dispersed bubbles in the plating solution, drawn through worn packing glands on circulating pumps; and (d) carbon dioxide formation during pH adjustments of acid plating baths with carbonates.

Harmful impurities are removed from a plating solution through a series of chemical, physical, and electrochemical methods. This may be illustrated by the following steps in the preparation of a bright nickel plating solution:

1. Weigh and dissolve the nickel salts in a clean rubber-lined purification tank (NOT in the plating tank) so that the impurities in the nickel salts can be removed before using the solution for plating. The purification tank, which should have a volume equal to or greater than the plating tank, is filled with hot water equivalent to about three quarters of the final solution volume, to allow room for the plating salts. While the required nickel salts are being stirred in, the temperature of the solution should be maintained at 65 to 70°C to dissolve the nickel salts quickly and completely. Typically, the nickel bath contains 330

g/liter nickel sulfate (heptahydrate) and 45 g/liter nickel chloride (hexahydrate).

2. Weigh and dissolve the boric acid in the hot nickel solution. Frequently, 45 g/liter boric acid is added as a buffering agent to stabilize the pH of the nickel plating bath during operation.

3. Raise the pH to 5.0 or above with a slurry of nickel carbonate, allowing ample time for the slurry to react before checking the pH. The hot solution will foam up at this stage because of the evolution of carbon dioxide from the chemical reaction of the nickel carbonate with the free acid in the nickel solution. To hasten this reaction, the temperature should be kept at 65 to 70°C and the solution should be well stirred until the pH is in the range of 5.0 to 5.5. This operation causes the precipitation of metallic impurities from the nickel salts, such as iron, copper, and aluminum.

4. Add hydrogen peroxide to the hot nickel solution at pH 5.0 to 5.5 to oxidize organic impurities and iron. A total of 2 ml 30% hydrogen peroxide per liter should be mixed in slowly since vigorous foaming will result if it is added too rapidly. The nickel solution should be kept hot and agitated for at least 1 hr.

5. Add activated carbon to remove organic impurities from the nickel solution by adsorption in the pores of the carbon. Because it is so finely divided, the carbon should be mixed separately into a slurry with some of the nickel solution so that it will not be blown into other plating tanks and cause rough deposits. Enough is made to contain 3 g of activated carbon per liter of solution and should be mixed in the solution for at least 2 he at 65 to 70°C. After mixing, the activated carbon and precipitated metallic impurities should be allowed to settle without agitation for several hours (preferably overnight) at 65 to 70°C. This will purge the liberated carbon dioxide, decompose any excess hydrogen peroxide, and make the filtration cycle easier.

6. Clean the plating tank and bag the nickel anodes to prevent organic contamination from the rubber tank lining (if the tank is new) and to prevent roughness caused by sludge from the anodes. The tank walls and bottom should be scrubbed with hot water and a wetting agent to remove dirt, grease, and oil. After thorough

rinsing, the tank should be filled with a 5% by volume solution of sulfuric acid containing 0.1 g wetting agent (such as Duponol ME) per liter and allowed to leach impurities from the tank lining overnight at 60 to 70°C. The anode bags, if not desized, should be soaked in boiling water containing 0.1 g/liter wetting agent to remove undesirable sizing fillers. After soaking, the anode bags should be thoroughly rinsed before being slipped over the nickel anodes. In the morning, the hot sulfuric acid is pumped out, the tank is thoroughly rinsed, and the anodes are put in place. Preferably, the anode area should be at least twice the cathode area. The tank is now ready for the plating solution.

7. Filter the plating solution from the purification tank into the clean plating tank. Since the metallic precipitates and activated carbon will be at the bottom of the purification tank after settling overnight, the intake hose from the filter should be kept about 15 cm below the solution level so that the maximum amount of clear solution can be filtered before the filter elements become coated with sludge. To facilitate removal of the filter cake at the end of the filtration cycle, the filter should be precoated with about 300 g of a paper pulp material (such as "Filterbestos") per square meter of filter surface. Then, the filter should be coated with about 600 g/m^2 of a filter aid of diatomaceous earth (such as Dicalite) to provide an incompressible filter cake for the removal of slimy metallic precipitates and finely divided activated carbon particles. As the bottom of the purification tank is appraached, some additional filter aid is sprinkled into the tank to help maintain an open filter cake as the heavy sludge is drawn up onto the filter elements.

8. Adjust the volume, temperature, pH, and surface tension of the plating solution. Sufficient water (preferably demineralized) should be added to the filtered solution in the plating tank to reach the working volume and the temperature should be adjusted to the recommended operating level (generally 50 to 60°C). Then the pH should be lowered gradually to the specified operating range (about pH 3.5 to 4.0 for bright nickel plating solutions), preferably with reagent-grade sulfuric acid. Enough time should be allowed to ensure thorough mixing of the sulfuric acid before the

pH of the plating solution is checked. Finally, the surface tension should be lowered with a wetting agent (such as filtered Duponol ME at pH 3.5) to 35 to 37 dynes/cm. At this surface tension, hydrogen pitting of nickel deposits is normally eliminated.

9. Electrolyze the solution at low current density to reduce the remaining metallic impurities to a harmless level. This nickel plating process, known as "dummying," involves the use of corrugated cathodes (dummies) to plate out impurities, such as iron, copper, lead, and zinc, at current densities in the range of 0.1 to 0.5 A/dm^2, along with the nickel. The corrugated faces should be about 5 cm wide and the included angle between faces should be about 60°. In most cases, the bath should be dummied for at least 15 A-hr/gal at low current density. The recesses of the corrugated dummies will show a light gray nickel deposit when the metallic impurities have been removed to an acceptable degree. To lessen the low current density electrolysis time, the maximum possible dummy area should be used together with good agitation of the nickel solution.

10. Add the brighteners to produce reflective nickel deposits, using the results of a Hull Cell plating test or a 4-liter sample of solution to establish the optimum concentrations of brighteners before adding them to the working solution. Since most of the bright nickel plating processes are of a proprietary nature, the directions of the supplier for control and maintenance of the solutions should be followed. The bright nickel plating solution is now ready for use.

The preparation of other plating solutions, such as chromium, copper, gold, tin, and zinc, is less complex than for nickel plating solutions, but still involves careful removal of impurities. As a rule, however, activated carbon treatment and electrolysis at low current density should be used when preparing a new plating solution.

C. Evaluation

Some of the factors requiring evaluation in both new and old plating solutions are the following: bright plating range, cathode efficiency, composition,

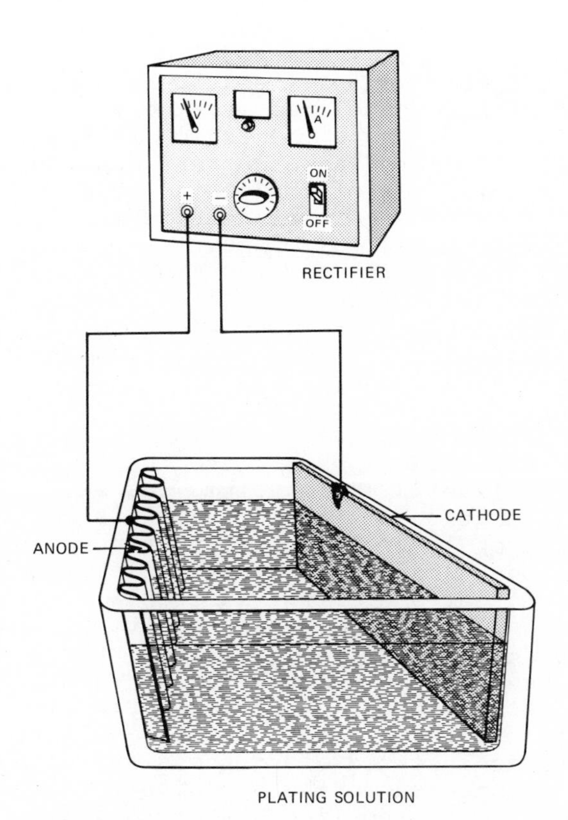

Fig. 1. Common 267-ml Hull cell for measuring the bright plating range of plating solutions.

pH, surface tension, and throwing power. Methods for evaluating these characteristics are as follows.

1. Bright Plating Ranges

Bright plating ranges of plating solutions are most readily determined by use of a Hull Cell. The basic Hull Cell, invented in 1936 by R. O. Hull, consists of a trapezoidal plastic cell, as shown in Fig. 1.

It is possible, in a single plating test in the Hull

Cell, to determine the charcteristics of deposits obtained over a 100-fold change in current density. Thus, for example, if the total current passing through the cell is 3A, the current density on the polished cathode panel nearest the anode will be around 15 A/dm^2, while the region farthest from the anode will be plated at a current density of only 0.15 A/dm^2. Examination of the plated panel will not only reveal the optimum current density range for bright plating, but it will also show any plating defects caused by the plating solution, such as pitting, highly stressed and cracked deposits, burning at the high current density area, or the presence of metallic impurities, which produces dark deposits in the low current density range.

In operation, the plating solution to be tested in preheated to a temperature about 5°C above the operating range and then poured into the Hull Cell up to the 267-ml level. (This volume was selected because a chemical addition of 2 g in 267 ml is equivalent to an adjustment of 1 oz/gal in the plating solution.) The nickel anode, which should cover the cross section of the rectangular end of the Hull Cell, is then put into position and connected to the positive terminal of a rectifier (or other controllable source of direct current). Next, the cleaned (free of water breaks when rinsed), acid dipped (5% hydrochloric acid at room temperature), and rinsed cathode panel (such as polished brass) is inserted along the slant edge of the Hull Cell and connected to the negative terminal of the rectifier. Care should be taken to avoid finger contact with the surface of the panel to be plated. Finally, the current should be adjusted to 3 A and the panel should be plated for 5 min with mild stirring of the solution. This will deposit approximately 0.5 μ of nickel in the low current density range of 0.5 A/dm^2 while the solution cools down over the optimum temperature range. The panel should then be removed, rinsed in hot water, dried, and examined.

Any required additions of brighteners, wetting agents, or other chemicals should be made to the sample of plating solution and the Hull Cell tests repeated until satisfactory nickel deposits are obtained. Then proportionate additions of chemicals should be made to the working bath.

For other plating solutions, the Hull Cell anodes should be of the metal under test, except in gold or rhodium solutions, where platinized titanium gauze anodes are preferred, and in chromium plating solutions where corrugated lead anodes are used. To minimize anode polarization (formation of a resistive oxide film at high current density, which inhibits electrochemical dissolution of the metal) corrugated anodes of brass, cadmium, copper, tin, and zinc should be used in the corresponding plating solutions. Two commonly used cathode surfaces are polished steel and polished brass. The polished steel cathodes are normally protected with a zinc plate that is stripped in 50% hydrochloric acid and rinsed immediately before use. The polished brass panels should be cleaned in a hot (70°C) brass soak cleaner for 3 min, to remove the protective plastic coating, and then cold water rinsed, dipped in 5% hydrochloric acid, cold water rinsed, and then inserted in the Hull Cell. The cell amperage employed will vary, depending upon the plating solution and the cathode current densities normally used in practice, but the amperage should be such that the normal operating current density falls around the middle of the Hull Cell panel. In the case of chromium plating, the Hull Cell test should preferably be run on the same source of direct current that is used in production (filtered to less than 5% ripple). For working with elevated-temperature electrolytes, modified Hull Cells are used, including thermostatically controlled, heated cells with agitation units. With experience, an operator will be able to use the Hull Cell to identify plating problems quickly and to evaluate corrective measures to restore the plating solution to optimum operating conditions.

2. Cathode Current Efficiency.

In some aqueous electroplating solutions, such as silver cyanide baths, all of the direct current flowing through the cathode is used to reduce positively charged ions (cations) in the plating solution to a metallic deposit on the cathode. Thus, in the case of silver plating from a cyanide bath, the weight of silver deposited is essentially equivalent to the quantity of electricity passed, in accordance with Faraday's law, and the plating solution is said to have a cathode current efficiency of 100%. For this reason, silver plating forms the basis of the definition of an ampere, namely, that current which will electrodeposit silver at the rate of 0.001118 g/sec.

On the other hand, in some aqueous acid electroplating baths, such as chromium plating solutions, only a fraction of the current flowing is used to electrochemically reduce metal ions. The balance of the current is consumed in overcoming the electrical resistance of the solution or in the reduction and discharge of hydrogen ions on the surface of the cathode in the form of hydrogen bubbles. Typically, the cathode current efficiency in hexavalent chromium plating baths is around 15%, while in trivalent baths it may be more than double that. Thus, at 30% cathode current efficiency it would take only one-half as long to deposit a given thickness of chromium from a trivalent bath as from a hexavalent bath. For most other plating solutions, the cathode current efficiency varies between 30% and 100%, as shown in Table 1.

In the evaluation of a plating solution, it is important to determine the cathode current efficiency under normal operating conditions so that the time required to deposit 25.4 μ of nickel at a current density of 0.205 A/dm^2 would be 60 min at a cathode current efficiency of 100%. But, if tests under actual plating conditions (at a given temperature, pH, and degree of agitation) indicated a plating efficiency of only 95% at a current density of 0.205 A/dm^2, the required plating time to deposit 25.4 μ of nickel would be

$$\frac{\text{Plating time at 100\% cathode efficiency}}{\text{Actual cathode efficiency}} = \frac{60 \text{ min}}{0.95} = 63.2 \text{ min}$$

Also, in the case of acid gold plating, where the gold is depositing from the trivalent state with insoluble anodes (such as platinized titanium), the cathode current efficiency decreases as the gold content of the plating solution is depleted. By measuring the actual plating efficiency at a given gold level in the bath, under normal operating conditions, it is possible to estimate both the time required to deposit a given thickness of gold and the required rate of replenishment of gold salts.

TABLE 1.

Typical Cathode Current Efficiencies of Plating Solutions

Metal	Valence	Type of Plating Solution	Range of Typical Cathode Current Efficiencies (%)
Cadmium	2	Cyanide	90-95
Chromium	3	Chloride	30-40
Chromium	6	Chromic acid	10-20
Copper	1	Cyanide, strike	40-60
Copper	1	Cyanide	100
Copper	2	Sulfate	95-100
Gold	1	Cyanide	100
Gold	3	Acid, soft gold	100
Gold	3	Acid, hard gold	10-40
Iron	2	Fluoborate	95-100
Lead	2	Fluoborate	100
Nickel	2	Sulfate-chloride	95-100
Palladium	2	Chloride	95-100
Platinum	4	Diammino dinitrite	15-65
Rhodium	3	Sulfate	60-75
Silver	1	Cyanide	100
Tin	2	Fluoborate	100
Tin	4	Stannate	60-90
Zinc	2	Cyanide	75-95

An approximate determination of the plating efficiency can be made by carefully weighing a light-gauge nickel sheet on an analytical balance before and after plating at the desired current density for, say, 5 A-min. At 100% cathode current efficiency in an acid gold bath, 0.2044 g of gold should be deposited in 5 A-min. If only 0.1226 g of gold were deposited in that time under actual plating conditions, the cathode current efficiency would be

$$\frac{\text{Actual weight gold in 5 A-min}}{\text{Theoretical weight gold in 5 A-min}} \times 100 =$$

$$\frac{0.1226\text{ g}}{0.2044\text{ g}} \times 100 = 60\%$$

Similarly, the required rate of gold replenishment would be estimated as follows:

$$\frac{\text{Actual weight of gold deposited}}{\text{5 A-min}} = \frac{0.1226\text{ g}}{\text{5 A-min}} =$$

$$0.0245\text{ g/A-min}$$

However, since the gold would be replenished in the form of gold chloride ($AuCl_3$), which has a theoretical metal content of 65.0%, the required rate of replenishment of gold chloride would be

$$\frac{\text{Weight of gold replenishment required}}{\text{Theoretical metal content of AuCl}_3} = \frac{0.0245\text{ g}}{0.65} =$$

$$0.0377\text{g Au}_3\text{Cl/A-min}$$

Although the approximate method described above would be adequate for routine production control, a more accurate procedure is required in research work for determining cathode current efficiencies. A laboratory device for measuring plating efficiencies is known as a coulometer; it measure the total coulombs (ampere-seconds) passed through the plating solution under test. Essentially, the coulometer consists simply of a plating bath known to have 100% cathode current efficiency under the conditions of test (such as the silver cyanide bath) connected in series with the bath of unknown plating efficiency at normal operating conditions. See Fig. 2. It should be noted that the cathode areas in the two plating cells are selected to provide the normally used cathode current densities.

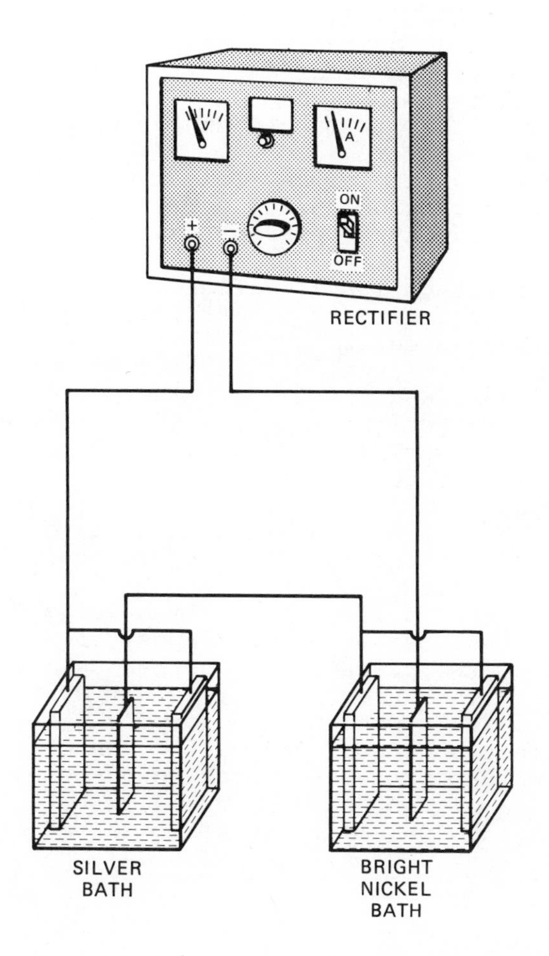

Fig. 2. Sketch of coulometer for measuring cathode current efficiency of a bright nickel plating solution.

From the gain in weight of the coulometer cathode (due to silver deposition) it is possible to calculate the number of coulombs passed through both the silver bath and the test bath. With this information, plus the corresponding weight gain of the cathode in the test bath, the cathode current efficiency of the test bath can be calculated as the ratio of the weight of test metal actually deposited to the theoretical weight that could be deposited by the flow of that quantity of electricity, in accordance with Faraday's law.

The fundamental principle of electrolysis is known as Faraday's law; it may be expressed by the following two statements:

1. The quantity of an element liberated at either the anode or cathode during electrolysis is proportional to the quantity of current that passes through the solution.

2. The quantities of different elements liberated by the same quantity of electricity are proportional to their equivalent weights.

In the silver cyanide bath, 1 C (1 A-sec) will deposit 0.001118 g of silver. Therefore, to deposit the equivalent weight in grams of silver, a total of (107.868 g silver)/(0.001118 g silver per coulomb) = 96,483 C is required. This quantity of electricity (often rounded off to 96,500 C) will theoretically deposit the equivalent weight in grams of any element and is known as one Faraday. It is equivalent to about 1608 A-min.

From Faraday's law, the maximum weight of a metal, in grams, that can be electrodeposited from a plating solution by the passage of one faraday (1608 A-min) of electricity is 1 g equivalent weight of that metal. Therefore, it is possible to compute a quantity known as the electrochemical equivalent, which is the theoretical weight of a metal that would be electrodeposited from a plating bath by the flow of 1 A-min of current if the sole cathode reaction was the reduction of metal ions. Thus,

$$\text{Electrochemical equivalent} = \frac{\text{Weight of metal deposited, g}}{\text{1 A-min}}$$

For example, in silver plating, where the gram equivalent weight of silver is:

$$\frac{\text{Atomic weight of silver}}{\text{Valence of silver}} = \frac{107.868}{1} = 107.868 \text{ g}$$

the electrochemical equivalent of silver is computed as follows:

$$\frac{\text{Gram equivalent weight of silver}}{\text{1 Faraday}} = \frac{107.868 \text{ g}}{1608 \text{ A-min}} =$$

$$0.06708 \text{ g/A-min}$$

Since it has been determined that all of the current flowing through the cathode of a silver plating cell is used to reduce silver ions to silver metal, the silver plating solution is said to have a cathode current efficiency of 100%. This means that the theoretical maximum weight of silver (one electrochemical equivalent, or 0.06708 g) will be electrodeposited by each ampere-minute of current flowing through a silver plating cell. Therefore, by accurately weighing the amount of silver deposited during a given plating cycle, it is possible to determine precisely how many ampere-minutes of current flowed through the plating cell:

$$\frac{\text{Weight of silver electrodeposited, g}}{\text{0.06708 g of silver/A-min}} = \text{A-min}$$

The advantage of using the silver coulometer is that it provides a convenient and accurate method of integrating the total ampere-minutes that passed through a plating circuit, regardless of fluctuations in the plating current or duration of the plating cycle.

A typical silver coulometer consists of a 1-liter beaker; two high-purity silver anodes, 5 cm wide by 15 cm long; a previously weighed silver foil (or silver-plated brass foil), 5 cm wide by 10 cm long, as the cathode; a small plating rectifier, having a range of 0 to 5 A; and 1 liter of a silver plating solution of the following composition:

Silver Cyanide Solution For Coulometer

Silver cyanide	40 g/liter
Potassium cyanide	60 g/liter
Potassium carbonate	30 g/liter
Free potassium cyanide	40 g/liter
Temperature	20-25°C

The silver cyanide solution is prepared by dissolving the potassium cyanide in 900 ml of demineralized water, stirring in the silver cyanide until it is completely dissolved. At this point, the solution is analyzed for free potassium cyanide (the amount in excess of the quantity required to complex the silver cyanide, forming potassium silver cyanide) and sufficient KCN is added to raise the free potassium cyanide to 40 g/liter. Then 3 g of activated carbon is added and stirred into the solution for 1 hr. After allowing the carbon to settle, the solution is filtered through filter paper, previously coated with a slurry of filter aid.

The clear silver plating solution is then poured into the 1-liter beaker. The cleaned and rinsed silver anodes are immersed in the solution, clamped vertically opposite each other, and connected to the positive terminal of the rectifier. The cleaned and rinsed silver cathode is suspended in the solution midway between the anodes, and facing them. (A silver cathode is used to prevent the formation of a nonadherent, chemical displacement film of silver, such as would form on a copper or brass cathode.) Then the electrical lead from the silver cathode is connected to the anodes of the next plating cell, whose solution is to be tested for cathode current efficiency. To complete the circuit, the electrical lead from the cathode of the test cell is connected to the negative terminal of the rectifier. A current of 1.0 A is then used in the silver coulometer circuit. This gives the recommended current density of 1 A/dm^2 on the silver cathode (2 sides x 0.5 dm^2/side).

Of course, the plating cell to be tested must be fitted with the proper anode material and the plating solution must be adjusted to the desired temperature, pH, composition, degree of agitation, and other plating variables corresponding to the intended use of the solution. Also, the size of the cathode must be

Table 2

Electrochemical Equivalents of Some Commonly Electrodeposited Metals[a]

Electrodeposited Metal	Chemical Symbol	Atomic Weight[b]	Valence In Solution	Equivalent Weight (g)	Electrochemical Equivalent (g/A-min)
Cadmium	Cd	112.40	2	56.20	0.03494
Chromium	Cr	51.996	3	17.332	0.01078
Chromium	Cr	51.996	6	8.666	0.00539
Copper	Cu	63.546	1	63.546	0.03952
Copper	Cu	63.546	2	31.773	0.01976
Gold	Au	196.967	1	196.967	0.02249
Gold	Au	196.967	3	65.656	0.04083
Iron	Fe	55.847	2	27.924	0.01737
Lead	Pb	207.19	2	103.595	0.06442
Nickel	Ni	58.71	2	29.355	0.01826
Palladium	Pd	106.4	2	53.20	0.03308
Platinum	Pt	195.09	4	48.773	0.03033
Rhodium	Rh	102.905	3	34.773	0.02133
Silver	Ag	107.868	1	107.868	0.06708
Tin	Sn	118.69	2	59.345	0.03691
Tin	Sn	118.69	4	29.673	0.01845
Zinc	Zn	65.37	2	32.685	0.02033

[a]The electrochemical equivalent of a metal is the theoretical weight of that metal that would be electrodeposited by the flow of 1 A-min of current in a plating solution having a cathode current efficiency of 100%. It is equal to the gram equivalent weight of a metal divided by 1608 A-min (1 Faraday).

[b]Values recommended by the International Commission on Atomic Weights (1967).

selected to give the desired current density under the intended plating conditions. For example, in a conventional hexavalent chromium plating solution, lead anodes are used, the bath temperature is 45°C, and the cathode current density is 15.5 A/dm^2. Thus, the total cathode area in the chromium solution (two sides) must be (1/15.5) dm^2 to give the desired current density when a current of 1.0 A is flowing through both the silver coulometer and the chromium plating solution. To avoid contamination of the silver cell with chromic acid spray, the top of the chromium plating cell should be covered during the plating cycle. After the plating cycle, both the silver and the chromium cathodes should be rinsed with hot distilled water and dried before weighing. The weight of metal electrodeposited from any plating solution onto a cathode is directly proportional to the current that flowed, the plating time, and the cathode current efficiency, with the proportionality constant being equal to the electrochemical equivalent of that metal:

$$\text{Weight of electrodeposit} = k \times A \times \text{min} \times \text{CCE}$$

where k is the electrochemical equivalent and CCE is cathode current efficiency. (The electrochemical equivalents of some of the commonly electrodeposited metals are shown in Table 2.

Thus, to determine the cathode current efficiency of any plating solution, under a specified set of operating conditions, it is necessary to know accurately the weight of metal deposited, the electrochemical equivalent of the metal being plated, and the number of ampere-minutes that flowed through the solution:

$$\text{Cathode current efficiency} = \frac{\text{Weight of metal electrodeposited} \times 100}{(\text{Electrochemical equivalent})\ (\text{A-min})}$$

For example, suppose that a silver coulometer is connected in series with a hexavalent chromium plating solution at the following operating conditions:

Chromic acid, CrO_3	250 g/liter
Chromic acid: sulfate ratio, $CrO_3:SO_4$ =	100:1
Plating temperature	45°C
Cathode current density	15.5 A/dm^2

If the weights of metal deposited during a plating efficiency test are 13.416 g of silver and 0.1585 g of chromium, what, then, is the cathode current efficiency of the chromium plating solution under the stated operating conditions?

From the gain in weight of silver on the coulometer cathode, the total ampere-minutes of current that flowed is computed as follows:

$$\frac{\text{Weight of silver electrodeposited}}{\text{Electrochemical equivalent of silver}} = \frac{13.416\text{ g}}{0.06708\text{ g/A-min}}$$

$$= 200\text{ A-min}$$

Then, using the electrochemical equivalent of hexavalent chromium from Table 2 (0.00539 g/A-min), the cathode current efficiency of the chromium plating solution is calculated:

Cathode current efficiency =

$$\frac{0.1585\text{ g chromium deposited}}{(0.00539\text{ g/A-min})(200\text{ A-min})} \times 100 = 14.7\%$$

Once the cathode current efficiency of a plating solution has been determined for a given set of operating conditions, the remaining problem is to determine how long it will take to electrodeposit a specified thickness of metal under the same operating conditions.

For example, if the cathode current efficiency of a hexavalent chromium plating solution is found to be 14.7% at a bath temperature of 45°C and at a cathode current density of 15.5 A/dm^2, how long would it take to deposit the normal decorative thickness of 0.254 µ of chromium over a bright nickel surface of 2 dm^2?

Since the density of chromium is 7.19 g/cm^3, the weight of electrodeposited chromium required is:

Weight of chromium required = Plated area x thickness x density

$= 2\ dm^2 \times 0.254\ \mu \times 719\ g/cm^3$

$= 200\ cm^2 \times 0.0000254\ cm \times 7.19\ g/cm^3$

$= 0.0365\ g$

The total current required for chromium plating is

Total current = Cathode current density x plated area

$= 15.5\ A/dm^2 \times 2\ dm^2$

$= 31.0\ A$

Since the weight of electrodeposited chromium required is also equal to k x A x min x CC, the plating time required to deposit 0.254 μ of chromium under the stated operating conditions will be

Plating time =

$$\frac{\text{Weight of electrodeposited metal required}}{\text{(Electrochemical equivalent)(A)(CCE)}}$$

$$= \frac{0.0365\ g\ \text{chromium}}{(0.00539\ g/A\text{-min})(31.0A)(0.147)}$$

$= 1.485\ min$

Several factors should be kept in mind about the relationship between cathode current efficiency and resulting thickness of electrodeposited coating. In the above example, for instance, it was assumed that the nickel plated object was being uniformly plated with chromium at a cathode current density of 15.5 A/dm^2. This would produce an average chromium plating thickness of 0.254 μ.

However, on most objects being plated, there is a range of current densities, with edges, corners, and projections drawing more current than at the center of flat areas or in recessed areas. Also, in the case of hexavalent chromium plating solutions, the cathode current efficiency tends to decrease when the bath temperature rises, when the chromic acid concentration rises, or when the cathode current density is lowered. Therefore, when plating chromium to a specified minimum thickness, on an irregular surface, it is advisable first to measure the actual thickness obtained in recessed areas during the calculated plating time and then to increase the plating time as required to obtain the specified minimum thickness.

For example, if the minimum specified thickness of chromium on an object is 0.254 μ and if the actual minimum thickness obtained in 1.485 min is only 0.200 μ, the required plating will be

$$\text{Required plating time} = \frac{\text{Specified minimum thickness}}{\text{Actual minimum thickness}} \times \text{calculated plating time}$$

$$= \frac{0.254\ \mu}{0.200\ \mu} \times 1.485$$

$$= 1.886\ \text{min}$$

For other electrodeposited metals, similar procedures should be followed to establish the plating time required to obtain a specified minimum thickness. This is especially important in the field of precious metal plating, where the intrinsic value of the electrodeposited metal is very high.

3. Composition

The plating solution is the sole source of metallic ions and additives available for electrodeposition on a cathode surface. Therefore, it is essential that the composition of the plating solution be established and maintained within the optimum concentration range for the type of electrodeposited metal required. The solution composition affects not only the operation of the plating bath, through such factors as buffering, conductance, current efficiency, and solubility, but it also determines many of the properties of the resulting electrodeposits, such as alloy composition, brightness, ductility, hardness, residual internal stress, solderability, strength, and structure. Of course, to obtain ideal electrodeposits, the composition of the plating solution must be carefully matched with other operating conditions, such as bath agitation, pH, surface tension, and temperature control. Data on the composition and operating conditions for standard electroplating solutions may be obtained from plating handbooks and technical articles. Corresponding data for proprietary processes are available through plating supply firms.

The following factors in electrodeposition are related to the composition of the plating solution.

a. Acidity. In terms of acidity, plating solutions may be classified into three categories: acid (pH below 2), neutral (pH between 2 and 8), and alkaline (pH above 8). Some metals, such as chromium and rhodium, require an acid electrolyte for metal solubility and bath stability. They are normally plated over a nickel undercoat. Typical neutral baths include the commercial bright nickel and acid-zinc solutions, which are operated at a pH between 3 and 5. The most important alkaline plating solutions are the copper, zinc, brass, cadmium, silver, and gold cyanide baths. In the case of silver or gold plating, it is necessary to avoid the formation of a loose, chemical displacement coating of noble metal on the substrate. This is done by first electrodepositing a very thin film of silver or gold from a low-metal-content "strike" solution.

b. Adhesion. Most plating cycles are designed to produce adherent electrodeposits on the basis metal. However, if a plating solution becomes contaminated with an organic material, such as oil or grease, the objects being plated may be partially or totally soiled and local blistering or general peeling of the electrodeposit may develop. Also, if a plating solution, such as a nickel bath, becomes contaminated with hexavalent chromium from improperly rinsed plating racks, an undesirable loss of adhesion of the nickel deposit may result because of passivation of the basis metal by hexavalent chromium ions. On the other hand, advantage is taken of this phenomenon in nickel electroforming, where the mandrel is intentionally dipped into a dilute solution of potassium dichromate and rinsed before being immersed in the nickel plating bath. This forms a parting film on the mandrel, which ensures easy separation of the electroformed nickel part.

c. Alloy Composition. Although it is generally desirable to avoid the introduction of foreign metal ions into a plating solution, it has been found that by suitable formulation it is possible to electrodeposit two or more metals simultaneously to produce alloy deposits having desirable properties. The trick in alloy deposition is to use a complexing agent in the plating bath that brings the deposition potential of each of the metallic ions to a common value for a given set of operating conditions and bath composition. Thus tin-lead alloys are readily plated from acid fluoborate solutions to produce easily solderable coatings on printed circuits. By closely maintaining the tin/lead ratio in the plating bath, it is possible to obtain a fairly constant alloy composition in the electrodeposited solder coating. Similarly, copper-zinc alloys may be plated from alkaline cyanide baths to produce attractive bright brass coatings on lamps and hardware. In gold plating, three or more metals may be dissolved in the plating solution to obtain alloy deposits having the desired color, hardness, electrical characteristics, or other properties. Some of the common alloying elements used in gold plating solutions are copper (pink gold), nickel (white gold), silver (green gold), and mixtures thereof. In general, alloy plating requires strict maintenance of the bath composition, by proper replenishment of plating salts in addition to using proper anodes, and close control of the operating variables, such as temperature, pH, solution agitation, and cathode current density.

d. Anode Depolarization. Most plating solutions are operated with soluble anodes of the metal or alloy being electrodeposited. Under ideal conditions, the anodes dissolve at 100% anode current efficiency and replenish the bath with metal at the same rate as it is being deposited on the cathode. However, if the plating rate is increased, by raising the current density, a point is reached at which the anodes become polarized, that is, they develop an insoluble oxide film at high current densities. This results in depletion of the metal content of the plating bath since the anodes are no longer able to replenish the solution with metallic ions, even though current is still flowing. To overcome this problem, high-speed baths are formulated with chemicals that tend to prevent or reduce the formation of oxide films on the anodes. Such chemicals are known as anode depolarizers. For example, most high-speed nickel plating baths contain some nickel chloride, since the chloride ion tends to eliminate or reduce polarization on the surface of nickel anodes. Similarly, high-speed alkaline copper cyanide baths employ potassium sodium tartrate (Rochelle salt) to reduce anode polarization, to permit lower free cyanide levels, and to obtain finer-grained copper deposits.

Alternatively, it is possible to manufacture depolarized metal anodes by incorporating small percentages of selected impurities in otherwise pure metal anodes. For example, nickel anodes may be depolarized by the addition of small amounts of carbon, oxygen, or sulfur, whereas copper anodes may be depolarized by the addition of small quantities of phosphorous. To prevent surface roughness on parts being plated, nickel anodes are almost always encased in cotton flannel or monofilament plastic anode bags to trap the sludge that forms as the nickel dissolves. Normally, copper anodes do not require bagging because they tend to dissolve smoothly.

e. *Brightness*. If a highly polished metal surface is electroplated with another metal, the first few layers of the electrodeposited coating will reproduce the bright surface of the basis metal. For instance, costume jewelry is frequently given a "gold flash" in which the gold deposit is so thin that it retains the bright finish of the underlying metal. However, as the thickness of the electrodeposit increases, the surface develops a dull, matte, frosty appearance due to the irregular growth of the electroplated metallic crystals. Originally, the normal procedure for obtaining a bright, reflective surface was to buff the electroplated coating with a buffing wheel coated with fine abrasive. As these manual operations became more costly, due to rising labor costs, automatic buffing equipment was developed. However, this process also became more expensive because of rising costs for materials, floor space, and handling. This led to the development of chemical additives to plating solutions, known as brighteners, which modified the structure of electrodeposited metals so that they had a lustrous, bright, reflective surface when removed from the plating tank. Most bright plating processes are based upon patented, proprietary plating solutions employing organic brighteners, metallic brighteners, or combinations thereof.

In bright nickel plating, for example, two classes of brighteners are frequently employed in a conventional Watts-type plating solution (containing nickel sulfate, nickel chloride, and boric acid). The first type of brightener is usually a sulfonate (such as the sodium salt of naphthalene trisulfonic acid), which produces a low-stressed, bright deposit over a polished substrate, but which does not build luster. Therefore, a secondary brightener of the metallic type (such as a cadmium or thallium salt) is used to build high luster, but without the first class of brightener in the plating solution, the nickel deposit would be excessively brittle, highly stressed, and too dark. In general, brighteners change the structure of electrodeposits from coarse, columnar grains to fine, laminar structures parallel to the basis metal.

f. Buffering Agents. Under ideal plating conditions, the rate of metal deposition from the cathode film is balanced by the rate of replenishment of metal ions from the bulk of the plating solution. Sometimes, in high current density areas on a part being plated, the rate of metal deposition will exceed the availability of metal ions in the cathode film. Then, another electrode reaction will take place, namely, the reduction of hydrogen ions and the evolution of hydrogen bubbles. This causes a localized increase of the pH in the high current density areas so that the next metallic ions entering that portion of the cathode film are precipitated to form a powdery, nonadherent deposit. Such high current density areas are said to be "burnt." Several techniques may be employed to prevent burning of electrodeposits, including the use of higher metal concentrations in the plating solution or higher agitation of the solution over the cathode surface, but the most effective method is to use a chemical buffering agent in the plating solution to stabilize the pH in the cathode film. Thus, for example, boric acid is used as a buffering agent in nickel plating solutions while sodium carbonate serves a similar purpose in alkaline cyanide plating solutions.

g. Cathode Current Efficiency. Ideally, the composition of every plating solution should be such that it will operate at 100% cathode current efficiency and permit the maximum rate of metal deposition per unit of plating power consumed. In some cases, however, other factors may be of even greater importance than simple economy of power costs. For example, a cyanide copper strike solution is intentionally formulated to produce a cathode current efficiency of only about 50% (by lowering the copper metal content and raising the free cyanide level). The resulting polarization at high current density areas, after copper deposition occurs, enables the plater to cover rapidly any remaining recessed areas on an irregular object with copper. The high amperage "strike," at 4 to 6 V, lasts for only 30 to 60 sec. As another example, when hard gold deposits are needed (having hardnesses from 100 to 450 Knoop), acid gold plating solutions are used to deposit alloys of gold with nickel, cobalt, indium, or manganese, even though the cathode current efficiency in such solutions may be below 40%. In most other cases, the composition of the plating solution is designed to yield the highest possible cathode current efficiency.

h. Dragout. When an object is removed from an electroplating bath, a certain amount of the plating solution is carried over into the rinse tank, either as a film of solution on the surface of the plated object or as entrapped liquid in the recessed areas of the object. The plating solution lost in this manner is known as "dragout." Naturally, the more concentrated the composition of the plating solution, the greater will be the dragout loss. However, the use of more concentrated plating solutions frequently makes it possible to electrodeposit metals at a faster rate and thereby increase the number of parts plated per day. For example, in the case of nickel electroforming, where heavy deposits may be required [say, up to 0.100 in. (2540 μ), versus 0.001 in. (25.4 μ) for decorative plating], concentrated solutions of the nickel sulfamate type or nickel fluoborate type are often used.

Plating rates may be doubled in some cases, because of the higher conductivity of concentrated solutions of these highly soluble nickel salts. On the other hand, the advantage of faster plating rates with such solutions is offset by the twofold expense of higher dragout losses, namely, greater loss of plating salts and higher waste treatment costs to stay within allowable effluent discharge limits. In some water treatment processes, such as the closed-loop type, the dragout metal is returned to the plating tank and the balance of the rinse water is recycled. Thus, it is possible to minimize the losses due to dragout by installing expensive waste treatment equipment.

i. Ductility. When a metal has the ability to be drawn through a die to form a wire of smaller diameter than before being drawn, the metal is said to be ductile. Conversely, if the metal cracks when deformed it is said to be brittle. In the case of electrodeposited metals, the composition of the plating solution has a marked effect upon the ductility of the plated coating. A purified plating solution having an adequate metal ion concentration will normally produce ductile deposits under the recommended operating conditions of temperature, pH, current density, and solution agitation. However, inorganic impurities (such as foreign metallic ions), organic impurities (such as improper stop-off lacquers), and gaseous impurities (such as carbon dioxide evolved during pH adjustment of a nickel bath with nickel carbonate) can cause brittle electrodeposits to be formed. A qualitative check of the ductility can be made by plating the metal to be tested on one side of a polished stainless steel sheet (whose back and edges have been masked with plater's tape) and then peeling off the non-adherent electrodeposit [say, 1 to 2 mils (25 to 50 μ) in thickness]. The test foil is then creased twice in succession: first, after being folded over from top to bottom and then, without unfolding, the foil is doubled over from left to right and creased flat again. If the foil is ductile, it will show no cracks along the creased lines when unfolded and no hole at the intersection of the creased lines. A quantitative measure of the ductility of an electrodeposit can be made by electroforming a thicker foil [say, 20 mils (508 μ) in thickness], cutting out a standard tensile specimen [with 0.5-in. (127-mm) reduced section], and measuring the percentage elongation in a 2-in. (50.8-mm) gage length after the pulled specimen fails in tension. Midway along the reduced section of the specimen, the width should be slightly tapered to cause the fracture to occur approximately midway between the two gage marks. Also, the crosshead speed of the tensile testing machine should be given when reporting the percentage elongation in 2 in. In general, it should be remembered that electrodeposits containing brighteners or codeposited metals (alloy deposits) will have lower ductility than pure metal electrodeposits.

j. Electrical Conductivity of Plating Solutions. In the process of electrodeposition, the major portion of the power consumed is for the electrode reactions (metal dissolution at the anode, gas evolution at both the anode and cathode, and metal deposition on the cathode). However, part of the energy is needed just to overcome the ohmic resistance of the plating solution. There are many ways to improve the electrical conductivity of plating solutions through adjustment of the bath composition. For example, the conductivity will be improved by a higher metal content (if the metal salt is sufficiently soluble), a higher free acid level (as in fluoborate baths), or a higher free cyanide level (as in silver plating baths). Frequently, the improvement in conductivity will also improve the "throwing power" of the bath, that is, its ability to deposit metal in recessed areas.

k. Hardness. Most of the commonly plated metals are relatively low in hardness, unless special additives are used in the plating solution. Notable exceptions are chromium and rhodium electrodeposits, which are inherently hard, scratch-resistant, and non-tarnishing. However, most chromium and rhodium deposits are very thin [say, 10 millionths of an inch (0.25 μ) in thickness] and, therefore, they are often used over a layer of hard nickel. In general, metallic and organic additives are used to produce bright, hard nickel deposits. Most such additives are proprietary compounds sold by plating supply firms. Similarly, there are additives that produce hard deposits of other metals, including cadmium, copper, silver, gold, and zinc. In measuring the hardness of electrodeposited coating, it is important to avoid errors caused by penetration of the diamond indenter through the coating and into the substrate. Thus, very light indentation loads should be employed (say, 10 to 25 g) or else the hardness should be measured on a polished cross-section of the electrodeposit. The hardness of plated metals varies widely, from about 80 Knoop hardness number (KHN) for pure silver, to about 500 Knoop for hard nickel, to about 800 Knoop for chromium.

l. Insoluble Anode Plating. Although most plating processes employ soluble anodes of the metal to be electrodeposited, as a source of replenishment for the metal ions plated on the cathode, it is often desirable or necessary to plate with insoluble anodes and to replenish the solution with metal salts. For example, insoluble stainless steel anodes may be used in cyanide gold baths to avoid the inventory cost and theft hazard of employing gold anodes. The gold content of the solution is maintained by regular additions of potassium gold cyanide. For nickel plating the inside of steel pipes, insoluble platinum-clad copper anodes are used with an all-sulfate nickel solution pumped through the pipe. This permits uniform plating with a constant anode-to-cathode distance and avoids the problem of dissolving away the internal anode before the end of the plating cycle. The depleted solution is replenished with nickel carbonate in an auxiliary tank and the evolved carbon dioxide is purged, by heating, before using the solution again for plating. In chromium plating from hexavalent baths, insoluble anodes must be used because metallic chromium will not dissolve in such solutions. Antimonial lead anodes are frequently used (sometimes cast over copper cores for improved conductivity and creep resistance) and the solution is replenished by additions of chromic acid.

m. Internal Stress in Electrodeposits. Most of the commonly electrodeposited metals have a tendency to shrink or contract as the deposit thickness increases. If the substrate is a thin metallic sheet, which is plated on one side only, the sheet will tend to curl along the edges and corners (high current density areas) in the direction of the anode. Many theories have been proposed to explain this behavior, but most agree that it is due to a residual tensile (contractile) stress in the structure of the electrodeposited metal. If a thin plated coating is applied to a thick substrate, the effect of the residual internal tensile stress can frequently be ignored.

In other cases, however, it causes serious problems. For example, in the electroforming of nickel phonograph record stampers the tensile stress in the nickel is often reduced by adding controlled amounts of a "stress reducer" to the plating solution. (Typical stress-reducing additives for nickel plating solutions are of the aromatic sulfonate type.) Thus, the nickel disks remain relatively flat when separated from the original silvered acetate recording. In the case of plated objects subjected to vibration during use, such as nickel-plated propeller blades, a residual tensile stress in the nickel coating was found to cause a serious lowering of the fatigue strength of the steel blades. This problem was overcome by using a proprietary stress reducer in the nickel plating bath to provide a residual compressive (expansive) stress in the nickel deposit at all current density levels used on the irregularly contoured blade surface. At the same time, unexpectedly, the nickel coating was found to possess a very high hardness (around 550 Vickers hardness number) combined with enough ductility to withstand stone impact damage without cracking.

Several methods may be used to measure the residual internal stress in electrodeposited metals. They employ the principle of measuring the deflection of metal strip or disk that has been plated on one side only. Such devices are called "contractometers." The simplest type is a "strip contractometer," in which the curvature of a spring steel strip is measured before and after plating one side only. If the plated side is concave, the electrodeposit is said to have a tensile stress. Conversely, a convex plated surface indicates a residual compressive stress in the electrodeposit. The magnitude of the residual stress may be calculated from the force required to deflect the unplated steel strip by the same amount (assuming that the plated coating is very thin versus the thickness of the steel strip).

In an ingenious variation of the simple strip device a "spiral contractometer" was developed by the National

Bureau of Standards in which a stainless steel strip was wound into a spiral coil. The upper end of the coil was clamped to the base of a dial indicator while the lower portion was immersed in the plating solution to be tested and a coating of the metal was plated only on the outer surface of the coil (the inside was masked). A clamp around the lower end of the coil held it against an insulating plug through which a stainless steel rod passed up to the dial indicator. By means of a sector gear connected to the upper end of the stainless steel rod, the rotation of the coil during the plating cycle was magnified on the dial indicator. If the coil tended to unwind, the deposit had a tensile stress proportional to the deflection of the dial indicator. On the other hand, a deflection in the opposite direction indicated a compressive stress in the deposit. The contractometer was calibrated by measuring the deflection caused by a known weight (1 oz) connected to the dial over a pulley. Then the internal stress could be calculated from the deflection caused by the electrodeposited metal on the outside of the spiral coil as follows (1):

$$S = \frac{2K}{pt} \times \frac{\bar{D}}{d} \text{ lb/sq in}$$

where K = deflection constant, in.-lb/degree
p = pitch of helix, in.
t = thickness of coil strip, in.
$\bar{D}$ = pointer deflection, degrees
d = deposit thickness, in.

By stripping the electrodeposited metal from the stainless steel helix, it was possible to make repeated measurements of stress with the same helix.

In general, impurities in plating solutions tend to increase the residual tensile stress, resulting in undesirable performance of the electrodeposited coatings. This is especially true in electroforming work where thick deposits tend to split open when the tensile stress exceeds the cohesive strength of the electrodeposited metal. On the other hand, a high tensile stress is desirable in the case of chromium plating, where a microcracked structure is wanted to provide maximum corrosion resistance over nickel undercoats. As the chromium coating splits open to relieve the internal tensile stress, more chromium is deposited in the crack until it, too, splits apart in a different place. Thus, it is possible to produce a chromium electrodeposit with a microdiscontinuous porosity that distributes any corrosive effect uniformly over the entire plated surface rather than concentrating it within a relatively few pits.

n. pH. In most electroplating baths the amount of free acidity or alkalinity will exert an important effect upon the quality of the electrodeposited metal. Thus, it is necessary to control this variable as closely as possible. To do so, the pH of the plating solution is measured and adjusted when necessary to stay within the limits specified for a given plating process. (The pH is equal to 1/log $[H^+]$, or to the negative logarithm of the hydrogen ion concentration. For example, hydrochloric acid concentrations of 0.1, 0.01, and 0.001 Normal would correspond to pH values of approximately 1, 2, and 3, respectively.)

Although the pH of a plating solution can be approximated by using paper strips impregnated with various indicators, it is more accurately measured electrometrically with a pH meter. However, if the solution is highly alkaline, or if it contains fluoride ions that might attack the glass electrode of the pH meter, then it is safer to use pH papers. Since most plating solutions contain buffering agents (such as boric acid in nickel baths), the fluctuations in pH are minimized and only occasional adjustments are needed.

o. Pitting. If the hydrogen bubbles formed on the cathode surface during metal electrodeposition are not removed, they prevent further plating at those spots. Then, when the plated object is removed from the plating bath, the bubbles burst and a series of pits, down to the basis metal, remain. Besides being unsightly, the pitted coating offers no corrosion resistance and, therefore, the plated part must be rejected. In some cases, the pitted coating may be stripped and the part replated, but normally it is too costly to reclaim pitted parts. Thus, it is important to prevent pitting in electroplating baths whenever possible. This may be done by ensuring that the surfaces to be plated are thoroughly cleaned and free of water breaks when rinsed before being placed in the plating solution. Also, the plating solution itself should be kept free of impurities and should contain controlled amounts of antipitting agents. In solutions operated with organic additives, such as brighteners or stress reducers, wetting agents are normally used an antipitting agents. They lower the surface tension to about 35 dynes/cm, and the hydrogen bubbles can no longer adhere to the cathode surface during the plating cycle. A simple method of checking the surface tension required to prevent pitting is to prepare several loops of stainless steel wire having loop diameters of, say, 3, 3.5, and 4 in. (75, 88, and 100 mm, approximately). When enough wetting agent has been added to stop the pitting in the plating bath, the stainless steel loops are dipped into the bath, one by one, and slowly withdrawn vertically until the loop size is found that will just support a film of the plating solution without popping. Then, future adjustments of the surface tension can be made by simply adding enough wetting agent to support a film of plating solution across the same wire loop.

However, when insoluble anodes are used in an electroplating bath, organic wetting agents should not be used to prevent pitting because they are decomposed by the strong oxidizing conditions on the insoluble anode surfaces. The decomposed wetting agents then act as impurities, causing high tensile stress in the electrodeposited metal and, possibly, dark, pitted deposits. In such cases, hydrogen peroxide may be added to the plating solution as an antipitting agent. The hydrogen bubbles are thereby oxidized and pitting stops. It is preferable to use acid-stabilized peroxide since organic stabilizers may contaminate the plating bath. The amount of peroxide required to stop pitting must be determined empirically for each plating bath and then maintained at that level by chemical analysis. An excess of hydrogen peroxide should be avoided since it would cause high tensile stresses in the electrodeposited metal, possibly leading to cracked deposits. Even if insoluble anodes are not employed, hydrogen peroxide may be used as an antipitting agent in plating baths that do not contain organic additives, as, for example, sulfate-chloride-type nickel baths.

p. Reflectivity. The ability of electrodeposited coatings to reflect light in the visible spectrum is one of the major reasons that such coatings are used for decorative applications, such as nickel-chromium deposits on appliances and on automotive hardware. These coatings also serve a functional purpose, namely, to provide resistance to tarnishing and corrosion at normal temperatures. However, when higher temperatures were encountered, as in searchlights operated with carbon arcs, thin, bright, nontarnishing, scratch-resistant rhodium electrodeposits were plated over highly reflective electroformed nickel reflector surfaces, up to 60 in. (1524 mm) in diameter. Although rhodium electrodeposits were also reflective into the infrared region, they were supplanted by other coatings that had a broader spectral range of reflectivity in the infrared. For example, searchlights that could operate with either visible or infrared light sources employed electroformed nickel reflectors that were subsequently vacuum coated with aluminum and silicon monoxide. Thus, electrodeposited coatings have provided a wide range of spectral reflectivity for various industrial and military applications.

q. Tensile Strength. By proper selection of the plating bath composition and operating conditions, it is possible to produce electrodeposits having a specified tensile strength (resistance to fracture under an applied stress; also, a measure of the cohesive strength of the metal). This property is particularly important in the field of electroforming, where structural parts are made by plating thick deposits over a mandrel or mold, followed by removal of the mold. Where high tensile strength is required, nickel is often used and tensile strengths can be obtained in the range of 345 MPa (50,000 lb/in.^{2}) from plain Watts type baths up to 11035 MPa (150,000 lb/in.^{2}) from Watts baths containing organic brighteners at carefully controlled concentrations. However, the brighteners will lead to embrittlement of the nickel at temperatures above 500°F (260°C) because of the formation of intergranular nickel sulfides. Thus, such high-strength nickel electrodeposits should be selected for use only at temperatures below about 450°F (232°C). When lower tensile strengths are acceptable, copper deposits may be used, either alone or as a backing for nickel electroforms. In the latter case, the copper may be built up rapidly in an acid copper sulfate bath, it has a low residual internal stress, and it is much cheaper than nickel.

r. Throwing Power In some plating solutions, the electrodeposited metal will not cover recessed areas of the parts being plated. Such a solution is said to have a poor throwing power. For example, it is normally difficult to electrodeposit chromium into corners, recesses, and around the base of protruding sections of a part. This problem may be overcome by suitable modification of the plating bath composition, combined with changes in the current density, bath temperature, agitation, and other operating conditions. A simple method of checking the throwing power of a plating solution is to run a "bent cathode test." In this test, both the throwing power and covering power may be observed. The cathode is bent 90° to provide a horizontal shelf extending out about 1 in. from the base of the cathode. After plating at the normal current density, the panel is examined to see if the recessed corner of the cathode has been completely covered. Often, there will be an unplated oval area in the center of the bend, extending from the horizontal shelf up onto the vertical section of the cathode. This indicates poor "macro throwing power" and calls for repeated process changes until the oval area is completely covered.

A second type of throwing power of a plating solution refers to its ability to deposit metal into narrow crevices on the surface of the object being plated. This is known as "micro throwing power." Normally, the electrodeposited metal will simply build up on each side of the crevice and finally bridge across, entrapping plating solution within the crevice. Later, this spot will blister open, especially if exposed to heat. For example, zinc die castings plated with a normal copper-nickel-chromium system exhibited "white rust" at such spots on exterior automotive parts, such as door handles. This problem was relieved by improvements in die casting technology combined with the use of copper strike solutions having good micro throwing power. Thus, copper pyrophosphate strike solutions were shown to deposit copper deep into crevices, protecting the zinc die casting from corrosion and blistering. Other plating solutions, especially precious metal baths, have been developed for jewelry and electronic applications where both macro and micro throwing power are required.

D. Control

To maintain plating solutions at optimum conditions requires regular control procedures. These procedures may vary from a simple temperature adjustment of a cleaning solution to a series of complicated analyses and chemical additions to a bright nickel plating solution.

In general, chemical additions to a plating solution should be made while the solution is in a spare tank. This will permit thorough dissolution of all additives, allow for purification of the chemical additions along with the original plating solution, and ensure that all of the insoluble particles are removed by the filter as the replenished solution is transferred back to the plating tank. Minor additions of liquid additives (such as wetting agents to lower the surface tension and thereby prevent pitting due to hydrogen bubbles, acids to lower the pH, brighteners to improve the reflectivity of the electrodeposited metal, and water to make up for evaporation losses) may be made directly to the plating tank, but only between plating cycles.

Thorough mixing of all liquid additives should occur before the next load of work enters the plating solution. To assist in dispersing the additives, they may first be stirred into a bucket full of the plating solution and then be added back to various parts of the plating tank. Alternatively, the liquid additives may be diluted with a portion of the plating solution and then be returned to the plating tank via the circulating pump and filter. Again, it must be emphasized that chemical additions to plating solutions should not be made while parts are being plated; otherwise, localized differences in concentration of additives could result in defective electrodeposits. Although the need for control will vary from job to job, the following basic schedule is recommended for daily and weekly control of plating solutions.

1. Daily Control

Before starting up a plating line, the plating solution levels in the various tanks should be adjusted with make-up water to replace that lost by evaporation. Next, the temperature of all heated solutions (such as cleaning solutions, pickling solutions, and plating solutions) should be measured and adjusted. (Thermostatically controlled electrical immersion heaters can help reduce the amount of temperature adjustment required.) The surface tension of plating baths, which use wetting agents to prevent pitting, should be checked along with the pH of the plating solutions. In cyanide plating processes, the free cyanide level should be analyzed and adjusted daily. When brighteners are used, a Hull Cell panel should be plated each day before starting production, to be sure that the plating solution contains sufficient brighteners. In special processes, such as electroforming, it may be desirable to measure the internal stress in electrodeposits from the plating bath on a daily basis. Continuous removal of organic impurities, by circulation of the plating solutions through a bed of activated carbon in a filter cartridge, is sometimes required. Similarly, continuous removal of metallic impurities, by low current density electrolysis (dummying) on a corrugated cathode sheet (dummy) in a separate compartment of the plating tank, is recommended for nickel plating solutions. Alternatively, the solution can be dummied overnight after production work has stopped. If careful attention is paid to the foregoing daily control procedures, high-quality, profitable plating production will result.

2. Weekly Control

When plating solutions are used intensively, that is, if they are operated continuously for two or three working shifts per day, for five days per week, a thorough purification treatment should be scheduled for the weekends. This includes removal of organic impurities by activated carbon in a separate tank, filtration to remove the carbon and other suspended solids as the solution is returned to the plating tank, and removal of metallic impurities by low current density electrolysis. In the case of nickel plating solutions, a high pH treatment with nickel carbonate and hydrogen peroxide may be required in addition to the activated carbon treatment. After filtration, it is necessary to adjust the nickel bath for pH, temperature, surface tension, and brighteners (if used). Cleaning solutions should be adjusted to the recommended levels of free alkalinity, or else they should be replaced with fresh solutions if the amount of soil in the solution prevents adequate cleaning of parts prior to plating. Similarly, acid dipping solutions should be analyzed and adjusted, if necessary, on a weekly basis. Other special solutions, such as metal activators, zincate solutions for aluminum surfaces, and metal stripping solutions, should be checked weekly.

While the plating solutions are removed for purification, the plating tank itself should be cleaned, dropped parts should be removed, and the tank lining should be inspected for any tears, breaks, or cracks (corrosion through the tank wall could lead to an expensive loss of plating solution). Also, the anode bags should be removed, inspected for holes, and then rinsed thoroughly, inside and out, before being replaced on the anodes. If the anodes have been consumed more than 50%, they should be replaced to ensure proper current distribution during plating production cycles. (The anode "spears" may sometimes be used in anode baskets for further plating; otherwise, they may be sent out to be recast into new anodes.) Finally, all electrical connections should be checked (including anode hooks, bus bars, and rectifier or generator contacts) and any mechanical plating equipment (cranes, racks, barrels, filters, hoses, etc.) should be inspected and repaired where needed. In general, the better the maintenance and control of plating solutions and equipment, the higher the quality of plated work that will be produced (with a minimum of rejected parts).

III. PREPARATION OF SURFACES TO BE PLATED

To obtain the optimum performance from an electrodeposited coating, the surface of the object to be plated must be finished to the desired degree of surface smoothness, must be thoroughly cleaned of foreign matter, and must be properly activated to ensure the desired degree of coating adherence. Therefore, it may be necessary to employ mechanical, chemical, and electrochemical processes in the preparation of surfaces to be plated.

A. Mechanical Processes

The three mechanical processes normally involved in metal finishing prior to plating are grinding, polishing, and buffing. However, a fourth process, barrel finishing or vibratory finishing, may be used when many parts of the same size require surface smoothness.

1. Grinding

Grinding is the coarsest mechanical finishing process and is essentially a metal removal operation. It is used when parting lines, stamping burrs, machining marks, or other gross defects must be eliminated. For fast metal removal, a coarse 60-grit wheel is used, followed by finer stone wheels up to, say, 200 grit. Large objects may be ground locally with a portable grinder fitted with a stone wheel or bonded abrasive disk. When worn or mismachined parts are to be salvaged by electrodeposition, they are first ground undersize to remove torn, pitted, grooved, or weakened surface layers; then they are plated oversize with a metal such as nickel; and, finally, they are ground or machined to finished dimensions.

2. Polishing

Polishing is a method of leveling or smoothing a metal surface to remove minor defects caused by prior fabrication processes and to provide a uniform surface appearance. Usually, a lathe with flexible abrasive belts or an abrasive-coated wheel is used for polishing. With flexible polishing wheels, surface speeds in the range of 20 to 30 m/sec are recommended to avoid excessive heat build-up on the surface of the part being polished. Generally, steels will tolerate higher polishing speeds than nonferrous metals. Where a nonreflective surface is required, or where bright finishing is difficult to obtain on small articles, a scratch brush finish is applied by using finer grit size greaseless compounds on unbleached muslin polishing wheels at surface speeds of 15 to 25 m/sec.

3. Buffing

Buffing involves the smoothing or lapping of a metal surface to provide a bright, smooth, scratch-free surface to ensure the clearest and brightest possible electrodeposit. Where decorative finishes are required, as on jewelry or electrical appliances, and where mirror-bright surfaces are needed, as on optical reflectors, it may be necessary to buff the electrodeposited coating as well as the basis metal. However, with the development of bright plating processes for most metals, it has been possible to eliminate much of the costly labor involved in buffing. In general, buffing is done with loose, full-disk buffing wheels of low-count cotton material, such as soft flannel. For softer metals, such as aluminum and brass, lime, unfused aluminum oxide or rouge compounds are used on 35-cm wheels at 1200 to 1500 revolutions per minute, to give surface speeds of 22 to 28 m/sec. Harder metals, such as steel and nickel, are buffed at higher speeds, in the range of 30 to 40 m/sec.

4. Barrel Finishing and Vibratory Finishing

Barrel finishing and vibratory finishing are bulk methods of improving the surfaces of parts prior to plating, at labor costs of only 10% to 40% of those required for hand polishing. Most metals and alloys can be processed by these methods to remove surface imperfections, such as flash, burrs, tool marks, and heat-treating scale, as well as to form radii on corners and sharp edges. The size and number of parts are limited only by the capacity of the equipment being used. Frequently, the machines are constructed of steel, with rubber or neoprene linings to dampen the noise and to provide corrosion resistance. In practice the machines are loaded with parts, abrasive media, lubricating and burnishing compounds, and water. Then, the speed of rotation or vibration is selected to provide enough impact and sliding motion between the parts and the abrasive media to give the desired degree of surface finish in a reasonable time. The speed and time cycle must be determined experimentally for each type of part. For normal cutting and deburring, the barrel speeds range from 0.6 to 0.9 m/sec (for example, a 30-cm-diameter barrel rotating between 38 and 57 RPM)

Where the parts are fragile, soft, or a high luster is desired, lower speeds are used any may range from 0.25 to 0.5 m/sec. In vibratory finishing, the machines may operate at frequencies up to 60 vibrations per second, depending upon the size of the equipment. Sometimes, combined rotary-vibratory equipment may be used at low rotational speeds to prevent settling of parts in the bottom of the mass, which might happen with vibratory finishing alone.

B. Chemical Processes

The two principal chemical processes involved in the preparation of surfaces for electroplating are the cleaning and activating steps. Unless both of these operations are properly employed, the plated parts will suffer defects such as blistering, peeling, pitting, and cracking.

1. Cleaning

After receiving the required degree of mechanical finishing, many parts may be covered with a residual film of dirt, buffing compound, or protective oil to prevent rusting or tarnishing. All of these foreign materials must be removed from the surface to be plated. In general, the nature of the basis metal and the type of soil to be removed determine the type of cleaning cycle required. The most practical way to test the effectiveness of a cleaning cycle is to examine the surface for "water breaks." If water rinses from the surface in a continuous film, the surface is considered clean. On the other hand, if breaks develop in the film of rinse water or beads of water form on the surface, the part is not sufficiently clean for electroplating.

The cleaning process is generally done in two steps, which may be called precleaning and final cleaning. The purpose of precleaning is to remove as much of the soil as possible or to prepare it for complete removal in the final cleaning process. Small parts are frequently precleaned by vapor degreasing in a chlorinated hydrocarbon, such as trichlorethylene. They are first soaked in the cool-solvent side of the degreaser to dissolve the majority of the soil and then are transferred to the hot section of the degreaser where they are flushed by condensation of the vapor on the cool parts. In some cases, small parts with deeply recessed areas and fine crevices are best cleaned in an ultrasonic cleaning unit with an organic solvent.

Large parts are often precleaned in hot (60 to 90°C), heavy-duty alkaline soak cleaners at concentrations of 100 to 200 g/liter. For every 10°C increase in cleaner temperature, the required cleaning time is approximately halved. Thus, if a soak time of 40 min were required at 60°C, only a 20 min soak would be needed at 70°C. Proprietary soak cleaners are compounded of sodium hydroxide, silicates, and carbonates, along with sequestering agents, soil dispersants, and various surface active agents. The parts to be cleaned may simply be suspended in the solution, but is preferable to agitate the solution or the parts and to use mild scrubbing. In the case of sensitive metals, such as aluminum and zinc, an inhibited alkaline soak cleaner should be used to prevent etching of the surface. Following soak cleaning, the parts should be thoroughly water rinsed to remove residual films of soil and cleaning solution.

Final cleaning employs direct current and is used to remove completely any remaining soils on the surface of parts to be plated. The solutions are basically hot, heavy-duty alkaline types, specially formulated with electrolytically stable organic additives. In reverse-current (anodic) electrocleaning, the parts are made positive at 3 to 12 V, to give current densities of 1 to 10 A/dm^2, for 0.5 to 2 min, with higher current densities being used for shorter cleaning times. Anodic electrocleaning is desirable for final cleaning whenever possible because oxygen evolution at the surface helps scrub off soils, metallic smuts are removed, deposition of nonadherent metallic films is prevented, and hydrogen embrittlement is avoided. However, it should not be used for cleaning metals that are soluble in alkaline electrocleaners, such as aluminum, brass, chromium, lead, magnesium, and tin.

In direct-current (cathodic) electrocleaning, the parts are made negative and then cleaned under the same operating conditions as in anodic electrocleaning. More gas scrubbing is achieved in cathodic electrocleaning because twice as much hydrogen gas is liberated on the cathode as compared to the volume of oxygen evolved on the anode at a given current density.

However, any metallic film deposited on the surface is usually nonadherent, difficult to detect, and hard to remove. Also, such a film could cause poor adhesion, roughness, or staining of electrodeposited metals. Finally, parts with hardness exceeding 40 Rockwell C, such as spring steel, may be subject to hydrogen embrittlement during cathodic electrocleaning and, therefore, should be baked for at least 1 hr at 200°C after plating. Two applications where cathodic electrocleaning is recommended are (a) for metals that are dissolved or etched by anodic electrocleaning, and (b) for buffed nickel prior to chromium plating (anodic electrocleaning would produce a passive film on the nickel, which would prevent the deposition of bright chromium). Of course, immediately after either anodic or cathodic electrocleaning the parts should be thoroughly water rinsed to remove any residual cleaning solution.

2. Activating

In most electroplating processes, the maximum possible degree of adhesion is desired between the electrodeposited metallic coating and the object being plated. On the other hand, in the process of electroforming on a permanent mandrel, the minimum degree of adhesion is desired so that the electroformed object may be easily and completely separated from the mandrel. Thus, two types of processing solutions may be required, one for obtaining completely adherent electrodeposits (activating solutions) and the other for completely nonadherent electrodeposits (passivating solutions).

In the first case, where completely adherent electrodeposits are required, the object to be plated is first cleaned until it is free of water breaks, rinsed thoroughly, dipped in a chemical activating solution, rinsed thoroughly, and then plated. The chemical activator may be either alkaline or acid, depending upon the nature of the substrate. For example, aluminum and its alloys are first given a series of acid dips (varying according to the alloy), rinsed, activated in an alkaline zincate solution (which forms an adherent displacement coating of zinc on the aluminum), rinsed, and then plated in an alkaline copper strike coating (about 10 to 15 μ thick). Another special case is encountered when plating on plastics. Here, a series of acid dips is used to roughen the surface prior to activation with a palladium salt, followed by an electroless (chemically reduced) coating of, say, copper. Details of these specialized activating processes are available from plating supply houses or from standard plating handbooks.

Normally, however, the chemical activating step simply involves the dipping of the thoroughly cleaned object in a dilute acid for about 10 sec at room temperature (20 to 25°C). This serves two purposes, namely, to neutralize any residual alkaline cleaning solution and to dissolve any residual film of oxide that would prevent adhesion of the electrodeposit. The type of acid to be used varies with the type of substrate to be plated. For example, a zinc-base die casting would be dipped in 0.5% sulfuric acid, a low-carbon steel object would be dipped in 5% hydrochloric acid, and a lead (or lead alloy) part would be dipped in 10% fluoboric acid. Immediately after acid dipping, the parts should be thoroughly rinsed with cold water and quickly transferred to the plating tank for coating with, say, a copper strike before the final topcoating system. Recommended plating cycles for various basis materials are given in most plating handbooks.

In the second case, where completely nonadherent electrodeposits are required, the surface is first cleaned and activated prior to passivation and plating. Thus, for example, when electroforming a replica of a nickel phonograph master stamper, the stamper is cleaned, rinsed, dipped in 5% sulfuric acid, rinsed, dipped 5 sec in a passivating solution (say, 4 g/liter potassium dichromate at 20 to 25°C), rinsed, and then placed to the desired thickness in a low-stress nickel plating solution (such as a sulfamate bath). After grinding the circumference, to remove the nickel that plated around the back of the master, the electroformed nickel "mother" can be pried away easily from the passivated surface of the master stamper. The "mother" can then be used to electroform many working stampers for use in phonograph record molding presses. The passivating sequence used for nickel mandrels is sometimes also used to ensure complete separation of electroformed nickel shells from stainless steel mandrels or chromium plated mandrels that are used repeatedly.

C. Electrochemical Processes

A number of electrochemical processes have been developed for treating metallic surfaces to ensure the adherence of electrodeposited coatings. Such processes include the electrocleaning procedures described above as well as various "strike" solutions, anodic etching methods, and alkaline descaling processes.

The purpose of a "strike" solution is to cover rapidly an an active substrate surface with an electrodeposited metallic coating to which subsequent electrodeposits will adhere strongly. In general, such solutions are operated with a low metal content, a high conductivity, and a high current density. Thus, as metal is deposited at high current density areas, the solution is depleted of metal, the resistance increases (because of cathode polarization), and the plating current tends to flow into low current density area (such as recessed areas and corners) where sufficient metal ions remain in solution to permit deposition. By this time, the high current density regions are replenished with fresh solution and the cycle is repeated. However, the structure of such "strike" deposits is very weak at thicknesses in excess of about 10 μ. Therefore, objects to be plated with heavier coatings are rinsed and transferred to conventional plating solutions having higher metal contents that produce stronger electrodeposits. To meet specific needs, both alkaline and acid strike solutions have been developed.

The most commonly used alkaline strike solutions contain copper, silver, or gold. Copper strike solutions, of the alkaline pyrophosphate or of the cyanide type, are frequently used to get rapid coverage of copper over active metal surfaces, such as zinc, which are to be plated later in an acid plating bath, such as a nickel solution. The thin copper layer protects the zinc recessed areas from the corrosive acid solution until the nickel deposit has completely covered such area. In the case of silver or gold strikes, from alkaline cyanide solutions, the purpose is to get rapid coverage of the substrate before a nonadherent chemical displacement film of silver or gold can be formed. The object to be plated is then rinsed and transferred quickly to a more concentrated conventional silver or gold plating solution.

For plating adherently over chromium and alloys containing significant amounts of chromium, several acid-nickel chloride strike procedures were developed. In a typical procedure, the parts are thoroughly cleaned, rinsed, and then plated for 5 min to 20 to 25°C with a current density of 5 A/dm^2 in a solution containing 240 g/liter nickel chloride and 120 g/liter hydrochloric acid. The parts are then rinsed quickly, to prevent passivation, and immersed in a low pH nickel bath, for instance, to electrodeposit an adherent coating of nickel.

An anodic etching cycle at high current density in cold surfuric acid was discovered as a suitable means of electrochemically treating low-carbon steels to obtain adherence of nickel deposits adequate to permit flame cutting and welding. For example, steel sheets were pickled in hot sulfuric acid to remove rust and scale, rinsed, cleaned anodically in a hot alkaline solution, rinsed, anodically etched at up to 20 A/dm^2 for 5 min in a 50% by volume sulfuric acid solution saturated with magnesium sulfate (to improve conducti-ity) at a temperature of 0 to 5°C (obtained by refrig-eration of the acid), rinsed, and then nickel plated to a thickness of about 250 μ. The resulting sheets were rolled into cylinders and welded to form nickel-lined tanks cars for shipping products that required protection from discoloration due to iron contamina-tion.

For alkaline descaling of steel, several proprietary solutions have been developed that function best with periodic-reverse current. In this method, the polarity of the steel is alternately positive (anodic) and nega-tive (cathodic) while the steel is immersed in the solution. This process offers an alternative to hot acid pickling, but its use is restricted because of its higher cost.

IV. EVALUATION OF ELECTRODEPOSITS

The evaluation of electrodeposited metals may encompass measurements of their chemical, electrical, magnetic, mechanical, metallugical, and physical pro-perties. The specific properties to be determined for a given electrodeposit may be related to the thickness of the coating as well as its function. Thus, for example, thin decorative coatings may be evaluated for their abrasion resistance, adhesion, color, corrosion resistance, ductility, hardness, reflectivity, surface appearance, tarnish resistance, and thickness. On the other hand, heavy electrodeposited metallic coatings for engineering purposes, such as electroforming and building-up of worn parts, may require evaluation of additional properties, such as composition, elongation, fatigue strength, internal stress, tensile strength, and yield strength. Finally, for special applications of electrodeposited metals, it may be necessary to measure their electrical, magnetic, oxidation, and thermal properties. The properties of electrodeposited metals and alloys have been described in detail by Safranek.

A. Thin Deposits

In general, thin electrodeposits may be characterized as those up to 50 μ in thickness. They are often evaluated for the following properties, which are especially important in decorative applications.

1. Abrasion Resistance

The ability of a plated coating to resist scratching and the tearing out of fragments from the surface determines its abrasion resistance. A qualitative measure can be obtained by observing the depth of scratch caused by running the corner of a file across the deposit. For a quantitative comparison of two different coatings, the depth of penetration of a fixed diamond shape, under a specific load, is measured after the diamond is drawn across the surface of each coating. Because of their hardness and toughness, very thin coatings of bright chromium or bright rhodium may be used as a top coating to provide abrasion resistance.

2. Adhesion

Several methods have been devised to measure the degree of adherence between a plated coating and the underlying substrate. One simple method is to simply run a coarse file across the edge of the plated object toward the coating. If the coating is nonadherent, it will peel up from the substrate. Another simple method is to rub the plated surface back and forth with a round object, such as a steel ball. If blisters develop, because of cold working and expansion of the plated coating, poor adhesion is indicated. A third method is to apply heat, either by baking in an oven at, say, 250°C, or by applying a torch to the plated object. Again, the appearance of blisters indicates a lack of adhesion. Other adhesion testing methods include the use of a chisel to lift the coating, repeated bending, and sudden stretching in a tensile machine. To measure the degree of adhesion quantitatively, rather complicated tensile-type tests have been developed. However, they are not suitable for routine quality control evaluations.

3. Color

The color of most electrodeposited coatings is a characteristic of the pure metal being plated. For example, copper deposits have a light salmon color, gold deposits are yellowish, and cadmium, chromium, nickel, silver, and rhodium all have a silver-white color. However, alloy deposits are also frequently employed for decorative and functional purposes. The color and hardness of such deposits varies with the alloy composition. Thus, pure gold is yellow and soft, copper-gold alloys have a pinkish cast and are somewhat harder, while nickel-gold alloys are silver-white and quite hard by comparison with pure gold. In the jewelry trade, it is quite important to match the color electroplated gold alloys from batch to batch. This is done by careful control of the plating bath composition, temperature, and current density.

4. Corrosion Resistance

One of the most important functions of a thin, decorative electrodeposit is to protect the basis metal from attack by the corrosive elements in the environment, especially water. For example, automobile bumpers frequently employ a system of copper, nickel, and chromium electrodeposits, in that order, to provide corrosion resistance again the salt used on highways to melt snow. The copper provides leveling of the polishing marks in the steel, the nickel imparts a high degree of corrosion resistance, and the chromium topcoat provides scratch resistance, tarnish resistance, and corrosion resistance. To measure the corrosion resistance afforded by a given plating system, tests may be run in a copper-accelerated acetic acid salt spray (Cass test). Long-term exposure tests are also run to verify the results of accelerated tests. Such exposure tests may be static (with panels mounted on test racks) or dynamic (the panels are mounted on moving vehicles, such as taxicabs). The quality of a plating system is measured by its ability to prevent the formation of rust spots due to pit-type corrosion of the steel substrate through the plated coating.

5. Ductility

Since one of the main purposes of electrodeposited metallic coatings is to protect underlying surfaces from corrosive attack, it is important that plated coatings have sufficient ductility to withstand impact and deformation without cracking. If the electrodeposited metal is brittle, it will develop numerous cracks when dented or bent, thereby exposing the substrate to corrosion. Accordingly, a simple method of checking the ductility of a plated coating is to bend or impact a test sample and examine it for cracks. In another method, a thin film (say, 50 μ thick) of the electrodeposited metal is plated nonadherently on a stainless steel panel, peeled off, and then creased. A ductile deposit will not crack when the creased foil is unfolded. To get a quantitative measure of the ductility of an electrodeposited coating, it is necessary to plate a thicker specimen (say, 250 to 500 μ in thickness) on a stainless steel panel, remove the coating from the stainless steel, cut a tensile specimen, and then measure the percentage elongation in a 50-mm gage length after pulling the specimen to fracture at a specified rate of stretching.

6. Hardness

One of the principal factors that contributes to the durability of an electrodeposited coating is its hardness. In general, the harder the coating the longer it will withstand removal through physical and mechanical wear, as in cleaning and polishing of electroplated objects, or during normal abrasion as with plated jewelry. A simple method of checking the hardness of a plated coating is to run the corner of a file lightly across the surface. A hard coating, such as chromium electroplate, will resist scratching whereas a soft coating, such as silver, will show a gouge mark. To obtain a more quantitative measure of the hardness of an electrodeposited coating, a diamond indentation under light loading (say, 10 to 25 g) is made on a polished, flat area of the plated object. (The smaller the size of the indentation, the higher the hardness of the coating.)

Since most decorative coatings are quite thin (on the order of 25 μ in thickness), it is normally necessary to measure the hardness on a cross section of the coating. This entails the cutting, mounting, and polishing a section of the plated object perpendicular to the plated surface. For thin deposits, a Knoop hardness tester is preferred because the indenter has an elongated diamond shape. This permits the long dimension of the diamond indenter to be positioned parallel to the surface of the plated coating and ensures that the narrow dimension of the indenter falls entirely within the cross-sectioned thickness of the coating. If the indentation has an irregular shape on any side or point, a faulty hardness reading will result because the indenter has penetrated beyond the coating in question. Either the mounting should be moved to expose a thicker section of the plated coating or else a lighter indentation loading should be used to ensure that the diamond penetrates only within the desired coating. A measuring eyepiece on the hardness tester is used to measure the size of the indentation and a conversion chart gives the equivalent Knoop hardness number.

For thicker deposits, a Vickers hardness tester may be used. In this case, the diamond indenter has equal axes so that all four sides of the indentation are of equal length. Similar precautions should be taken to ensure that the indentation is made entirely within the coating to be tested and that there are no irregularities in the indentation pattern. Measurements of the length of the indentation axes (between opposite corners) are made with a filar eyepiece and converted to the corresponding Vickers hardness number. It is important to specify the indentation loading used when reporting the hardness of a plated coating (on either the Knoop or the Vickers scale) since the data vary with lighter loads.

7. Reflectivity

For most decorative applications, the reflectivity of electrodeposited coatings is important only within the visible spectrum. Thus, the reflectivity of bright chromium deposits on appliances, automotive hardware, and jewelry is one of the major reasons for its selection as a decorative topcoat (in addition to its resistance to abrasion, corrosion, and tarnishing). Most bright electroplated metal surfaces reflect between 50 and 70% of white light. Typical reflectivity values for brass, chromium, copper, gold, nickel, platinum, tin, and zinc fall within this range. Notable exceptions are bright electrodeposited rhodium (72%) and silver (95% reflectivity). For this reason silver and rhodium coatings are preferred for reflectors. In other parts of the spectrum, such as the infrared, gold and rhodium provide a high degree of reflectivity combined with tarnish resistance.

8. Surface Appearance

Although most decorative electrodeposited coatings are applied over smooth substrates, to obtain maximum reflectivity, it is sometimes desirable to produce nonreflective finishes for utilitarian or styling effects. For example, objects subject to frequent handling, such as lipstick cases, may have a slightly knurled or scratch-brushed finish to improve their nonslipping qualities. Also, such finishes improve the surface appearance by minimizing the effect of fingerprint stains, which are quite apparent on highly polished objects. Interesting styling effects can be obtained by plating a dull finish over an irregular object and then polishing the projecting areas to highlight them against the dull background. Conversely, a brightly plated object such as a silver-plated candy tray, can be scratch brushed on a wire wheel to provide a contrast between the polished rim and dull center. By selecting metals with different colors, it is also possible to vary the surface appearance of electroplated objects. For example, the inside of silver-plated cream pitchers may be gold plated to provide an attractive color contrast between the silver and gold surfaces. In another variation of this technique, an irregular surface may be plated with two layers of metal having contrasting colors, such as nickel followed by a thin layer of copper. Then, by polishing through the copper deposit on projecting areas, it is possible to obtain bright nickel highlights against a copper background. In the case of automotive hardware, selected areas of chromium plated parts may be decorated with baked-on paint finishes to produce objects such as nameplates. Thus, it is evident that wide variations of surface appearance are possible with electrodeposited coatings.

9. Tarnish Resistance

Since most electrodeposited metals react with gaseous and liquid materials in the environment, they tend to form dull, dark surface coatings, such as oxides or sulfides, known as tarnishing. For example, silver-plated objects rapidly tarnish when used to eat hard boiled eggs, because of the formation of a film of silver sulfide. On the other hand, nickel-plated objects tarnish slowly when exposed to the atmosphere. Although metal polishes may be used to restore the original metallic luster of electrodeposited metals, it is usually desirable to provide tarnish resistance to the original coating, if possible. This may be done by applying suitable lacquers, as on brass-plated lamps or by electrodepositing a tarnish-resistant coating, such as chromium over nickel deposits. To evaluate the tarnish resistance of an electroplated object, it may be placed in a simulated environment, such as hydrogen sulfide. By evaluating various thicknesses and types of topcoats, it is possible to specify a coating system that provides the necessary degree of tarnish resistance.

10. Thickness

Probably the major factor affecting the performance of a decorative electrodeposit is its thickness. For this reason, it is important to be able to measure the thickness of an electrodeposited coating with a high degree of reliability. However, since most electroplated objects have irregular contours, the coating thickness in recessed areas and corners will be less than the average thickness while projecting areas and corners will have a greater than average coating thickness. Therefore, most plating specifications are based on the minimum thickness required on significant surfaces (excluding those areas that are concealed, functionally unimportant, or economically impractical to electroplate to the specified thickness). It may be necessary to test several plating cycles to find the average number of A-hr/dm^2 required to meet the minimum thickness specification for a given object. Furthermore, in some cases, as on threaded parts, both a minimum and a maximum coating thickness may be specified. When a precious metal, such as gold, is to be plated to a specified minimum thickness, it is essential for economic reasons to determine the lowest possible average thickness that will provide the required minimum thickness in recessed areas. For these technical and economic reasons, two general methods have been developed to measure the thickness of electrodeposited coatings, namely, nondestructuve and destructive test methods.

In nondestructive thickness testing methods, the integrity of the electrodeposited metal is kept intact. The simplest test method of this type would be to measure the overall thickness at designated areas on the object before and after plating, using a micrometer caliper. The difference in dimensions would represent the coating thickness (or double the coating thickness if the object were plated on both sides). However, this method is not very accurate for measuring deposit thicknesses below 25 μ.

In some cases, a calibrated magnetic gauge may be used to measure the local thickness of electrodeposited coatings. This method is applicable when a magnetic coating is deposited over a nonmagnetic base (such as nickel, plated over brass) or when a nonmagnetic coating is deposited on a magnetic substrate (such as copper, plated on steel). In the first case, the attractive force of the magnet is proportional to the thickness of the magnetic coating, whereas in the second instance the force of magnetic attraction is reduced as the thickness of the nonmagnetic coating is increased. Although magnetic thickness testing methods are rapid and nondestructive, they are sensitive to vibration and surface roughness, have an accuracy of about 90% depending upon the type of deposit and base metal, and are limited to appropriate combinations of magnetic materials. There are a number of commercial magnetic testing devices available.

Another nondestructive thickness testing method for electrodeposited coatings is based upon the difference in intensity of eddy currents flowing through a coating and through the basis metal. In this method, probes are used to circulate minute alternating electrical currents in short, closed paths in the skin of a conductive material. As the coating thickness decreases, more of the eddy currents flow through the basis metal and vice versa. Standard coating thickness samples are used to calibrate such eddy current devices, which are sold commercially.

In still another nondestructive method for measuring the thickness of electrodeposits, especially precious metal coatings, the intensity of beta-ray backscattering may be used if the atomic number of the coating and the basis metal differ by a sufficiently large interval. In this method, a sealed container of radioactive isotope emits beta particles (fast electrons) into the plated coating where the particles suffer loss of energy and deflection. Some of the beta particles are reversed in direction and pass back out through the coating (backscattered), where they are detected and counted with a Geiger counter. Since thicker coatings cause greater backscattering, this method can be calibrated to measure the thickness of various electrodeposited coatings. Commercial direct- reading digital beta backscattering instruments are available.

Other specific nondestructive thickness testing methods have been developed, including a method based upon X-ray fluorescence, but such methods are generally too expensive for routine thickness measurements. They are, however, very useful for checking the thickness and composition of precious metal electrodeposits, such as gold alloys.

In destructive thickness testing methods, the electrodeposited coating is either cut, chemically dissolved, electrochemically stripped, or otherwise destroyed. Perhaps the most accurate method of thickness measurement is to cut through the electroplated object, at the area to be tested, and to measure the coating thickness on the exposed cross section under a microscope. This requires great care, a trained technician, and a well-equipped metallurgical laboratory.

First of all, the object to be measured is cleaned, activated, and plated with about 25 to 50 μ of a metal topcoat having a contrasting color, hardness, or structure. Frequently, bright nickel is used to back up the final coating on the test object because, when etched, the bright nickel has a contrasting banded structure parallel to the substrate and its high hardness prevents the edge of the test coating from being rounded during the subsequent polishing of the cross section. Next, the test object is cut transversely (perpendicular to the surface) in the area where the thickness is to be measured.

After rough polishing of the cut surface, one or more sections are mounted in a plastic holder with the rough polished surfaces exposed. Then the exposed cross sections are polished on successively finer emery paper and, finally, on wet polishing wheels with alumina slurries of finer and finer grades until a scratch-free surface is obtained. To distinguish the plated coating under test from the back-up deposit and the substrate, a chemical etchant is swabbed across the polished surface to show the structure of the desired coating in contrast to the other coatings and the substrate (appropriate etchants are listed in metallography handbooks). The etched surface is then rinsed with distilled water and quickly dried in a stream of warm air (as from a portable hair dryer). Finally, by means of a calibrated filar eyepiece, it is possible to measure directly the thickness of the etched electrodeposited coating under a microscope.

In the chemical stripping method, the plated object is weighed before and after the electrodeposited coating is dissolved in a solution that selectively removes the coating without attacking the basis metal. The average thickness is then computed as follows:

$$\text{Average thickness} = k \frac{(\text{Weight of deposit})}{(\text{Area of deposit})(\text{Density of deposit})}$$

For example, if the weight of a nickel deposit (which has a density of 8.9 g/cm^3), stripped from an object having a total area of 100 cm^2, is 2.225 g, what is the average thickness?

Average thickness of nickel =

$$10,000 \frac{(2.255\ g)}{(100\ cm^2)(8.9\ g/cm^3)} = 25\ \mu$$

(Note: k= 10,000 to convert centimeters to microns.)

It should be remembered that the properties of an electrodeposited coating are usually related to the minimum thickness, rather than the average thickness. Thus, the stripping method should not be used to measure coating thicknesses where a critical performance requirement must be met.

In the electrochemical stripping method for measuring the thickness of electrodeposited coatings, the time required to dissolve the coating anodically from a fixed area is directly proportional to the thickness of the coating, in accordance with Faraday's law. An accurately defined area of the coating, on a flat surface of the object to be tested, is made anodic at a current density that will dissolve the coating at 100% anode efficiency, in an electrolyte that will selectively strip the coating without attacking the basis metal. In general, a special electrolyte is required for each type of coating. This makes it possible to measure the separate thicknesses of composite coatings, such as copper-nickel-chromium. The deposit dissolves anodically at a rate directly related to the current density and, when the basis metal is exposed, the endpoint of the test is indicated by a sharp change in voltage. From the current used and the time required to dissolve the deposit, the thickness of the coating can be determined. Automated instruments employing this coulometric method of thickness measurement are commercially available.

B. Thick Deposits

In the electrodeposition of thick metallic coatings for engineering purposes, it may be necessary to evaluate a number of properties in addition to the thickness and adherence of the electrodeposits. Such properties include composition, elongation, fatique strength, internal stress, tensile strength, and yield strength.

1. Composition

When thick electrodeposits (over, say, 50 μ) are used for engineering purposes, the chemical composition may be critical. For example, if nickel deposits are to be exposed to high temperatures, as in welding, they must be essentially free of sulfur and lead, which cause intergranular embrittlement upon heating. A qualitative check for embrittlement may be made by electroforming a strip of nickel from the same bath, heating it to redness, allowing it to cool, and then bending the strip back on itself. If the bent strip can be opened without cracking, it is acceptable. In electronic applications, it is often necessary to use heavy gold deposits for contact surfaces and such deposits must meet specifications with regard to gold content (say 18 karat, or 75% gold) and impurity levels. A convenient method of checking such deposits is by X-ray fluorescence, using standard alloys as reference materials. When heavy electrodeposits are made with successive layers, as in electroforms having a nickel face and copper backing, the composition may be checked by polishing an exposed edge and measuring the relative thicknesses of the two metals. (Acid copper is frequently used as a backing material in electroforming processes because of its lower internal stress and lower cost relative to nickel.)

2. Elongation

One method of determining the ability of an electrodeposited metal to be deformed without cracking is by measuring its tensile elongation, as described earlier (p.397). This property is important, for example, when an electroformed object must be bent during use, as when a nickel phonograph record stamper is clamped into the platen of the molding machine. Another method of measuring the elongation of an electrodeposit is to plate one side of a spring steel strip (say, 1 cm x 15 cm x 250 μ thick) to a specific thickness of deposit (say, 250 μ). Then the number of degrees which the plated strip can be bent around a 2-cm-diameter mandrel without cracking (with the electrodeposit in tension) is measured. A bend of 180° indicates acceptable ductility. This test was developed for hard nickel coatings used to protect steel propeller blades from erosion and corrosion.

3. Fatigue Strength

The fatigue strength of a metal may be defined as the limiting stress below which it may be flexed for an indefinite number of stress cycles without fracture. When steel, for example, is electroplated with nickel, the fatigue strength of the steel is lowered by an amount roughly proportional to the residual internal tensile stress in the nickel electrodeposit. Thus, when nickel-plated steel objects are subjected to vibration in use, such as propeller blades (in which the nickel coating is alternately in tension and compression during each revolution of the blade), the plating process should be controlled so as to produce a residual compressive stress in the electrodeposited nickel at all current densities employed on the irregular blade surface. In this way, the fatigue strength of the steel will be maintained at or near its original value. In addition to employing compressively stressed nickel coatings, another factor of safety against fatigue failure (via propagation of a crack developed in the vibrating nickel coating through the steel substrate) can be obtained by electrodepositing a relatively thick intermediate coating of a metal having a low modulus of elasticity, such as zinc.

4. Internal Stress

Most electrodeposited metals have a residual tensile (contractile) stress in the as-deposited condition, even when the plating solution is purified and operated under optimum conditions. If impurities, whether inorganic, organic, or metallic, are introduced into the plating bath, the resulting electrodeposits normally become so highly stressed that they spontaneously crack and peel away from the substrate. On the other hand, certain organic brighteners, such as those used in nickel plating baths, have the property of converting the residual tensile stress to a residual compressive (expansive) stress within a selected range of current densities (at higher plating rates, the stress once again becomes tensile). For example, aromatic sulfonates and sulfonamides (such as naphthalene trisulfonic acid, sodium salt, and para-toluenesulfonamide) have been employed as "stress reducers" in nickel plating solutions. Often, it is necessary to make a stress survey, including several current densities, to be sure that the nickel deposits have a compressive stress over the proposed plating range. Methods for measuring residual stresses in electrodeposited metals, using contractometers, were discussed earlier (p. 400).

5. Tensile Strength

By proper selection of a plating process it is possible to obtain electrodeposited metals having a wide range of tensile strengths to meet specific engineering requirements, as in electroformed components. For example, copper may be deposited with a tensile strength in the range of 100 to 200 MPa (about 15,000 to 30,000 lb/in^2), whereas nickel deposits may be obtained in the range of 400 to 700 MPa (without organic additives) and up to 1000 MPa when organic stress reducers are employed in the plating solution. The procedure for determining the tensile strength of an electrodeposited metal was discussed earlier (p. 419), in connection with the measurement of ductility of deposits.

6. Yield Strength

In designing an object for an engineering application in which the part is to be subjected to tensile stresses, it is essential to know the minimum stress at which plastic flow will occur, namely, its yield strength. This property may not be important for thin electrodeposited coatings since the tensile strength of the substrate will dominate. However, in a free-standing object, such as a plastic mold made of electroformed nickel, it is important to select the plating process such that the yield strength of the nickel will be higher than the applied stresses during the plastic molding operation. Otherwise, the mold will be stretched out of shape. Normally, the yield strength is measured by plotting applied stress versus resulting strain during a tensile strength measurement (see above) and using an offset of 0.2% to determine the elastic limit (the stress at which plastic flow begins).

V. SPECIAL PLATING TECHNIQUES

Some of the special techniques of importance to electroplaters are barrel plating, brush plating, electroforming, industrial safety and hygiene, periodic-reverse current plating, rinsing, stripping of metallic coatings, and waste water treatment. Each of these techniques is described briefly below, but further references should be consulted by the reader in the areas of specific interest before using such techniques.

A. Barrel Plating

When many small parts are to be electroplated, from the size of a pin or a small ball bearing up to about 0.5 kg in weight, it is often possible to avoid the expense of racks, racking of parts, unracking, and rack maintenance by using rotating plating barrels. Also, barrel plating is compatible with other bulk-finishing operations such as barrel polishing and vibratory finishing. The most commonly used plating barrel is of the horizontal type, 30 to 40 cm in diameter by 60 to 80 cm in length, made of plastics that withstand acids, alkalies, and high temperatures (up to 80°C) without distortion, and having perforated side panels assembled to form a hexagonal cross section. A suitable drive mechanism above the plating barrel allows it to be rotated at speeds of 3 to 6 revolutions per minute so that the parts will slide toward the center of the barrel. The least troublesome cathode contact is a flexible dangler consisting of an insulated cable terminating in a chromium-plated contact weight from which successive layers of plated metals can be easily removed with a hammerblow. Anodes should conform to the shape of the rotating barrel so as to minimize internal resistance in the plating solution. Parts that tend to "nest," such as flat plates or cup-shaped parts, do not move freely relative to each other while the barrel rotates and, therefore, will have wide variations in deposit thickness. The current density used in barrel plating should be adjusted to avoid exceedingly low ranges (whereby the coating might be removed as fast as it was deposited) and to avoid "burning" at very high rates. In general, plating barrels can be processed through the entire cycle of cleaning, activation, and plating to produce parts having fairly uniform average plating thicknesses at low unit costs.

B. Brush Plating

Brush plating is a technique whereby an insoluble, insulated, conductive stylus (such as graphite) is tipped with an absorbent material, saturated with plating solution, and then connected to the positive terminal of a direct-current source (such as a battery or rectifier) to form the anode. The part to be plated is connected to the negative terminal, thereby becoming the cathode. After proper masking, cleaning, and activation, selected localized areas may be electroplated by brushing the stylus against the cathode and replenishing the absorbent tip with fresh plating solution. Metals that can be brush plated include cadmium, zinc, tin, copper, nickel, gold, and silver. The obvious advantage of brush plating is that a damaged area can be replated without disassembling the part and immersing it in a series of plating solutions. However, since this process is more expensive than tank plating, it is recommended for use only where the cost of disassembly or other reasons, such as accessibility, make it competitive.

C. Electroforming

In the process of electroforming, an article is produced or reproduced by the electrodeposition of a relatively thick coating of one or more metals on a conductive mandrel or mold, which is subsequently removed from the electrodeposit. The thickness of electroformed articles may vary from 250 μ up to 10 mm, or more, depending upon the structural requirements of the finished part. Among the advantages of electroforming are the following: (a) reproduction of fine detail, as in printing plates for paper currency or in phonograph record stampers; (b) reproduction of mirror-like surfaces, as in searchlight reflectors; (c) ability to produce intricate internal shapes within close dimensional tolerances, as in electronic wave guides; and (d) ability to produce multilayered parts to combine function, properties, and cost, such as gold-plated nickel reflectors backed with lower-cost copper.

The techniques employed in electroforming are quite similar to those used for conventional electroplating, with two major exceptions: the electrodeposit must be nonadherent so that it can be separated from the mandrel, and the deposit thickness is normally far greater than the 25 to 50 μ thickness of decorative coatings. Mandrels may be "permanent," such as stainless steel or metallized plastics which may be reused, or "expendable," such as fusible metals or wax, soluble metals, such as aluminum alloys that are dissolved in hot 10% sodium hydroxide, or plaster and wood that are destroyed during separation from the electroform. The most commonly electroformed metals are copper, iron, and nickel, but these may be combined with thin, initial coatings of precious metals, such as silver and gold. By proper selection of plating baths and operating conditions, it is possible to produce large quantities of identical parts that have close dimensional tolerances, low internal stresses, and high surface finishes. Although electroforming is an expensive operation, it is especially suitable for tooling, required for short production runs, that would be very expensive to produce by conventional machining methods.

D. Industrial Safety and Hygiene

Because of the many chemical, electrical, mechanical, and thermal hazards that exist in a plating shop, every effort should be made to ensure that safe and hygienic procedures are followed. To protect workers chemical burns, whether acidic or alkaline, goggles, respirators, rubber gloves, rubber aprons, acid-proof clothing, and rubber boots with steel toe inserts should be worn when handling, measuring, mixing, or transferring plating solutions. In addition, deluge showers and eye wash fountains should be provided for use in the event of accidental splashing. When adding chemicals to plating baths, many small additions should be made with thorough stirring rather than dumping one large quantity into the tank. Adequate ventilation should be provided at the tank surfaces and in the general working area to prevent the accumulation of noxious fumes. To prevent the formation of toxic hydrocyanic acid fumes, all cyanide compounds and solutions should be prevented from reaching acids or liquids that are acidic in reaction. Even cyanide plating rinse waters should be ventilated if acid rinse waters follow in the plating cycle. Similarly, adequate ventilation should be provided in the area of vapor degreasing equipment to prevent the accumulation of harmful vapors.

To prevent electrical shock, it is recommended that wooden duckboards be provided around the plating tanks and that all electric immersion heater, electric motors, and motor-equipped devices, be properly grounded. Also, safety guards should be erected around the tops of all open tanks if they are less than 1 m above the floor or platform level, to prevent workers from accidentally falling into them. Of course, all local, state, and federal safety regulations should be obeyed with regard to the labeling, handling, storage, transfer, and disposal of plating chemicals. In addition, workers should be trained in safe working procedures and first aid measures. Finally, arrangements for emergency medical services should be made in case of accident, with employees advised in advance on how to secure such services promptly.

E. Periodic Reverse Current Plating

Around 1950, a number of patents were granted to Westinghouse Electric Corporation for a novel method of electrodeposition known as periodic-reverse-current (PR) plating. In this process, the plating current is periodically reversed for appreciable lengths of time to remove a substantial amount of previously deposited metal, especially from high-current density areas. Besides provided a controlled anodic film on the plated work, the PR plating process improves the degree of leveling so that less buffing is required, greatly improved uniformity of deposit thickness is obtained on irregular surfaces, higher plating current densities are permissible so that shorter plating cycles are needed in spite of intermittent deplating cycle, and thick, uniform, high-quality deposits are obtainable. Although a number of metals could be plated by the PR process, it was developed primarily for copper plating from cyanide baths containing organic additives to produce bright, decorative finishes.

F. Rinsing

Perhaps one of the most frequent sources of trouble in the electroplating industry is improper rinsing following each plating step. Since the purpose of rinsing is to dilute the contaminants on the cathode surface to the point where they will not interfere with the next plating operation, good design practice is necessary. Of course, the source of rinse water should be clean, free of excessive mineral content, and low in metallic contaminants from previous rinse cycles. Ideally, the rinse water should be violently agitated with air (by means of a blower, not an air compressor which would introduce an oil mist into the rinse water), fresh water should be introduced at the bottom of the tank, and (optionally) spray rinses may be provided as the work leaves the rinse tank. In severe cases, multiple counterflow rinsing should be provided, with the final rinse tank containing the fresh water inlet. To conserve water, reduce costs, and minimize waste treatment, it is sometimes possible to reclaim the rinse waters, concentrate them, and return them to the plating tank. This is especially true in the case of precious metal plating.

G. Stripping of Metallic Coatings

Electrodeposited coatings are sometimes unacceptable for reasons such as blistering, cracking, peeling, pitting, or staining, and it may be desirable to strip the coating and then to replate the object. This is especially true when a lot of expensive machining was performed before the part was plated or when a precious metal coating was deposited. In other cases, it may cost less to scrap the part than to strip and replate it. Careful planning is required to remove the electrodeposited metal without unduly damaging the substrate. Several techniques are available for removing plated coatings, including immersion in chemical stripping solutions, anodic etching in a solution that will selectively strip the plated coating, or dissolving the undercoat to permit the plated layer to fall off (as in stripping a thin rhodium layer from nickel-plated brass).

If possible, a simulated part should be processed through the proposed stripping cycle before deciding on the method to be used. Thicker deposits require a longer time to remove and care must be taken to prevent uneven attack of areas that are stripped before all of the deposited metal has been stripped. Frequently, stripped parts must be polished to remove surface etching or pitting before replating. Immersion methods are preferred because of their simplicity, but the stripping solutions must be replenished to maintain the rate of metal removal. For example, chromium deposits may be stripped from brass, copper, nickel, or steel by simple immersion in 10% hydrochloric acid, but the part should be watched carefully for complete stripping so that the substrate will not be attacked unduly. In the case of cyanide soluble metals (such as brass, cadmium, copper, gold, and silver) on steel or nickel substrates, the metallic deposits may be removed anodically at 6 V in a solution of 90 g/liter sodium cyanide plus 15 g/liter sodium hydroxide at room temperature. Other stripping solutions are described in plating handbooks, and a number of proprietary processes are available from plating supply firms.

H. Waste Water Treatment

In recent years, increasingly stringent controls have been imposed on the types and amounts of effluents that may be discharged from electroplating plants into local sanitary sewage systems or into flowing water systems. For example, effluent limits for copper, nickel, chromium, and zinc electroplating were published in the Federal Register, Vol. 39, No. 61, Part II, on March 28, 1974 in compliance with the Federal Water Pollution Control Act. Standards governing effluents from other electroplating processes remain to be promulgated, but they undoubtedly will be of equivalent stringency. In general, the Environmental Protection Agency (EPA) has established effluent limits based upon the area of processed parts requiring rinsing, the number of rinsing steps in the plating cycles, and a rinse water rate of 160 liters/m^2. State or municipal agencies may impose stricter requirements, but may not permit a reduction in the quality of waste water treatment required by the EPA.

Among the effluents to be controlled are the following: cadmium, chromium, copper, cyanide, fluoride, iron, lead, nickel, phosphate, silver, suspended solids, tin, and zinc. Over the years, a number of waste treatment processes have been developed, including alkaline chlorination (for cyanide decomposition), oxidation (for conversion of alkali metal cyanides to cyanate precipitates), reduction and neutralization (for conversion of hexavalent chromium to trivalent chromium followed by precipitation and filtration), ion exchange (especially for precious metal recovery on resins used to treat rinse waters), dialysis and electrodialysis (for copper removal from a nitric acid pickling solution), reverse osmosis (for concentration of nickel plating dragout rinses), and evaporation (for simultaneous recycling of metal dragout to the plating tank and revoery of pure rinse water). The selection of waste water treatment processes is governed by the types and quantities of effluents to be treated, allowable effluent limits, cost of treatment chemicals, and value of recycled or recovered metals, among others. To comply with all local, state, and federal waste water treatment regulations at minimum cost, the amount of toxic contaminants reaching the effluent should be kept to a minimum by proper design and operation of the plating plant.

ADDITIONAL READING

A. Kenneth Graham, Ed., Electroplating Engineering Handbook, 3d Ed. Reinhold, New York, 1971.

Metal Finishing Guidebook and Directory, 43rd Annual Edition, Metals and Plastics Publications, Inc., Hackensack, N.J., 1975.

F. A. Lowenheim, Ed., Modern Electroplating, 2d Ed. John Wiley, New York, 1963.

REFERENCES

1. A. Brenner and S. Senderoff, Proceedings of the 35th Annual Convention of the American Electroplaters' Society, pp. 74-75, 1948.
2. W. H. Safranek, The Properties of Electrodeposited Metals and Alloys, A Handbook, American Elsevier Publishign Company, Inc., New York, 1974.

VII. ELECTRODIALYSIS OF AQUEOUS SOLUTIONS*

Irving F. Miller

University of Illinois at Chicago Circle
Chicago, Illinois

I. INTRODUCTION

Electrodialysis is a process in which the components of an ionic solution may be separated by means of an electric current. This is accomplished by interposing across the current flux lines a particular type of electrically conductive membrane. The membrane, an ion exchanger, is so designed that the fraction of the current carried by ions of a particular sign (the transport number) within the membrane is substantially

*Completed 9/70. The reader is referred to the current literature for later work.

different from that fraction in free solution. By proper design of the electrodialysis cell, this property can be exploited to achieve a substantial degree of separation of the solution components.

As a method of desalting brackish water for industrial and consumer purposes, electrodialysis falls into the category of such selective transport processes as reverse osmosis (hyperfiltration) and solvent extraction. In these processes, advantage is taken of the different rates of movement of the various species in different parts of the system, rather than the exclusion of certain species from different phases within the system. Typical phase-change systems include distillation and partial freezing processes.

From the point of view of energy consumption, particularly for low-salinity feed stocks, selective transport processes have certain inherent advantages. From a thermodynamic point of view, the minimum energy required to separate the components of a solution can be calculated from the partial molal enthalpies of these components at the various concentrations involved In principle, one need only supply the heat of mixing of a solution to completely separate it. Of course, the second law of thermodynamics prevents one from transferring this heat at a finite rate with 100% efficiency. In fact, to maintain reasonable production rates, approximately 10 to 20 times this minimum energy is typically required.

The energy requirement can be considerably greater with phase-change processes. In such processes, in addition to the heat of mixing, one must also supply either the heat of vaporization or the heat of fusion of water. Although substantial portions of this heat are recoverable, again the second law of thermodynamics prevents complete recovery.

However, this apparent advantage tends to disappear as the solute content of the feed increases. While the energy required to separate salt from water by distillation is, in practice a very weak function of salt concentration (heat-transfer inefficiencies are determining), the energy required to separate the same components by electrodialysis is a very strong function of concentration, for reasons that will become evident. When one also considers the capital investment required to build a distillation plant, as opposed to an electrodialysis plant, it usually turns out that electrodialysis is superior for smaller-scale plants involving lower-salinity feeds.

The engineering of electrodialysis plants, and the economics involved, have been discussed in detail by such authors as Tribus (1), Schaffer and Mintz(2), Lacey et al (3), and others, and is not repeated here. Rather, we devote our attention to more fundamental aspects of the process, emphasizing material that has not appeared elsewhere.

II. FUNDAMENTALS OF THE PROCESS

Electrodialysis depends on the fact that the transport number for ions of a given sign is different in an ion-exchange membrane than it is in free solution. To see how this can result in desalination, consider Fig. 1. For example, two membranes - one whose transport number for cations is 0.9 (marked "C"), and the other whose transport number for anions is 0.9 (marked "A") - are placed in a solution containing a salt in which the transport number of cation is 0.4. Suppose one Faraday of current is passed across the stack, as indicated. Across the cation-exchange membrane, 0.9 equivalents of cation pass out of the center compartment. Across the anion-exchange membrane 0.1 equivalents of cation enter the center compartment, while 0.9 equivalents of anion leave. Thus, the

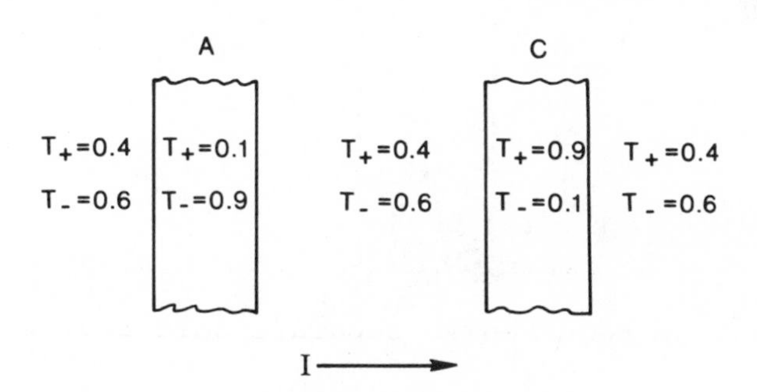

Fig. 1. Unit electrodialysis cell. A, anion selective membrane. C, cation selective membrane.

passage of one Faraday of current results in a net depletion of 0.8 equivalents of salt from the center compartment. (It should be pointed out that commercial membranes are available whose transport numbers are greater than 0.99. The numbers above were chosen for illustrative purposes only.)

This basic concept has been adapted for practical electrodialysis units in several ways, some of which are described below.

A. Conventional Electrodialysis

Figure 2 represents a typical design for a conventional electrodialysis unit. In this design an alternating series of cation-permeable and anion-permeable membranes is placed between two electrodes.

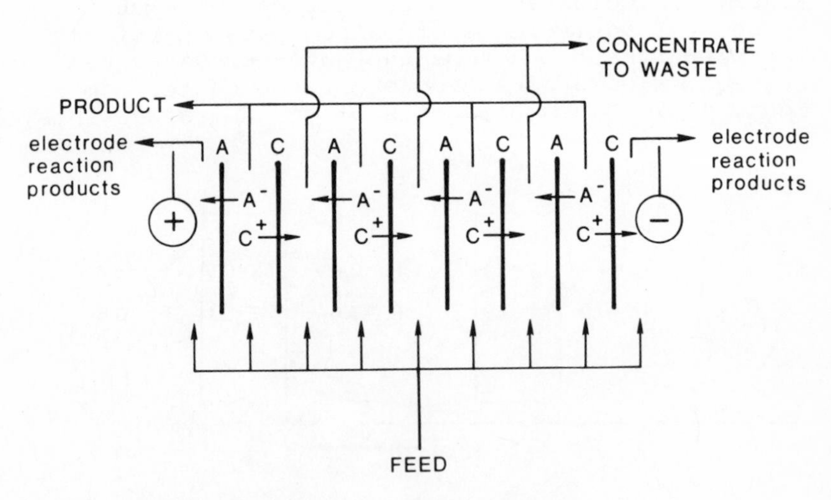

Fig. 2. Conventional electrodialysis.

Feed water flows through the stack parallel to the membranes, and a current is passed normal to the plane of the membranes. As a result of the selective transport of cations and anions across the membranes, alternate cells are depleted of solute, while the cells adjacent to these are concentrated in solute. The depleted solution is taken off as product, while the concentrate and the electrode reaction products flow to waste.

In principle, it would appear that any desired degree of desalination at any production rate can be achieved by the simple expedient of drawing sufficient current. Actually, several phenomena occur to defeat this possibility.

The first obvious limitation lies in the fact that the stack has a finite electrical resistance. Thus, as the current is increased, a larger and larger fraction of the usable current is dissipated as heat.

However, another phenomenon, membrane polarization, occurs at considerably lower current levels to prevent efficient desalination long before significant amounts of Joule heating can occur. This phenomenon is closely tied to the presence of hydrogen and hydroxyl ions in the solution. At low current levels, the current is carried almost exclusively by solute ions, because of the low concentration of the more mobile hydrogen and hydroxyl ions in neutral solutions (10^{-7} M). At these current densities one observes a small transport of hydrogen ions across the cation exchange membrane, and a small migration of hydroxyl ions away from this membrane. The reverse effect occurs at the anion-exchange membrane.

As the current density is increased, this effect increases until, after a certain critical current density is reached, the hydrogen-ion flux at the cation-exchange membrane (or the hydroxyl-ion flux at the anion-exchange membrane) becomes a substantial fraction of the total current. This effect, first observed by Bethe and Toropoff (4), not only results in decreased efficiency of desalting, but also in highly undesirable pH changes in the solutions. This effect has been studied extensively in recent years, and some results of these investigations are presented in Section IV.

The phenomenon of polarization also contributes to the Joule heating problem. As the current flows, counterions in the boundary layer upstream adjacent to the membrane are drawn into the membrane. These ions are replenished by diffusive and coulombic flow from the bulk solution. As the current increases, eventually a condition is reached where the finite flow from the bulk phase to the interface can no longer keep up with the current demand. As a result, the boundary layer adjacent to the membrane becomes depleted of ions, and its electrical resistance increases. This boundary-layer depletion is thought to be the trigger for the polarization phenomena described above. These conditions are illustrated in Fig. 3.

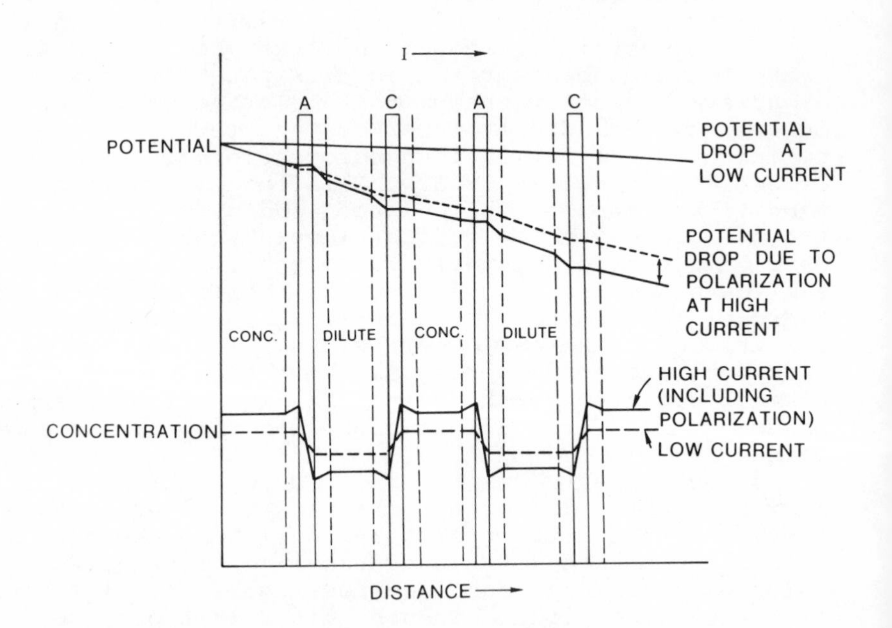

Fig. 3. Concentration polarization.

The increased stack resistance brought about by concentration polarization is an important contributing factor to decreasing desalination efficiency. Because it is essentially a diffusive effect, it can be alleviated somewhat by inducing turbulence within the cells. Turbulence promoters of various designs have been used; they are usually made of plastic screening material. Unfortunately, these promoters mask portions of the current path and therefore contribute somewhat to increasing the stack resistance.

Another phenomenon that results in the decreased efficiency of the electrodialysis process is electro-osmosis. At all current densities a flux of water approximately proportional to the flux of ions is observed to occur across all ion-exchange membranes. This flux, which varies with the ion carrying the current, and with the porosity of the membrane, can be significant, and is not solely attributable to ion hydration. Some results of recent studies of electro-osmosis are presented in Section V.

B. Transport Depletion

In this process, somewhat less common that electrodialysis, the anion-exchange membrane in the electrodialysis stack is replaced by a neutral membrane. To see how such an arrangement can result in desalting, consider Fig. 4. Suppose one Faraday of current is passed across the unit cell, as shown. Across the cation-exchange membrane, one equivalent of cation leaves the center compartment and enters the compartment on the right. Across the neutral membrane, 0.4 equivalents of cation enter the center compartment, and 0.6 equivalents of anion leave the center compartment. Thus the passage of one Faraday of current has resulted in the depletion of solute in the center compartment by 0.6 equivalents of salt.

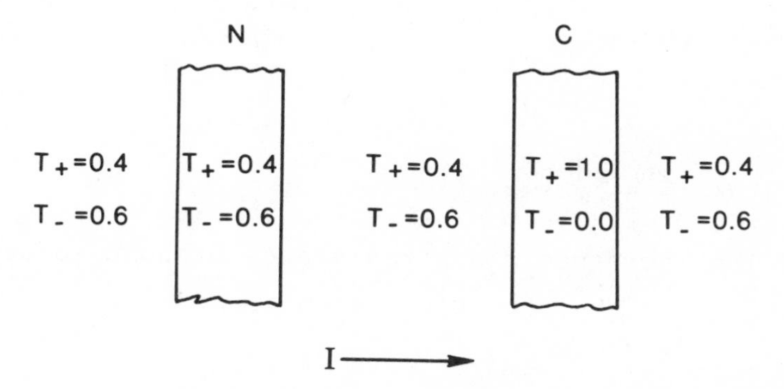

Fig. 4. Unit transport depletion cell. N, neutral membrane. C, cation-exchange membrane.

This process has several advantages over conventional electrodialysis. First of all, possible polarization effects on the surface of the anion-exchange membrane are avoided by eliminating this membrane. Since the critical current density for an anion-exchange membrane is considerably lower than for an equivalent cation-exchange membrane, this allows efficient operation of the stack at considerably higher current densities than is possible with conventional electrodialysis. Secondly, neutral membranes are considerably cheaper than are anion-exchange membranes, so significant savings are possible.

The chief disadvantage, of course, lies in the reduced desalination efficiency brought about by the presence of a membrane that transports in both directions. Other things being equal, the transport depletion process is approximately half as efficient, in terms of current utilization, as the equivalent conventional electrodialysis process.

There are two different types of transport depletion processes that have been under study at the Southern Research Institute (5) and elsewhere. The two processes both involve a cation-exchange membrane to produce ion-depleted and ion-enriched solutions, but differ in the way that the two product streams are kept separated so that they can be withdrawn separately.

One of these, the cation-neutral membrane transport depletion process, is shown in Fig. 5. This design is identical in concept to the example shown in Fig. 4. The neutral membrane, in this case, serves to separate the depleted and concentrated solutions, so that they can be withdrawn separately. A laboratory stack based on this design operated at the Southern Research Institute a number of years ago at current densities as high as 106 Ma/cm^2 without encountering the usual indications of excessive polarization. For comparison, conventional electrodialysis units are normally limited to current densities below about 25 Ma/cm^2.

The other type of transport depletion process depends on the fact that concentrated salt solutions are denser than dilute solutions. This process is referred to as electrogravitation, and is shown in Fig. 6. In this process the membranes are all cation-exchange membranes. The feed water is introduced midway between the vertical membranes at various heights, and current is passed. The passage of cations across

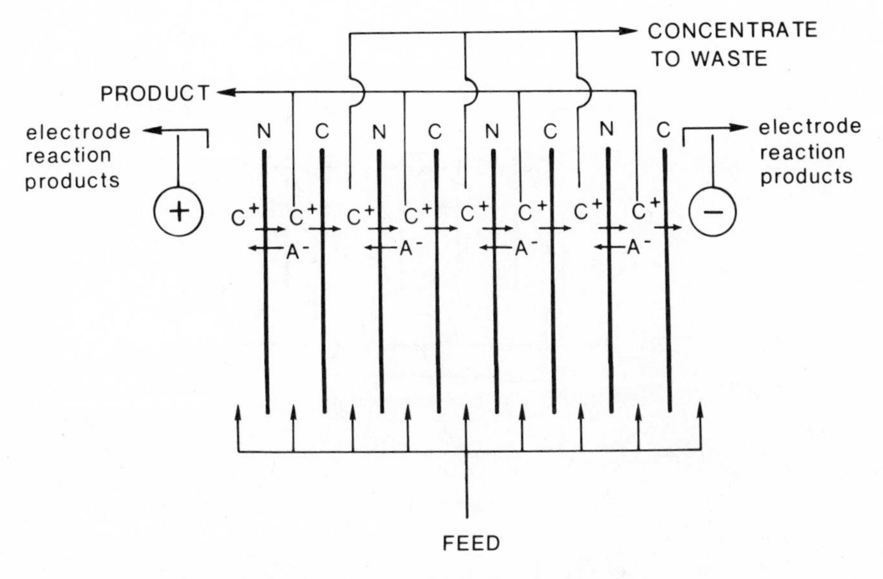

Fig. 5. Cation-neutral membrane transport depletion process.

the membranes and the migration of anions within the compartments result in the establishment of a dense layer that flows down the downstream side of the membranes to waste, and a lighter product layer that flows up the upstream side of the membranes to a collecting reservoir.

Preliminary comparisons of these two different types of transport depletion processes indicate that the cation-neutral process would cost about half as much per unit product as electrogravitation. This difference is caused by the greater coulomb efficiencies (90% of theoretical, as compared to 33 to 67% of theoretical) and higher production rates per unit membrane area for the cation-neutral process. However, electrogravitation has the advantages of simplicity of design, ease of maintenance, and low capital cost.

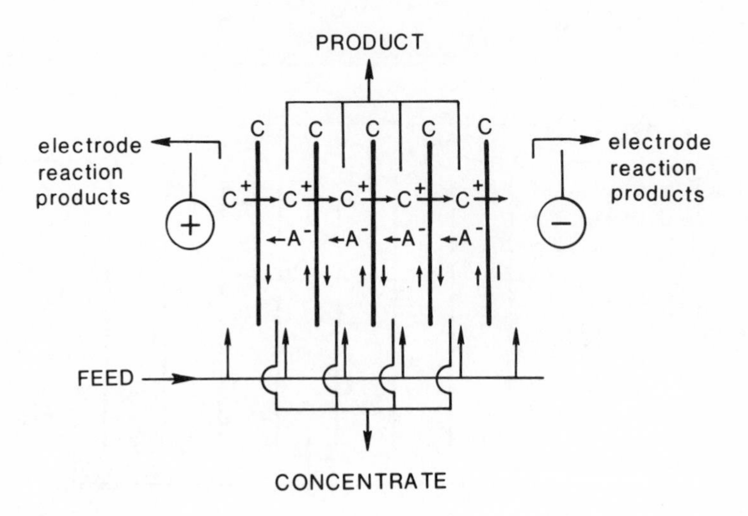

Fig. 6. Electrogravitation transport depletion process.

III. MEMBRANES

A. General Considerations

The function of the membrane in electrodialysis is to provide a selective barrier for ions of a particular sign, while still maintaining reasonable production rates. For efficient, economic operation, the membrane should have the following properties:

1. High degree of selectivity for ions of a given sign. In fact, the transport number for the ion in question should approach unity.
2. Low electrical resistance. This can be achieved by using a membrane with a large water content, coupled with a high concentration of fixed charges. This requirement conflicts somewhat with the requirement for high selectivity. However, it is currently possible to meet both requirements quite adequately.
3. Low degree of water transport. Electroosmosis, the movement of water as a result of current flow,

tends to reduce the electrodialytic efficiency of the system, and should be avoided. This can be almost achieved by keeping the porosity low, but cannot be avoided altogether.

4. Inertness. The membrane should not deteriorate in the presence of whatever chemical and biological agents with which it may come in contact. This requirement does not seriously limit the range of suitable membrane materials.

5. Physical strength. The membrane should be able to withstand whatever physical stress it encounters. For materials without intrinsic physical strength, methods have been developed to improve such properties. One method is to cast the membrane over a woven backing material that does not affect its electrical properties. Another method is to use a heterogeneous membrane made up of the ion-exchange materials, and a "film-former," which is an inert polymer that provides suitable physical properties.

6. Resistance to fouling. With many feed stocks, electrodialytic purification is accompanied by precipitation of pH- and thermally-sensitive solutes that tend to clog the membranes. Although no method has yet been devised to completely eliminate this problem, both pretreatment of feed stocks and periodic reversal of current seem to help.

7. Low cost. The ion-exchange membranes represent the most important single item of capital investment for electrodialysis. The high cost of membrane replacement, and the limitations inherent in the properties of available membranes represent the most important obstacle to widespread usage of electrodialysis to desalt brackish waters.

8. Resistance to deterioration at high temperature. It has been found that the effects of polarization tend to decrease as the temperature is raised. Studies sponsored by the Office of Saline Water in Israel (6) have indicated that the optimum operational temperature for electrodialysis plants desalting sea water appears to be about 70°C. If high-temperature operation becomes more widespread, the ability of the membrane to withstand such temperatures without deterioration will become important.

B. Methods of Preparation

An extensive literature exists on general methods of preparation of ion exchangers, although specific details of formulations are well kept trade secrets. A good general reference is the book by Helfferich (7).

The synthesis of an ion exchange membrane must yield a film that is superficially uniform and that, on a microscopic level, consists of a three-dimensional crosslinked matrix of hydrocarbon chains carrying fixed ionic groups. The membrane must be insoluble, but it must be able to swell in water to a limited extent. This property is determined by controlling the degree of crosslinking in the final structure. An example of how the degree of crosslinking can be controlled is shown in Fig. 7.

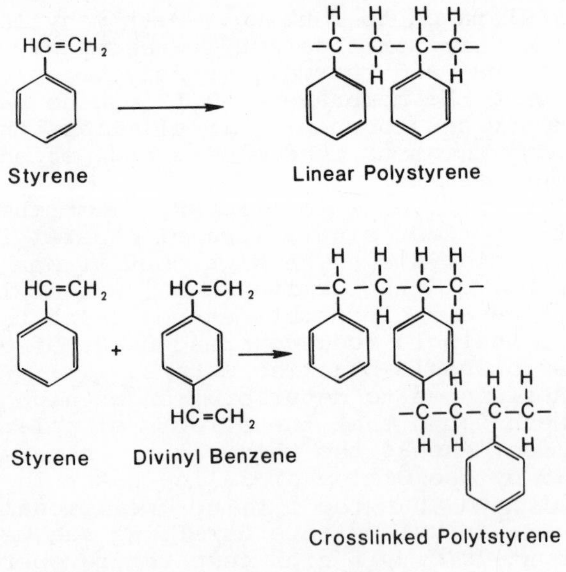

Fig. 7. Polymerization of styrene.

If styrene is polymerized, linear polystyrene is formed, which is soluble and which cannot be used as a membrane material. If a crosslinking agent, such as divinylbenzene, is added, polymerization yields a three-dimensional matrix. The degree of crosslinking is determined by the proportion of divinylbenzene used.

An example of how such an approach can be used to produce a commercial ion-exchange membrane is provided by the method used by Asahi Chemical Industries of Japan (ACI). In this method membranes are prepared by dissolving linear polystyrene (or a copolymer with styrene as the principle ingredient) in a mixture of styrene and divinylbenzene, which is then polymerized to form a suitable three-dimensional cage structure containing linear polystyrene trapped in a crosslinked polystyrene matrix. Blocks of this material are then microtomed to form sheets of suitable thickness.

If these membranes are treated with concentrated sulfuric acid or chlorosulfonic acid, sulfonate groups are introduced on the aromatic rings to form a cation-exchange membrane. If, on the other hand, they are treated with chloromethylether, followed by trimethylamine, an anion-exchange membrane is formed. These reactions are illustrated in Fig. 8.

Membranes such as these, where a homogeneous polymer matrix is first formed and the ion-exchange groups are added subsequently, are referred to as homogeneous membranes of the addition type. Homogeneous membranes may also be formed by addition on a polyethylene or a polyethylene-polystyrene copolymer base, or by condensation of such organic electrolyte monomers as phenolsulfonic acid with formaldehyde, or polyethyleneimine with epichlorohydrin. While the intrinsic strength of such membranes is dictated by the polymers used, they may be reinforced by polymerizing on wide-mesh plastic screen supports.

Heterogeneous ion-exchange membranes are formed by combining the ion-exchange material with an inert binder that provides mechanical strength. These may be formed either by embedding colloidal ion-exchanger particles in such binders as polyethylene, polystyrene, fluorocarbons, and so on, or by impregnating microporous membranes with soluble polyelectrolytes, subsequently drying the membranes to reduce their ability to swell, and thus trapping the polyelectrolyte.

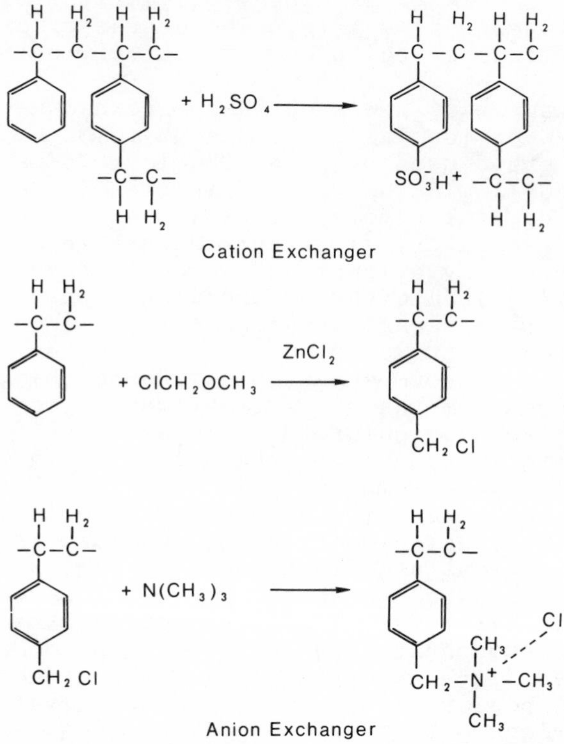

Cation Exchanger

Anion Exchanger

Fig. 8. Preparation of ion exchangers.

Another method of forming a heterogeneous ion-exchange membrane is to crosslink a polyacid such as linear polystyrenesulfonic acid in the presence of an inert polymer such as linear polyvinylidene fluoride. The three-dimensional polyacid structure traps the inert polymer and creates a membrane with quite good chemical and physical properties. The same technique may be used with such poly bases as polyethyleneimine as well.

Other techniques for forming ion-exchange membranes include "interpolymer" membranes, in which a solution containing a linear polyelectrolyte and a linear inert polymer is evaporated. Apparently, the chains of the water-soluble polyelectrolyte and the water-insoluble film former become so entangled that the polyelectrolyte cannot be leached out by water even though no crosslinking is used.

Graft polymerization has also produced ion-exchange membranes. In this technique, polyethylene films are impregnated with styrene and divinylbenzene and exposed to γ-irradiation. The radiation causes the styrene and divinylbenzene to be grafted onto the polyethylene base, and the resulting film is then either sulfonated or chloromethylated and aminated to form the ion-exchange membrane.

C. Methods of Characterization

The important properties of ion-exchange membranes that must be measured to determine their value for electrodialysis include:

1. Ion-Exchange Capacity. A sample of cation-exchange membrane in the hydrogen form is treated with an excess of standard 0.1N KOH and allowed to equilibrate. The solution is then separated and titrated. A sample of anion-exchange membrane in the hydroxyl form is similarly treated with HCl. The membrane capacity is normally reported as meq/dry gram and requires a measure of the water content. Typical values for commercially available membranes run from 0.5 to 3.0 meq/dry gram.

2. Water Content. A sample of cation-exchange membrane in the sodium form or anion-exchange membrane in the chloride form is soaked in distilled water, carefully blotted to remove surface moisture, and weighed. It is then vacuum dried at 70°C until a constant weight is obtained. The water content is reported on a weight percent basis, and typically runs from 20 to 50%.

3. Permselectivity. A sample of membrane in the appropriate ionic form (sodium or chloride, depending on whether the membrane is cationic or anionic) is placed in a cell with one concentration of NaCl on one side, and another concentration of NaCl on the other. The electrical potential developed across the membrane is measured. If it is assumed that the solutions

behave ideally, this concentration potential can be related to the transport number of cation through the membrane by the equation

$$\Delta\phi = \frac{-RT}{F} (2t_+ - 1) \ln \frac{C'}{C''} \quad (1)$$

where ϕ is the potential in volts, t_+ is the transport number for cation, F is Faraday's constant, and C' and C" refer to the concentrations on the two sides of the membrane. All commercially available membranes completely exclude co-ion, for all practical purposes, over the range of conditions of interest for desalination.

4. Membrane Resistance. A sample of membrane, equilibrated with the appropriate concentration of salt solution, is placed in a conductance cell, and its resistance is measured with a Wheatstone bridge arrangement. The membrane resistance is reported as ohm-cm^2, with acceptable values less than 10 ohm-cm^2.

5. Electroosmotic Coefficient. A sample of membrane, equilibrated with the appropriate concentration of salt solution, is placed in a constant-volume cell equipped with a horizontal capillary, and a constant current is drawn. The volume flux is measured, corrected for ion flux, and the results are reported in terms of the electroosmotic coefficient, defined as moles water transported per Faraday of current. To ensure precise measurements, a number of important precautions must be taken, but their description is beyond the scope of this report. Acceptable values would be less than about 5 to 10 moles/Faraday.

6. Dimensional Stability. Here the dimensions of a membrane sample in the proper state are measured initially, upon vacuum drying, and upon rewetting. Good membranes will be reversible and will swell 10 to 20% in linear dimensions on rewetting.

7. Strength. A Mullen burst-strength test is performed on the membrane in the wet state, and the results are reported in psi. A typical commercially available homogeneous electrodialysis membrane might have a burst strength of about 50 psi. Reinforcement would bring this value to over 100 psi.

Other membrane properties that might be measured include hydraulic permeability, streaming potential, self-diffusion coefficient of ions of particular interest, diffusion coefficients of neutral species, co-ion concentration, and so on. Information on the properties of the membrane might also be obtained by measuring any of the above properties in the presence of different counterions. In the case of cation-exchange membranes, the tetra-alkylammonium ions have proven particularly useful because their size, shape, and degree of hydration are accurately known.

IV. MEMBRANE POLARIZATION

A. Experimental Observations

When a direct electric current is passed across most types of porous membranes interposed between neutral electrolytic solutions, a pH change is produced in the solutions, which is independent of any electrode reaction. This effect, first observed by Bethe and Toropoff (4), results in the anolyte becoming alkaline and the catholyte acidic, which is just the reverse of what happens in ordinary electrolysis of neutral solutions between platinum electrodes. This effect is identifiable with the presence of a fixed charge in the membrane, and correlates with production and transport of hydrogen and hydroxyl ions.

Because of its importance to the practical application of the electrodialysis process, this effect has been studied extensively in recent years. In general, what one observes is that, at low current densities (in neutral salt solutions), the solute ion carries the current almost completely, with negligible transport of hydrogen ion (for cation-exchange membranes) or hydroxyl ion (for anion exchangers). As the current density is increased across a cation-exchange membrane, the production of hydrogen and hydroxyl ion at the interface increases, with the hydrogen ion migrating into the membrane, and the hydroxyl ion migrating away from the membrane. When a certain critical current density is reached, the observed hydrogen-ion flux increases sharply and becomes a substantial fraction of the total current. This effect depends on the degree of stirring, the concentration and nature of the salts present, the properties of the membrane, and so on.

Since the mobility of hydrogen ion is different from sodium ion (as hydroxyl is from chloride), one would expect that this phenomenon would result in a change in the normal current-voltage curve for the system. This has indeed been observed by such workers as Peers (8), Rosenberg and Tirrell (9), and Cowan and Brown (10). A good example of such an observation is shown in Fig. 9, from data presented in Peers (8). In this experiment, a pair of AgCl probe electrodes were placed on opposite sides of a Permutit C-10 cation-exchange membrane bathed on both sides with flowing 0.01N NaCl solutions. The sharp break in the curve indicating the critical current density is apparent.

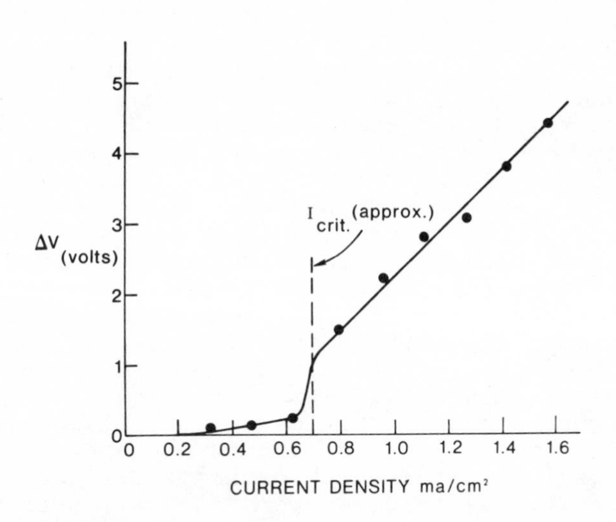

Fig. 9. Current voltage curve for Permutit C-10 cation-exchange membrane in 0.01N NaCl solution [Data of Peers (8)].

Observations of this phenomenon in terms of pH changes have been made by Uchino et al (11) and Gregor and Peterson (12). The results of Gregor and Peterson are typical. Figure 10 is a schematic representation of the apparatus used. In this experiment a Nalfilm 1 (Nalco Chemical Company) cation-exchange membrane of well-defined area was interposed between two compartments containing KCl solution of various concentrations. The compartments were connected to electrode compartments containing Ag-AgCl electrodes, so that a current could be drawn. The complete cell can thus be written as:

$$\mathrm{Ag}\,|\,\mathrm{AgCl(s)}\,|\,\mathrm{KCl}(\alpha)\,|\,\mathrm{Membrane}\,|\,\mathrm{KCl}(\beta)\,|\,\mathrm{AgCl(s)}\,|\,\mathrm{Ag}$$

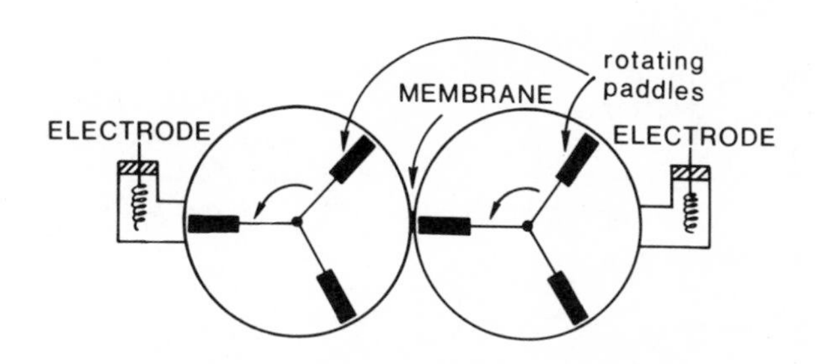

Fig. 10. Schematic of apparatus of Gregor and Peterson.

The solution compartments were fitted with paddles that give a high rate of stirring (peripheral velocities of about 100 cm/sec with the paddle coming within 1 to 2 mm of the membrane) so that the hydrodynamic conditions could be well defined. The compartments were also fitted with cooling coils to maintain constant temperature, and inlet and outlet ports so that flow experiments could be run in which the entire system could be maintained at steady state.

The hydrodynamic conditions in the system were defined by performing a diffusion exchange experiment employing potassium and ammonium ions in the absence of current. For such an experiment, Fick's Law for each ion, given as

$$J_i = -D_i \frac{dC_i}{dx} \tag{2}$$

where J_i is observed ion flux (moles/sec-cm^2), D_i is diffusivity (cm^2/sec), C_i is ion concentration (moles/cc), and x is distance (cm), can be integrated from one bulk phase to the other to give an apparent concentration boundary-layer thickness, δ, adjacent to the membrane on both sides, which is a function of the rate of stirring.

After the hydrodynamic conditions were defined, experiments were run in which current was drawn at a fixed solution concentration and rate of stirring. Observations were made of pH changes after enough time had elapsed to convert the membrane to the mixed potassium-hydrogen state, but before the solution bulk concentration changed significantly. The results were put in terms of hydrogen-ion flux, or in terms of transport number of hydrogen ion. One typical result is shown in Fig. 11, taken at a concentration of KCl of 0.0005M, and a calculated boundary-layer thickness of 4.3 μ. The circles represent the experimental points. The theoretical curve refers to the analysis presented below.

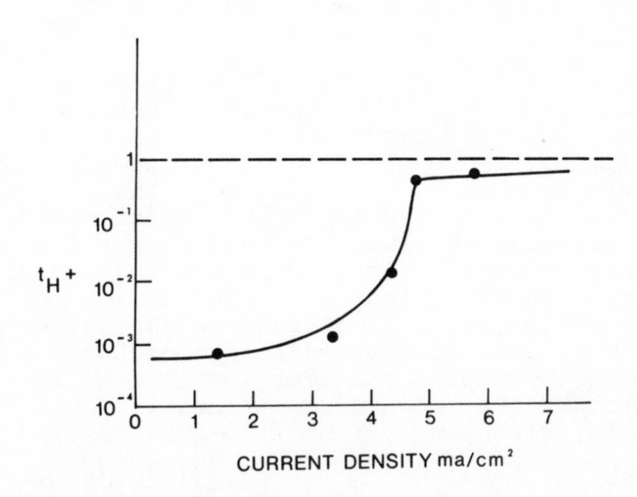

Fig. 11. Typical result of Gregor and Peterson. The circles are experimental data, the line is the theoretically predicted curve.

The "critical current density" phenomenon is clearly visible here. At current densities below about 4 mA/cm^2, a negligible flux of hydrogen ions is observed, amounting only to about 0.001 of the total current. However, as the current is increased to about 5 mA/cm^2, the hydrogen-ion flux suddenly jumps to over 0.5 of the total current.

B. One-Dimensional Theory

Gregor and Peterson identified the sharp increase in acid and base production with the point where the concentration of electrolyte at the solution-membrane fell to zero when solving the Nernst-Planck equation considering only the salt present. The equation for this case can be written

$$J_i = -D_i \frac{dC_i}{dx} - \frac{Z_i F D_i C_i}{RT} \frac{d\phi}{dx} \quad (3)$$

where Z_i is the valence. The other symbols are as previously defined.

Defining $\phi^* = Z F \phi / R T$, and using subscript 1 to denote K^+ ion, and subscript 2 to denote Cl^- ion,

$$-\frac{J_1}{D_1} = \frac{dC_1}{dx} + C_1 \frac{d\phi^*}{dx} \quad (4)$$

$$-\frac{J_2}{D_2} = \frac{dC_2}{dx} - C_2 \frac{d\phi^*}{dx} \quad (5)$$

For an ideally selective cation-exchange membrane, $J_2 = 0$ at the membrane surface and, in the steady state, everywhere. For this case, then, if it is assumed that $D_1 = D_2$ (reasonable for KCl), and $C_1 = C_2$ (any space charge present is neglected), Eqs. 4 and 5 may be added or subtracted to give

$$-\frac{J_1}{D} = 2C \frac{d\phi^*}{dx} = 2 \frac{dC}{dx} = -\frac{I}{D} \quad (6)$$

where I is the total current density in equiv./sec-cm^2.

If this equation is integrated from the bulk phase to the membrane surface, a linear concentration gradient is obtained, given by

$$C_{bulk} - C_{membrane} = \frac{I\delta}{2D} \tag{7}$$

The potential gradient across the boundary layer is given by

$$\left(\frac{d\phi^*}{dx}\right)_{boundary\ layer} = -\frac{I}{(2DC_{bulk} - I\delta)} \tag{8}$$

Thus, as $I \longrightarrow 2DC_{bulk}/\delta$, the field goes to infinity. This is defined as the critical current density, and corresponds to the point where $C_{membrane} \longrightarrow 0$.

More recently, Gregor and Miller (13) presented a more detailed analysis of this phenomenon. In this analysis, the system is assumed one dimensional (flows parallel to the membrane face are neglected) isothermal and at steady state, and the membrane is assumed ideally selective. Here, subscripts 3 and 4 refer to hydrogen and hydroxyl ion, respectively, and an overbar refers to membrane properties.

In the boundary layers adjacent to the membrane, the following flux equations may be written

$$\frac{J_1}{D_1} = -\frac{dC_1}{dx} - C_1\frac{d\phi^*}{dx} \tag{9}$$

$$\frac{J_2}{D_2} = 0 = -\frac{dC_2}{dx} + C_2\frac{d\phi^*}{dx} \tag{10}$$

$$\frac{J_3}{D_3} = -\frac{dC_3}{dx} - C_3\frac{d\phi^*}{dx} \tag{11}$$

$$\frac{J_4}{D_4} = = \frac{dC_4}{dx} + C_4\frac{d\phi^*}{dx} \tag{12}$$

In addition, the total current density, I, is given by

$$I = J_1 + J_3 - J_4 \tag{13}$$

The space charge is given by Poisson's equation:

$$\frac{d^2\phi^*}{dx^2} = -\frac{4\pi}{\varepsilon}(C_2 + C_4 - C_1 - C_3) \qquad (14)$$

where ε is the electric permittivity in equiv/cm.
The dissociation constant for water is given by

$$C_3C_4 = K_w B \qquad (15)$$

where K_w is the ordinary dissociation constant for water 10^{-20} ($mole^2/cm^6$) and B is the second Wein effect term, wherein the rate of dissociation of water is influenced by the field (14).
In addition, the equal rate of production of hydrogen and hydroxyl ions results in

$$\frac{dJ_3}{dx} = \frac{dJ_4}{dx} \qquad (16)$$

At the membrane surface,

$$\frac{C_3}{C_1} = \sigma \frac{\bar{C}_3}{\bar{C}_1} \qquad (17)$$

where σ is the equilibrium distribution coefficient.

Within the membrane

$$\frac{\bar{J}_1}{\bar{D}_1} = -\frac{d\bar{C}_1}{dx} - \bar{C}_1\frac{d\bar{\phi}^*}{dx} \qquad (18)$$

$$\frac{\bar{J}_3}{\bar{D}_3} = -\frac{d\bar{C}_3}{dx} - \bar{C}_3\frac{d\bar{\phi}^*}{dx} \qquad (19)$$

$$\frac{d^2\bar{\phi}^*}{dx^2} = -\frac{4\pi}{\varepsilon}(\bar{C} - \bar{C}_1 - \bar{C}_3) \qquad (20)$$

$$\bar{J}_1 + \bar{J}_3 = I \qquad (21)$$

$$\bar{J}_1 = J_1 \qquad (22)$$

where $\bar{C}$ is the total membrane capacity in moles/cm^3.

The entire system of equations has been solved numerically and tested with the data of Gregor and Peterson. The results indicate the presence of a phenomenon that has not been observed previously.

In 1934, Onsager (14) presented a detailed analysis of the second Wein effect. In this phenomenon it has been observed that the presence of an electric potential gradient tends to increase the dissociation constant for water (the factor B presented in Eq. 15). Onsager's analysis, limited to homogeneous solutions, showed that this effect is due to the increased rate of dissociation of the water molecule caused by distortion of the ion cloud surrounding the molecule, but that the rate of recombination of hydrogen and hydroxyl ions is independent of field. This independence is due to the homogeneity of the system.

However, the surface of an ion-exchange membrane is not homogeneous. The membrane phase is forbidden region to mobile co-ions (the hydroxyl ion, in the case of a cation exchanger). Thus, while in free solution, the field may draw hydrogen and hydroxyl ions closer together or further away, depending on their orientation, on the surface of the ion exchanger, the field can only draw the ions further away. Therefore, the rate of recombination of hydrogen and hydroyxl ions must be decreased by the presence of a field at the ion-exchange surface, and the dissociation constant for water must be increased. This effect can be presented as

$$(C_3\ C_4)_{\text{membrane surface}} = K_W BA \tag{23}$$

where A is this additional effect. Gregor and Miller attempted to fit the coefficient A from the data of Gregor and Peterson under a variety of hydrodynamic conditions and solution concentrations. The results, as a function of potential gradient at the membrane solution interface, are presented in Fig. 12. All experiments correlated with a single curve for A. The theoretical curve presented in Fig. 11 was obtained by solving Eqs. 9 to 22, with Eq. 23 and Fig. 12 used at the membrane-solution interface

A further refinement of this approach is presented by Lang (15). In this work, a narrow "repulsion zone" is defined in the solutions immediately adjacent to the membrane surfaces, in which forces exerted by the fixed membrane charges on co-ions dominate the situation. The effect of these repulsion zones is that the rate of recomination of hydrogen and hydroxyl ions in these zones is strongly affected by the local electric field. A consequence is that polarization is associated with the buildup of high concentrations of H^+ ion adjacent to a cation exchange membrane (or OH^- ion adjacent to an anion exchange membrane), thus helping to explain the observed pH changes.

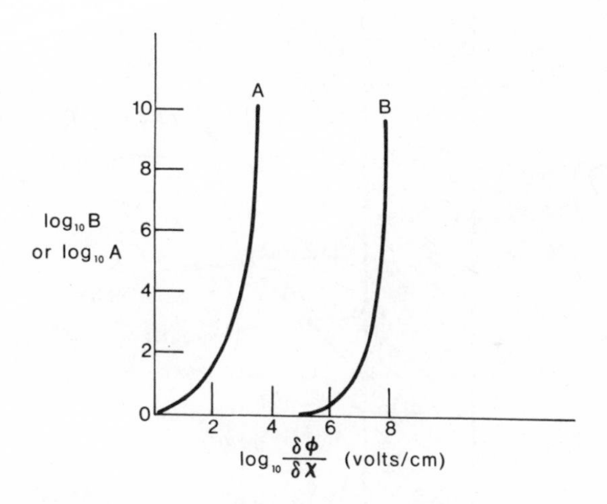

Fig. 12. Effect of field on dissociation constant of water. A, the empirical effect of Gregor and Miller. B, the second Wein effect.

C. Effect of Hydrodynamic Conditions on Polarization

The model described above, while accounting for the data of Gregor and Peterson reasonably well, still has little to say about the effect of hydrodynamic conditions on polarization. These hydrodynamic conditions were indirectly taken into account in the model, in that the assumed boundary-layer thickness adjacent to the membrane is a function of the rate of stirring within the cell.

However, there is a profound difference between the Gregor and Peterson experiment and commercial electrodialysis. In the Gregor and Peterson experiment the batch nature of the experiment, combined with the high degree of mixing and the small membrane area used, resulted in a condition in which the concentration of ion at the interface was not a function of position on the membrane. Thus a one-dimensional model was adequate to explain the experimental results.

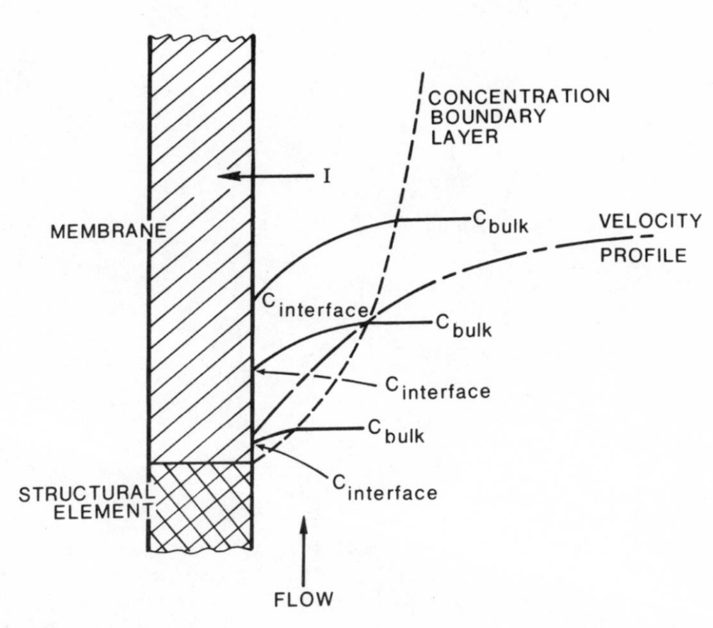

Fig. 13. Concentration polarization as influenced by convection.

In commercial electrodialysis, however, the ionic concentration in the solution at the interface is a strong function of location on the membrane surface. For example, consider Fig. 13. As the flow passes upward and enters the region across which current is passed (the "membrane window"), a concentration boundary layer develops. As this boundary layer develops, the concentration of transporting ion at the interface drops from its value in the bulk fluid to an equilibrium value calculable from the Nernst-Planck equation. Because of these flow processes parallel to the membrane face, the average concentration of transporting ion within the boundary layer would thus be larger than could occur in a nonconvective system. Also, since the liquid enters the system considerably upstream of the membrane window, the velocity profile within the unit cell is more fully developed than the concentration profile; one should therefore not expect the two profiles to be similar.

Under the circumstances, one should expect that the observed critical current density should be a function of bulk concentration of solute and of flow rate, and this, indeed, turns out to be the case. In fact, Cowan (16) has proposed the following empirical expression for critical current density as a function of these two variables:

$$I_c = \frac{0.04C_b}{y} W^{1/2}(W + 1.9)^{1/2} \qquad (24)$$

where I_c is the critical current density in mA/cm^2, C_b is the bulk salt concentration in equiv/liter, y is the cell width in inches, and W is the flow in gal/min-in. of cell width.

It is well established [see, for example, Schlichting (17), Levich (18), or Bird et al. (19)] that for both velocity boundary layers in laminar flow past a flat plate and concentration boundary layers in laminar flow past a flat dissolving plate, the boundary layer-thickness is proportional to the inverse square root of the Reynolds number. This fact, coupled with the Cowan equation, is additional evidence of the strong connection between critical current density and convective boundary layer thickness.

Both Espelosin (20) and Salzarulo (21) have studied this phenomenon in considerable detail, theoretically as well as experimentally. In the theoretical development, the phenomenon is assumed to be analogous to the case of the development of a laminar boundary layer when a liquid approaches a semi-infinite flat plate with a certain velocity, V_∞. This approach is used for both the non-fully developed flow case described above and, also, for the fully developed flow case (a semi-infinite membrane window).

Following the method described by Wranglen and Nilsson (22), the function $C_{bulk}-C_{interface}$ is approximated by a polynomial in the variable y, (the distance from the membrane surface), which, after evaluation of constants from appropriate boundary conditions, yields:

$$C_{bulk}-C_{int.} = \frac{2}{3}\frac{J}{D}\,\delta\left[1 - \frac{3}{2}\,\frac{y}{\delta} + \frac{1}{2}\left(\frac{y}{\delta}\right)^3\right] \qquad (25)$$

where J is the total diffusive flux.

For the case of non-fully developed flow, a velocity profile described by Schlichting (17) for the case of a semi-infinite flat plate may be taken:

$$\frac{\upsilon}{\upsilon_{bulk}} = \frac{3}{2}\,\frac{y}{\delta_h} - \frac{1}{2}\left(\frac{y}{\delta_h}\right)^3 \qquad (26)$$

where δ_h is the hydrodynamic boundary-layer thickness.

If a material balance is made across a control volume within the concentration boundary layer and standard procedures for analyzing boundary layers, such as are described by Schlichting, are applied, one can determine the concentration boundary-layer thickness as a function of the various parameters of the system:

$$\delta = 3.59\ (S_c)^{-1/3}\ (R_e)^{-1/2} \qquad (27)$$

where S_c is the Schmidt Number, given by $S_c = \nu/D$, dimensionless; R_e is the Reynolds Number, given by $R_e = \upsilon_{bulk}\,x/\nu$, dimensionless; X is the distance to the point of interest, cm; and ν is the kinematic viscosity, cm^2/sec.

If the diffusive flux is written as a function of current, and if the assumption is made (as it was by Gregor and Peterson) that the critical current density corresponds to the condition where $C_{interface} = 0$, then the following equation is obtained:

$$C_{bulk} = \frac{2.39}{D}\,(S_c)^{-1/3}\ (R_e)^{-1/2}\ I_c\left(\frac{1-t}{F}\right) \qquad (28)$$

where t is the transport number for the ion of interest in the solution.

For the case of fully developed flow, consideration of the system as flow through a narrow slit of some fixed half-width yields

$$C_{bulk} = \frac{4}{3D}\sqrt{\frac{DX}{\upsilon_{bulk}}}\, I_c\, \frac{1-t}{F} \tag{29}$$

In both cases, the critical current density is proportional to the bulk electrolyte concentration, and to the square root of the bulk linear velocity of the fluid, an obvious consequence of diffusive boundary-layer control.

Both Espelosin and Salzarulo verified this relationship experimentally, by studying transport of ions from dilute KCl solutions across various types of commercial ion-exchange membranes under a variety of flow conditions. In these experiments, Reynolds' numbers ranging between 30 and 3500 were employed, and turbulence-producing spacers were used to modify flow conditions. In the presence of the spacers, the flow was considered fully developed. In the absence of the spacers, the flow was non-fully developed.

The critical current density was determined by several methods. In one set of experiments, current-voltage curves were obtained, with the critical current density being indicated by the current at which the curve changed slope.

In another set of experiments the pH in the downstream compartment was monitored, with the critical current density being indicated by the current at which a rapid increase occurred in the hydrogen-ion flux (in the case of cation-exchange membranes), or the hydroxyl-ion flux (in the case of anion-exchange membranes).

A third set of experiments involved the placement of a set of fine wire potential probes in the compartment of interest. At various currents these probes accurately measured the potential gradient in the compartment and indicated the critical current density by indicating a sharp break in the gradient.

In all these experiments, under all the conditions chose, the theoretical behavior of critical current density with bulk concentration and flow rate was well verified.

V. THERMODYNAMIC MODELS FOR MEMBRANE PROCESSES

For a number of reasons, several attempts have been made to characterize ion transport across membranes by use of irreversible thermodynamics. Chief among these reasons is the fact that such an approach is relatively model independent; that is, the transport equations for membrane flow can be written without resort to assumptions regarding the microscopic structure of the membrane.

In principle, such a purely phenomenological approach should prove quite powerful in its ability to predict membrane behavior. In practice, however, approaches involving membrane models (for example, hydrodynamic models involving Stokes' law, or capillary flow models) have proved more successful in their ability to predict how membranes will function under a variety of conditions. Why this should be so will become clear as we proceed.

To develop a thermodynamic model for membrane processes, let us assume a system in which a membrane is used to separate two compartments containing electrolyte solution. Let us further assume that the compartments contain electrodes that undergo a reversible reaction with one of the ions in solution (for example, $Ag + Cl^- \rightleftharpoons Ag\,Cl + e^-$), so that no additional components need be introduced.

At equilibrium (in the absence of all gradients), no flows will occur between the two compartments. If an external force of some sort, such as a pressure difference is applied across the membrane, a flow in response to the force will occur. In the system assumed, three possible flows can occur, consisting of counterion (subscript 1), co-ion (subscript 2), or solvent water (subscript w). In a thermodynamic sense, the forces establishing these flows consist of gradients in the electrochemical potential for the flowing species across the membrane.

The electrochemical potential gradient is defined by

$$\Delta\tilde{\mu}_i = \Delta\mu_i - Z_i F \Delta \phi \tag{30}$$

where $\Delta\mu_i$ is the chemical potential gradient; Z_i is the valence, F is the Faraday constant, and $\Delta\phi$ is the electrical potential gradient across the membrane.

Following the methods of irreversible thermodynamics, as formulated by Onsager and others (23-25), one may write a "dissipation function" that is proportional to the rate of production of entropy (or the rate of dissipation of free energy), as a sum of products of fluxes and their corresponding forces:

$$\Phi = J_1 \Delta\tilde{\mu}_1 + J_2 \Delta\tilde{\mu}_2 + J_w \Delta\mu_w \tag{31}$$

Following the approach used by Katchalsky and coworkers (26, 27), Eq. 31 is transformed into one written in terms of volume flow (v), salt flow (s), and electric current (I). This is accomplished by a series of linear transformations as follows:

Define salt flow by

$$J_s = J_i/\gamma_i \tag{32}$$

where γ_i is the stoichiometric number,

so

$$J_1 = \gamma_1 J_s \tag{33}$$

Define total electric current by

$$I = Z_1 F J_1 + Z_2 F J_2 = \gamma_1 Z_1 F J_s + Z_2 F J_2 \tag{34}$$

Also,

$$\Delta\mu_s = \gamma_1 \Delta\tilde{\mu}_1 + \gamma_2 \Delta\tilde{\mu}_2 \tag{35}$$

The condition of electroneutrality yields

$$Z_1 \gamma_1 + Z_2 \gamma_2 = 0 \tag{36}$$

Define

$$E = \frac{\Delta\tilde{\mu}_i}{Z_i F} \tag{37}$$

Proper combination of Eqs. 31 - 37 yields

$$\Phi = J_s \Delta\mu_s + IE + J_w \Delta\mu_w \tag{38}$$

For convenience, define the volume flow, J_v, by

$$J_v = J_s \bar{V}_s + J_w \bar{V}_w \tag{39}$$

where $\bar{V}_i$ is the partial molal volume of component i in the flowing solution.

The chemical potentials for salt and water are given by

$$\Delta\mu_s = \bar{V}_s \; \Delta P + \frac{\Delta\pi}{\bar{C}_s} \tag{40}$$

and

$$\Delta\mu_w = \bar{V}_w \; (\Delta P - \Delta\pi) \tag{41}$$

where ΔP is the pressure difference between the two compartments, $\Delta\pi$ is the osmotic pressure difference, and $\bar{C}_s$ is the mean salt concentration.

Combining Eqs. 38 - 41 yields

$$\Phi = J_v \; (\Delta P - \Delta\pi) + J_s \frac{\Delta\pi}{\bar{C}_s} (1 + X_s) + IE \tag{42}$$

where X_s is the average volume fraction of solute in free solution ($X_s \ll 1$). So, to a good approximation,

$$\Phi = J_v(\Delta P - \Delta\pi) + J_s \frac{\Delta\pi}{\bar{C}_s} + I\,E \tag{43}$$

Once a suitable dissipation function is obtained, the methods of irreversible thermodynamics allow one to write a linear set of equations relating flows to forces [see Kedem and Katchalsky (26)], as

$$J_v = L_{11} \; (\Delta P - \Delta\pi) + L_{12} \; E + L_{13} \frac{\Delta\pi}{\bar{C}_s} \tag{44a}$$

$$I = L_{21}(\Delta P - \Delta\pi) + L_{22} \; E + L_{23}\frac{\Delta\pi}{\bar{C}_s} \tag{44b}$$

$$J_s = L_{31} \; (\Delta P - \Delta\pi) + L_{32} \; E + L_{33}\frac{\Delta\pi}{\bar{C}_s} \tag{44c}$$

Provided that the system is close enough to equilibrium so that the dissipation function as written is valid (a condition easily met for systems of interest here), Onsager has shown that the matrix of coefficients L_{ij}, defined by Eqs. 44 is symmetric; that is,

$$L_{ij} = L_{ji} \tag{45}$$

and, rather than nine phenomenological coefficients, one need only evaluate six coefficients to completely define the system.

Inspection of Eq. 44 suggests a number of relationships among membrane phenomena that can yield significant information and, at the same time, also suggests experiments that can be done to obtain this information.

For example, consider the flux equations in the absence of an electrical potential gradient. With $E = 0$, the volume flow equation can be rearranged to give

$$J_V = L_{11}\left[\Delta P - \left(1 - \frac{L_{13}}{L_{11}\overline{C}_3}\right)\Delta\pi\right] \tag{46}$$

From this equation the reflection coefficient of the salt, ε, can be defined as

$$\varepsilon = 1 - \frac{L_{13}}{L_{11}\overline{C}_S} \tag{47}$$

This coefficient represents the degree to which the membrane acts as an osmotic barrier to salt. A value of $\varepsilon = 1$ means the membrane is ideally semipermeable. A value of $\varepsilon = 0$ means the membrane is freely permeable to both water and salt. A knowledge of the reflection coefficient of a given membrane is central to its prospective usefulness for such desalination processes as reverse osmosis. The coefficient L_{11} is usually referred to as the "filtration coefficient" at zero potential for the membrane, and may be obtained by measuring the hydraulic permeability of the membrane to pure water when the membrane is short-circuited.

Consider the flux equations in the absence of pressure gradients ($\Delta P = 0$), or concentration gradients ($\Delta\pi = 0$). The flux equations then become

$$J_V = L_{12}E \tag{48a}$$

$$I = L_{22}E \tag{48b}$$

$$J_S = L_{32}E \tag{48c}$$

Obviously, L_{22} is the overall electrical conductance of the system. If Eqs. 39, 48a, and 48b are combined, an equation for water flow can be obtained as

$$J_w = \left(\frac{L_{12} - L_{32} \bar{V}_s}{L_{22} \bar{V}_w} \right) I \tag{49}$$

This coefficient, $(L_{12} - L_{32}\bar{V}_s)/L_{22}\bar{V}_w$, is the electroosmotic coefficient of the membrane, and is a measure of the amount of water transported per unit of electric current.

Consider the flux equations when a concentration difference is applied between the two compartments in the absence of electric current ($I = 0$), or pressure gradient ($\Delta P = 0$). In such a situation, a concentration potential develops, given by Eq. 44b:

$$E = \frac{L_{21} - L_{23}/\bar{C}_s}{L_{22}} \Delta\pi \tag{50}$$

From the Onsager reciprocal relations, this coefficient is seen to be intimately related to the electroosmotic coefficient. Both of these coefficients depend on the ion-exchange character of the membrane since, without such character, the coupling coefficients L_{12} ($=L_{21}$) and L_{32} ($=L_{23}$) would both be zero.

The membrane selectivity can be determined by combining Eqs. 48b and 48c ($\Delta P = 0$, $\Delta\pi = 0$) to give

$$J_s = \frac{L_{32}}{L_{22}} I \tag{51}$$

If the membrane is ideally permeselective,

$$J_s = I \tag{52}$$

That is, the current is carried by movement of ions in a single direction only, and

$$L_{32} = L_{22} \tag{53}$$

As the selectivity of the membrane decreases, L_{32} decreases, and for a membrane that has no ion-exchange character,

$$L_{32} = 0 \tag{54}$$

This ratio, L_{32}/L_{22}, is termed the transport number for the membrane, and, for good electrodialysis membranes, approaches 1.0.

A. "Friction Coefficient" Approach

The phenomenological equations represented in Eq. 44, in principal, characterize the system completely. A knowledge of the six phenomenological coefficients, L_{ij}, is sufficient to develop any membrane property. However, a serious drawback in the use of these coefficients is that their physical meaning is not clear and they cannot be readily interpreted in terms of molecular properties.

In an attempt to get around this problem, Spiegler (28), following the work of Onsager (29), Klemm (30), and Laity (31), proposed the use of "friction coefficients," r_{ik}, instead of L_{ik}. These coefficients, essentially the inverse of the matrix L_{ik}, are defined by Laity as

$$F_i = \sum_{k=1}^{n} X_k r_{ik}(V_i - V_k) \qquad (55)$$

where F_i is the total force exerted by all other species on component i, X_k is the mole fraction of component k in the membrane, and V_i and V_k are the average velocities of components i and k, respectively.

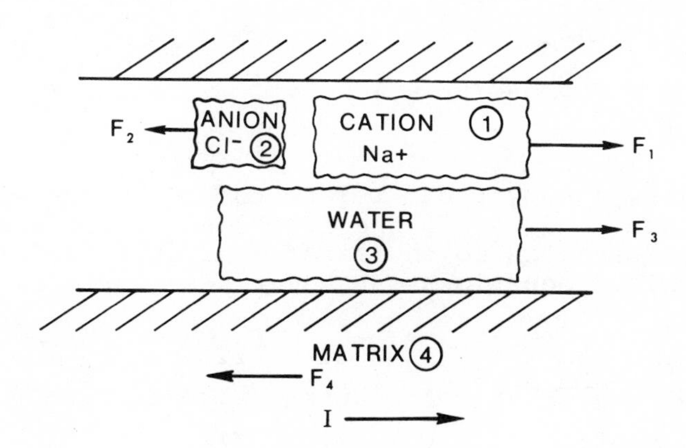

Fig. 14. Phenomenological picture of cation-exchange membrane.

Spiegler's picture of the membrane, for the case of a cation exchange membrane, is shown in Fig. 14. In this case, an electric current applies across the membrane exerts a force directly on the moving cation (F_1), on the stationary membrane poly anion (F_4), and on the small amount of intruding solution anion present (F_2). The transported cation applied a drag force (F_3) on the water present in the pores.

As before, the force exerted on the moving ions is given by the difference in electrochemical potential for the ion across the membrane. In the absence of concentration and pressure gradients, this force is proportional to the electrical potential difference (Eq. 30).

Consider, now, an experiment in which the self-diffusion coefficients of the various mobile species present are independently measured. Making use of Eq. 55, Caramazza et al. (32) point out that Laity has shown that, for such experiments in the membrane,

$$\frac{RT}{\bar{D}_{11}} = \bar{X}_1 r_{11} + \bar{X}_2 r_{12} + \bar{X}_3 r_{13} + f_{14} \tag{56a}$$

$$\frac{RT}{\bar{D}_{22}} = \bar{X}_1 r_{12} + \bar{X}_2 r_{22} + \bar{X}_3 r_{23} + f_{24} \tag{56b}$$

$$\frac{RT}{\bar{D}_{33}} = \bar{X}_1 r_{13} + \bar{X}_2 r_{23} + \bar{X}_3 r_{33} + f_{34} \tag{56c}$$

Here, the over bar refers to conditions within the membrane, and f_{i4} is the frictional force exerted by the membrane on molecules of i at unit velocity.

Similar equations can be written for free solution as

$$\frac{RT}{D_{11}} = X_1 r_{11} + X_2 r_{12} + X_3 r_{13} \tag{57a}$$

$$\frac{RT}{D_{22}} = X_1 r_{12} + X_2 r_{22} + X_3 r_{23} \tag{57b}$$

$$\frac{RT}{D_{33}} = X_1 r_{13} + X_2 r_{23} + X_3 r_{33} \tag{57c}$$

Here, the Onsager reciprocal relations have been assumed; that is, the basic flux equations derived by Onsager are

$$J_i = \sum_{j=1}^{n} L_{ij}X_j \tag{58}$$

with

$$L_{ij} = L_{ji} \tag{59}$$

Equation 55 represents an inverse of Eq. 58, and the inverse of a symmetric matrix is symmetric. Therefore,

$$r_{ij} = r_{ji} \tag{60}$$

From Ohm's law the velocity of the mobile cation and anion (V_1 and V_2) is proportional to the product of the ionic conductance in the medium (λ) and the electrical potential difference. The velocity of the water (V_3) is proportional to the electroosmotic coefficent (E), and to the electrical potential difference. The velocity of the matrix (V_4) is taken to be zero. On this basis, Eq. 55 may be used to calculate the total membrane conductance ($\bar{\lambda}=\bar{\lambda}_1 + \bar{\lambda}_2$) as

$$\gamma = \frac{F^2[\bar{X}_3\ (r_{13}+r_{23}) + f_{14}+f_{24}] - (\bar{X}_3E)\ (r_{23}f_{14}-r_{13}f_{24})}{\bar{X}_3[r_{12}r_{13}\bar{X}_1 + r_{12}r_{23}\bar{X}_2+r_{13}r_{23}\bar{X}_3] + f_{14}[r_{12}\bar{X}_1+r_{23}\bar{X}_3]+ f_{24}[r_{12}\bar{X}_2+r_{13}\bar{X}_3] +f_{14}f_{24}} \tag{61}$$

and the membrane selectivity, defined as

$$\eta + \frac{J_1}{J_2} \tag{62}$$

$$= \frac{r_{12}(\bar{X}_1-\bar{X}_2)+r_{23}\bar{X}_3+f_{24}+(\bar{X}_3E/F^2)[r_{12}r_{13}\bar{X}_1+r_{12}r_{23}\bar{X}_2+r_{13}r_{23}\bar{X}_3+r_{13}f_{24}]}{r_{12}(\bar{X}_2-\bar{X}_1)+r_{13}\bar{X}_3+f_{14}+(\bar{X}_3E/F^2)[r_{12}r_{13}\bar{X}_1+r_{12}r_{23}\bar{X}_2+r_{13}r_{23}\bar{X}_3+r_{23}f_{14}]} \tag{63}$$

The system represented by Eqs. 55 to 63 contain nine unknown friction coefficients (all the mole fractions are assumed to be known), given by r_{11}, r_{12}, r_{13}, r_{22}, r_{23}, r_{33}, f_{14}, f_{24}, and f_{34}. Experiments can be performed to measure the following quantities: D_{11}, D_{22}, D_{33}, $\bar{D}_{11}$, $\bar{D}_{22}$, $\bar{D}_{33}$, η, $\bar{\lambda}$, and E. Thus, in principal, all the frictional coefficients can be calculated, and the system can be completely characterized.

Lippe (33) performed just such a calculation for a typical electrodialysis membrane, the American Machine & Foundry Co. membrane AMF C-103. Data used in this calculation are given in Table 1. The molality referred to in the table is determined by measuring the membrane strong acid capacity per unit imbibed water.

Table I

AMF C-103 Membrane in 0.1M KCl @ 25°C

$D_{11} = 1.96 \times 10^{-5}$ cm²/sec ± 10%

$D_{22} = 1.89 \times 10^{-5}$ cm²/sec ± 10%

$D_{33} = 2.50 \times 10^{-5}$ cm²/sec ± 10%

$\bar{D}_{11} = 1.27 \times 10^{-6}$ cm²/sec ± 10%

$\bar{D}_{33} = 3.53 \times 10^{-5}$ cm²/sec ± 10%

$\lambda° = 149.86$ ohm^{-1}-cm² equiv^{-1} (infinite dilution)

λ (0.1M KCl) = 128.96 ohm^{-1} -cm² equiv^{-1}

$\bar{\lambda}/\lambda° = 0.065$

E = 7.65 moles/Faraday

Water content = 13.4 wt%

$\bar{m}_1$ = 8.17 equiv./(1000 g H_2O)

$\bar{m}_2$ = 0.009 equiv./(1000 g H_2O)

$\eta > 100$

An examination of the data before any calculation presents certain problems at the outset. It may be noted that the amount of co-ion present in the membrane is extremely small ($\bar{X}_2$ = ~0.0001). Thus, it becomes extremely difficult to evaluate any friction coefficients from terms in the equations containing $\bar{X}_2$. In fact, the large amount of water present results in imprecise determinations of friction coefficients from terms containing $\bar{X}_1$ also.

The determination of friction coefficients from measurements in free solution also runs into trouble, because in 0.1M KCl solution, X_1=X_2=0.0018, X_3=0.9964.

Self-diffusion coefficients in free solution are functions of concentration, and are usually measured at infinite dilution in order to eliminate the effects of non-idealities. In practice, then, Eq. 57 becomes

$$\frac{RT}{D_{11}^{\circ}} = r_{13} \qquad (64a)$$

$$\frac{RT}{D_{22}^{\circ}} = r_{23} \qquad (64b)$$

$$\frac{RT}{D_{33}^{\circ}} = r_{33} \qquad (64c)$$

For a given membrane, the charge density is fixed by electroneutrality considerations, and one cannot vary $\bar{X}_1$ arbitrarily.

Considering these limitations, a set of friction coefficients was calculated from the data in Table 1. These results are given in Table 2. Of the calculated friction coefficients, only r_{13}, r_{23}, r_{33}, and f_{14} are known with sufficient accuracy to allow confident prediction. Coefficients r_{11}, r_{12}, and f_{24} are known to within a factor of 2, and r_{22} and f_{34} are known only to within an order of magnitude. This loss of precision is due largely to the nature of Eqs. 56 to 63, which involve taking differences between large terms whose values are very close, and in which important terms disappear because coefficients differ vastly in size.

Table 2

Friction Coefficients From Data of Table 1

$r_{11} = 6.0 \times 10^9 \pm 100\%$
$r_{12} = 5.7 \times 10^9 \pm 100\%$
$r_{13} = 1.27 \times 10^8 \pm 100\%$
$r_{22} = 0\ (5 \times 10^9)$
$r_{23} = 1.22 \times 10^8 \pm 10\%$
$r_{33} = 9.91 \times 10^7 \pm 10\%$
$f_{14} = 1.70 \times 10^9 \pm 10\%$
$f_{24} = 3.0 \times 10^9 \pm 100\%$
$f_{34} = 0\ (3 \times 10^6)$

From these coefficients, a value of η was predicted to be

$$\eta > 300$$

While this value is consistent with the experimental value, this consistency is fortuitous. Very minor changes in the experimental values of self-diffusion coefficients, well within the experimental error, would have allowed η to take on almost any value.

It would be interesting to perform this calculation on other data taken under a wider variety of conditions but, unfortunately, a complete set of data on a given membrane system is not available at this time.

B. General Considerations

All thermodynamic models based on the Spiegler scheme suffer from the problems encountered above; that is, the lack of precision inherent in the equations makes it extremely difficult to predict membrane transport properties to within any reasonable degree of confidence. This is at least true for the highly selective membranes of interest for electrodialysis.

Perhaps a quantitative approach to the problem, based on the Kedem and Katchalsky model, might prove more fruitful.

VI. HYDRODYNAMIC MODELS FOR MEMBRANE PROCESSES

Another approach to the problem of correlating membrane structure with function is through the use of a hydrodynamic model. This apprach differs from a purely thermodynamic approach in that a homogeneous membrane phase is not assumed. Instead of developing a dissipation function and writing flux equations as functions of thermodynamic forces, a particular simplified membrane model is chosen and macroscopic hydrodynamic theory is applied to calculate functional properties. Examples of models that have been investigated include capillary flow models and models based on calculations of drag on submerged objects.

A good example of this approach is given by Kawabe et al. (34). In this paper an attempt is made to estimate the apparent pore diameters of certain cation-exchange membranes of interest for electrodialysis. The method is based on the work of Heymann and O'Donnell (35), and subsequent workers (36, 37, 38), which showed that the conductivity of cation-exchange membranes in different ionic states was directly proportional to the limiting equivalent conductivity of the exchange cation in free solution for small cations (H^+, Li^+, Na^+, K^+, NH_4^+), but that this proportionality constant decreases for larger cations (tetra-alkyl ammonium ions). Kawabe et al. investigated this phenomenon for a number of commercially important cation-exchange membranes, and the results obtained for one of them (AMF C-103) are shown in Table 3.

Based on these data, and such other data as pore volume fraction, membrane thickness, counterion concentration, coion concentration, and so on, in these various ionic states, a model was assumed. In this model, the ion was assumed to be a sphere moving through an irregularly shaped hole filled with homogeneous liquid water. Stokes hydrodynamics was assumed and a force balance was made on the ion. The force pulling the ion through the pore is simply the electrical potential gradient. In the absence of

Table 3

AMF C-103 Membrane in 0.1N[a] Salt Solution at 25°C

Membrane State	Resistance (ohm-cm)	$\bar{\lambda}/\lambda°$
H^+	0.6	0.085
Li^+	6.0	0.074
Na^+	4.8	0.071
K^+	3.5	0.065
NH_4^+	3.3	0.069
$(CH_3)_4N^+$	6.6	0.059
$(C_2H_5)_4N^+$	19.1	0.032
$(C_3H_7)_4N^+$	48.7	0.017
$(C_4H_9)_4N^+$	296.0	0.0041

[a]These values did not change significantly as the salt concentration was decreased to 0.001N.

accelerations, this force is balanced by a drag force given by

$$F = \frac{-6\pi\mu\ r\ v}{k\ f} \tag{66}$$

where F is the drag force, μ is the viscosity of the medium, r is the ionic radius, v is the ionic velocity relative to the membrane, and k is a correction factor to account for the fact that the ionic radius is close to the radius of a water molecule, and, so, the medium is not really homogeneous [see Robinson and Stokes (39)]. The factor f, referred to as the "drag factor," is based on the work of Faxen (40), and accounts for the effect of the pore walls on the drag force. If the pore is assumed cylindrical in shape, the drag factor is given by

$$f = 1-2.014\ (\frac{r}{a}) + 2.09\ (\frac{r}{a})^3 - 0.95\ (\frac{r}{a})^5 + \ldots \tag{67}$$

where a is the pore radius.

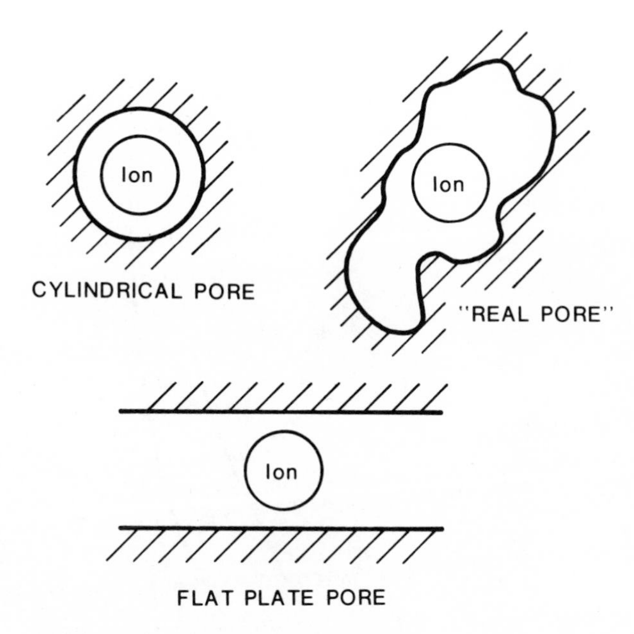

Fig. 15. Models of membrane pores.

Most work on hydrodynamic models in the literature [see Pappenheimer et al. (41), Renkin (42), and Solomon (43)] assumes such a cylindrical pore, but Kawabe et al. point out that this is a poor assumption. The reason for this is illustrated in Fig. 15. If the pore were really a macroscopic opening, and the ion were really a sphere, then the irregularity of the pore walls would result in its effect being felt at only one or two points on the periphery of the sphere. Since the factor f is such a strong function of r/a (see Fig. 16), the particle will not really see any part of the wall further away than the points of closest approach.

If the pore were cylindrical, the drag force would be evenly spread around the particle. On the other hand, if the pore were represented as a pair of parallel flat plates, the resulting drag force would closely approach that which must actually be present for an irregular pore.

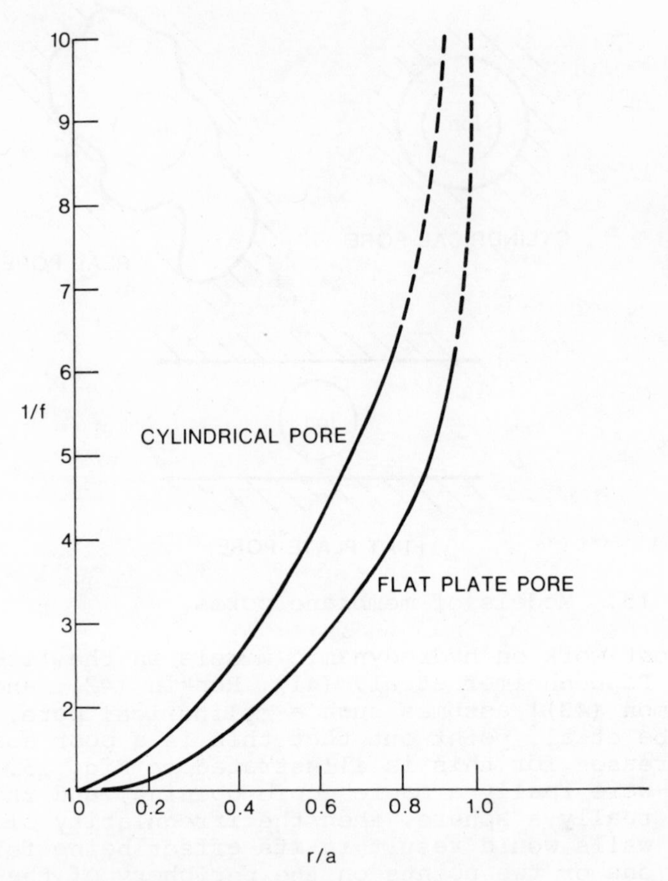

Fig. 16. Drag factor versus r/a.

Faxen has presented an equation for the drag factor for the case of parallel flat plates as

$$1 = 1-1.004\left(\frac{r}{a}\right) +0.418\left(\frac{r}{a}\right)^3-0.169\left(\frac{r}{a}\right)^5+\ldots \quad (68)$$

where a is the pore half width.

This equation is also shown in Fig. 16.

In order to satisfy the no-slip condition at the wall, Kawabe et al. have added a term to force f to go to zero as r/a goes to one. In addition, Kawabe et al. have added a term suggested by Ferry (44), to account for entry of the sphere into the pore from free solution. With these two changes, the drag factor becomes

$$f=(1-\frac{r}{a})[1-1.004(\frac{r}{a})+0.418(\frac{r}{a})^3-0.169(\frac{r}{a})^5-0.245(\frac{r}{a})^7] \tag{69}$$

Using these considerations, the equivalent conductances of the membrane in different ionic states were related by

$$\frac{(\bar{\lambda}/\lambda^\circ)_1}{(\bar{\lambda}/\lambda^\circ)_2} = \frac{(\phi h\ A\ f)_1}{(\phi h\ A\ f)_2} \tag{70}$$

where ϕ is the pore volume fraction, h is the tortuosity, defined as the membrane thickness divided by the pore length, and A is related to the activation energy for diffusion. The pore volume fraction was measured. hA was assumed the same for all ionic states, and Eq. 70 was used to calculate a for a given pair of ions of known radius. For this purpose, pairs of the tetra-alkyl ammonium ions were chosen, since they are non-hydrated, of known size, and enough bigger than the water molecule, so that the assumption of a water continuum is reasonable.

One the pore widths in the tetra-alkyl ammonium ion states were calculated, the measured pore volume fractions were used to calculate pore widths for membranes containing the other hydrated ions. From these values of $\hat{a}$, Eq. 70 was used to calculate effective ion size for these ions within the membrane. The results of this calculation are shown in Table 4.

A number of interesting observations may be made with regard to Table 4. First of all, it should be noted that calculated pore widths vary from about 6 Å to about 10 Å. This value is consistent with data in the literature on the permeability of these membranes to uncharged solutes. Measured permeability coefficients indicate that molecules larger than about 10 Å diffuse with great difficulty through these membranes.

Table 4. AMF C-103 membrane at 25°C

Ionic state	Pore half-width (Å)	Ionic radius in pore (Å)	Hydrated ionic radius in free solution (Å) Robinson and Stokes (38)	Crystal radius (Å)
H^+	3.52	2.12		
Li^+	3.77	2.51	3.7	0.68
Na^+	3.25	1.97	3.3	0.97
K^+	2.95	1.71	2.3	1.33
NH_4^+	3.43	2.19	2.3	1.43
$(CH_3)_4N^+$	4.52	3.47	3.47	3.47
$(C_2H_5)_4N^+$	4.52	4.00	4.00	4.00
$(C_3H_7)_4N^+$	5.52	4.52	4.52	4.52
$(C_4H_9)_4N^+$	5.38	4.94	4.94	4.94

Secondly, calculated ion sizes within the pores indicate the possibility of partial dehydration of the alkali metal cations. This dehydration is probably caused by competition for the available water (pore liquid is about 5 molal) and has been discussed in the literature by such authors as Ambard and Trautman (45) and others.

Similar results were obtained on the other membranes investigated (DK-1 of Asahi Chemical Industries; Nalfilm-1, and Nalfilm-3, of Nalco Chemical).

The use of hydrodynamic and molecular models has also produced useful results for the analysis of other membrane processes as well. Two good examples are the work of Schmid (46) and of Breslau and Miller (47) on electroosmosis.

In particular, the work of Breslau and Miller is a direct extension of the work of Kawabe et al. This hydrodynamic model for electroosmosis was based on the unidirectional flow of macroscopic spheres in a homogeneous bounded medium. Corrections were made to account for the influence of the matrix, viscosity changes due to ion-solvent interaction, and the molecular size of the ionic sphere. Electroosmotic coefficients calculated from measured membrane properties were found to agree with experimentally determined values to within ±5%. As with Kawabe et al., the results indicate that membrane pore structure is well approximated by a model employing parallel-plate pores, with ions within the membrane being partially stripped of their hydration shells.

Another example is the work of Miller et al. (48) on ion-exchange selectivity in which the membrane polyion is approximated by a charged cylinder in the center of an annular region containing the counterions.

Based on the current literature, there is considerable effort under way in this area, and excellent prospects of fruitful results yet to be obtained.

NOMENCLATURE

A	Defined by Eq. 23, dimensionless; or defined by Eq. 70, dimensionless
a	Pore radius or half width, Å
B	Second Wein effect term, dimensionless
c	Concentration, moles/cm^3
D	Molecular diffusivity, cm^2/sec.
E	Potential difference, V (defined by Eq. 37); or electroosmotic coefficient, moles water/Farady
F	Faraday's constant = 96,500 C/equiv or, if subscripted, total force, joules/mole-cm.
f	Friction coefficient between species in solution and matrix, joules-sec/mole-cm^2 or drag factor, dimensionless
h	Tortuosity, dimensionless
I	Total current density in mA/cm^2, or total flux in equiv/sec-cm^2
J	Ion flux, equiv/sec, or ion flux density, equiv/sec-cm^2

K_w	Dissociation constant for water = 10^{-20} mole^2/cm^6
k	Defined by Eq. 66, dimensionless
L	Onsager coupling coefficient, consistent units
M	Molarity, moles/liter solution
m	Molality, moles/1000 g H_2O
P	Pressure, consistent units
R	Universal gas constant = 8.3144 joules/mole-°K
r	Friction coefficient between two species in solution, or ionic radius, consistent units
T	Absolute temperature, °K
V	Partial molal volume, cm^3/mole
v	Velocity, cm/sec
w	Flow rate, gal/min-in. of cell width
X	Volume fraction or mole fraction, dimensionless
x	Distance, cm
y	Cell width, in.
Z	Valence, dimensionless
ϕ	Electrical potential, V
ϕ^*	Dimensionless electrical potential
δ	Boundary-layer thickness, cm or Å
ε	Electric permittivity, equiv/cm
σ	Equilibrium distribution coefficient, dimensionless
ν	Kinematic viscosity, cm^2/sec
μ	Chemical potential, joules/equiv, or viscosity, poise
$\tilde{\mu}$	Electrochemical potential, joules/equiv
Δ	Gradient
γ	Defined by Eq. 32
π	Osmotic pressure, consistent units
ξ	Reflection coefficient, dimensionless
λ	Ionic conductance, ohm^{-1}-cm^2/equiv.
η	Membrane selectivity, dimensionless
θ	Pore volume fraction, dimensionless
$\underline{\Phi}$	Dissipation function, defined by Eq. 31

Nomenclature

Subscripts

+	Cationic
-	Anionic
i	Refers to a particular ion
1	K^+ or counterion
2	Cl^- or co-ion
3	H^+ or water
4	OH^- or matrix
bulk, b	Mixed average property
membrane, int.	Property at interface between solution and membrane
c	Critical
h	Hydrodynamic
s	Salt
v	Volume
w	Water

Superscripts

' "	Refer to different sides of the membrane
*	Dimensionless unit
-	Within the membrane, mean value, partial molal property
~	Electro
o	Infinite dilution

Dimensionless Groups

Re	Reynolds number = $\mu_{bulk} X/\nu$
Sc	Schmidt number = ν/D

REFERENCES

1. M. Tribus, Süsswasser aus dem Meer, Dechema Monographien, Verlag-Chemie, Weinheim/Bergstrasse, 1962.
2. L. H. Schaffer and M.S. Mintz, Electrodialysis, in "Principles of Desalination", K. S. Spiegler, Ed., Academic Press, New York, 1966.
3. R. E. Lacey, E. W. Lang, and E. L. Huffman, Advan. Chem. Ser. 38, 1962.
4. A. Bethe and T. Toropoff, Z. Physik. Chem. 88, 686 (1914); 89, 597 (1915).

5. Office of Saline Water, 1966 Saline Water Conversion Report, U. S. Dept. of the Interior, 1967, p. 224.
6. Office of Saline Water, 1966, Saline Water Conversion Report, U. S. Dept. of the Interior, 1967, p. 222.
7. F. G. Helfferich, Ion Exchange, McGraw-Hill, New York, 1962.
8. A. M. Peers, Disc. Faraday Soc. 21, 124 (1956).
9. N. W. Rosenberg and C. E. Tirrell, Ind. Eng. Chem. 49, 780 (1957).
10. D. A. Cowan and J. H. Brown, Ind. Eng. Chem. 51, 1445 (1959).
11. T. Uchino, S. Nakooka, H. Hani, and T. Yawatawa, J. Electrochem. Soc. Japan 26, 366 (1958).
12. H. P. Gregor and M. A. Peterson, J. Phys. Chem. 68, 2201 (1964).
13. H. P. Gregor and I. F. Miller, J. Am. Chem. Soc. 86, 5689 (1964).
14. L. Onsager, J. Chem. Phys. 2, 599 (1934).
15. K. C. Lang, Ph.D. Dissertation, Poly. Inst. New York, 1974.
16. D. Cowan, Advan. Chem. Ser. 27, 224 (1960).
17. H. Schlichting, Boundary Layer Theory, McGraw-Hill, New York, 1955.
18. V. G. Levich, Physiochemical Hydrodynamics, Prentice-Hall, Englewood Cliffs, N.J., 1962.
19. R. B. Bird, W. E. Stewart, and E. N. Lightfoot, Transport Phenomena, John Wiley, New York, 1960.
20. M. E. Espelosin, Ph.D. Dissertation, Poly. Inst. Bklyn., 1965.
21. L. M. Salzarulo, Ph.D. Dissertation, Poly. Inst. Bklyn., 1966.
22. G. Wranglen and O. Nilsson, Electrochim. Acta 7, 124 (1962).
23. L. Onsager, Phys. Rev. 37, 405 (1931); 38, 2265 (1931).
24. S. R. DeGroot, Thermodynamics of Irreversible Processes, North-Holland, Amsterdam, 1959.
25. I. Prigogine, Introduction to Thermodynamics of Irreversible Processes, Interscience, New York, 1961.
26. A. Katchalsky and O. Kedem, Biophys. J. 2, Suppl. 53 (1962); O. Kedem and A. Katchalsky, Trans. Faraday Soc. 59, 1918 (1963).

27. A. Katchalsky and P. F. Curran, Non-equilibrium Thermodynamics in Biophysics, Harvard Univ. Press, Cambridge, Mass., 1965.
28. K. S. Spiegler, Trans. Faraday Soc. 54, 1408 (1958).
29. L. Onsager, Ann. N.Y. Acad. Sci. 46, 241 (1945).
30. A. Klemm, Z. Naturforsch. 8a, 397 (1953).
31. R. Laity, J. Phys. Chem. 63, 80 (1959).
32. R. Caramazza, W. Dorst, A. J. C. Hoeve, and A. J. Staverman, Trans. Faraday Soc. 59, 2415 (1963).
33. A. Lippe, M. S. Project Report, Poly. Inst. Bklyn., 1966.
34. H. Kawabe, H. Jacobson, I. F. Miller, and H. P. Gregor, J. Coll. Int. Sci. 21, 79 (1966).
35. E. Heymann and J. O'Donnell, J. Coll. Sci. 4, 405 (1949).
36. J. S. Mackie and P. Mears, Disc. Faraday Soc. 4, 111 (1956).
37. G. Manecke and H. Heller, Disc. Faraday Soc. 21, 101 (1956).
38. M. A. Peterson and H. P. Gregor, J. Electrochem. Soc. 106, 1051 (1959).
39. R. A. Robinson and R. H. Stokes, Electrolyte Solutions, 2nd ed. rev., Butterworths, London, 1965.
40. H. Faxen, Arkiv. f. Math., Astron. Phys. 17, No. 27 (1922).
41. J. R. Pappenheimer, E. M. Renkin, and L. M. Borrero, Am. J. Physiol. 167, 13 (1951).
42. E. M. Renkin, J. Gen. Physiol. 38, 225 (1954).
43. A. K. Solomon, J. Gen. Physiol. 43, Suppl., 1 (1960).
44. J. D. Ferry, J. Gen. Physiol. 20, 95 (1936).
45. L. Ambard and S. Trautmann, Ultrafiltration, Charles C. Thomas, Springfield, Ill., 1960.
46. G. Schmid, Chemie-Ing.-Tech. 37, No. 6, 616 (1965).
47. B. R. Breslau and I. F. Miller, I.E.C. Fund. 10, 554 (1971).
48. I. F. Miller, F. Bernstein, and H. P. Gregor, J. Chem. Phys. 43, 1783 (1965).

Author Index

Author Index

Author Index

Subject Index

Subject Index

Subject Index